T0400074

Multidimensional Stationary Time Series

Multidimensional Stationary Time Series

Dimension Reduction and Prediction

Marianna Bolla
Tamás Szabados

CRC Press
Taylor & Francis Group
Boca Raton London New York

CRC Press is an imprint of the
Taylor & Francis Group, an **informa** business
A CHAPMAN & HALL BOOK

First edition published 2021
by CRC Press
6000 Broken Sound Parkway NW, Suite 300, Boca Raton, FL 33487-2742

and by CRC Press
2 Park Square, Milton Park, Abingdon, Oxon, OX14 4RN

Library of Congress Cataloging-in-Publication Data

ISBN: 9780367569327 (hbk)
ISBN: 9781003107293 (ebk)

Typeset in CMR10 font
by KnowledgeWorks Global Ltd.

To our families.

Contents

Foreword

The purpose of this book is to give a brief survey of the theory of multidimensional (multivariate) weakly stationary time series, with emphasis on dimension reduction and prediction. We restricted ourselves to the case of discrete time, since this is a natural model of measurements in practice, and also, it is a good viewpoint to extend these techniques to the theory of continuous time as well.

Understanding the covered material requires a certain mathematical maturity, a degree of knowledge in probability theory and statistics, and also in linear algebra and real, complex and functional analysis. It is advantageous if the reader is familiar with some notions of abstract algebra, but the corresponding details may also be skipped. The main tools include harmonic analysis, state space methods of linear time-invariant systems, and methods that reduce the rank of the spectral density matrix.

We made an effort to give an almost self-contained version of the theory, with proofs that seemed the easiest to grasp. Also, we collected important results from a rather wide range of topics that are relevant in the theory of stationary time series. We hope that this book can be used as a text for graduate courses and also as a reference material for researchers working with time series.

The applications of multidimensional stationary time series cover a very wide range of fields nowadays; there are important applications in statistics, engineering, economics, finance, natural and social sciences.

Preface

"Deep is the well of the past.
Should we not call it bottomless?"
(Thomas Mann: Joseph and his brothers)

To understand the remote past is indeed crucial if we want to classify the weakly stationary (briefly, stationary) time series. The presence of remote past causes different types of singularities, that themselves make the future predictable with zero error. We use the beautiful spectral theory of stationary time series, developed mainly for one-dimensional processes a century ago by H. Cramér, H. Wold, A.N. Kolmogorov, and N. Wiener, just to mention the most prominent ones. Singular, or in other words, deterministic processes in 1D may arise due to spectral distribution that is singular with respect to the Lebesgue measure (we call it Type (0) singularity), but even if an absolutely continuous spectral measure exists, the process can be singular (we call those Type (1) and Type (2) singularities), e.g. the sliding summation (two-sided moving average) that cannot be written as a one-sided moving average. The famous Wold decomposition is able to separate the singular and regular part of a non-singular process, where the regular part is, in fact, a one-sided moving average, and the singular part is of Type (0) in 1D. Here it is important that the Wold decomposition is not applicable to singular processes, as the possible Type (1) or Type (2) singularities cannot coexist with a regular process. Unfortunately, less attention has been paid to this distinction, however, understanding the down side makes us capable of better understanding the regular part, where prediction makes sense with positive error. The theory extends to multidimensional processes where more complicated situations can occur, e.g. regular and singular parts may coexist when concentrated on different components. Regular processes are also called purely non-deterministic or causal, as probabilistic predictions can be made for the future on the basis of the past, of course, with positive error. Those are the innovations (called fundamental shocks in finance) that contain the added value of the future observations, and so, they provide a driving force of the whole process, and establish the method of dynamic factor analysis too.

We shall work with multidimensional stationary time series of discrete time, both in the frequency and in the time domain, and find analogies between these two approaches as for dimension reduction. Our main objects are (weakly) stationary time series with a spectral density matrix that has constant rank (almost everywhere in the frequency domain). Those correspond to the sliding summation, which is in general singular, but still, its spectral density can be factorized, and the (two-sided) power series expansion of the transfer function provides the coefficient matrices, in terms of which the two-sided moving average can be written. Important special case of a constant rank process is the one-sided moving average that (by the Wold's theorem) is always a regular process. Here the (one-sided) power series expansion of the transfer function gives the coefficient matrices that are called impulse responses, and the regular process can be expanded with them in terms of the innovations. Further subclass of the regular processes is the set of processes that have a rational spectral density matrix (its entries are complex rational functions of $e^{-i\omega}$) and also a rational transfer function. When factorizing rational spectral densities, usual row and column operations of the classical matrix decomposition algorithms should be adapted to modules over the ring of complex polynomials, to obtain e.g. the Smith–McMillan form.

Under some additional conditions, there is a one-to-one correspondence between the spectral density matrix (frequency domain) and the pair of the transfer function and the innovation covariance matrix (time domain). Under the same conditions, there is also an (infinite) matrix relation between the block Hankel matrix of the impulse response matrices and that of the autocovariance matrices. However, by the Kronecker's theorem, in the possession of a rational spectral density, these block Hankel matrices have bounded rank, so we can confine ourselves to finite segments of them.

The processes with a rational spectral density matrix have either a stable VARMA (vector autoregressive plus moving average), or an MFD (matrix fractional description), or a state space representation; latter ones are linear dynamical systems with hidden state variables and observable variables, further, they contain white noises as error terms. All these representations have a finite number of parameter matrices that are sometimes overparametrized, and are not always uniquely determined by the transfer function, but can be estimated under some conditions. We do not want to go into details of the methods of estimation, but will discuss the Yule–Walker equations that emerge in different situations.

In practice, we have a finite time series observed from a starting time, and infinite past can only be imitated if we go farther and farther to the future. Ergodicity under general conditions guarantees that we can make estimations based on a single trajectory, observed for long enough time. Luckily, the theory of finite past predictions is well elaborated since Gauss, with the projection principle that is widely used in the theory of multivariate regression (now our sample is the set of the finitely many, say n, past observations). As $n \to \infty$, we approach the infinite past situation. The innovations, obtained by the Gram–Schmidt orthogonalization and, in the multidimensional case, by the block Cholesky decomposition, will better and better approach the innovations based on the infinite past prediction; whereas, the coefficient matrices will better and better approach the impulse responses. In this way, we assign a bottom to the well and ignore the remote past.

When the spectral density matrix has a lower rank than its size, then it is important that its rank is equal to the dimension of the innovation subspaces in the multidimensional Wold decomposition. In this case, the impulse response matrices are not quadratic, but rectangular. In practice, the rank of the spectral density matrix is only estimated from a sample, and hence, we can speak of an essential rank of it. The theory of the General Dynamic Factor Models (GDFM) makes it possible to separate the structural (large absolute value) eigenvalues of the spectral density matrix and to look for lower dimensional innovations. Actually, this is the task of the dynamic factor analysis, and it can be realized with different techniques. As for the dimension reduction, we also introduce the "spectra of spectra" technique, i.e. in the possession of a d-dimensional time series observed at n consecutive time instances, we take the spectral decomposition of the $d \times d$ spectral density matrices at the n Fourier frequencies, and keep only the structural eigenvalues with corresponding eigenvectors. The computational complexity is of order nd^3 that is smaller than the block Cholesky decomposition of the $nd \times nd$ large block Toeplitz matrix. However, it is proved that there is an asymptotic correspondence between the nd eigenvalues of this large matrix and the union of the d eigenvalues of the spectral density matrices at the n Fourier frequencies. Of course, both the autocovariance matrices and these spectral density matrices are only estimated by means of multivariate periodograms. We also touch upon the Kálmán's filtering. This method is able to predict the state variables in a state space model based on finitely many observed variables (past ones, current ones, or future ones, corresponding to prediction, filtering, or smoothing problems, respectively). As a byproduct, it is able to find the innovations as well.

The organization of the chapters is as follows. In Chapter 1, we introduce the equivalent views of a weakly stationary, multidimensional time series (with discrete time and complex state space): they have a non-negative definite autocovariance matrix function and a non-negative definite matrix valued spectral measure on $[-\pi, \pi]$ (the two are related via Fourier transformation). We deal with complex valued multivariate time series, though most of the real-life time series (e.g. financial ones) have real coordinates. However, by the spreading of quantum computers and quantum information theory, there are emerging telecommunication systems that are complex valued by nature. Our tools are well applicable to this situation, since the frequency domain calculations need complex harmonic analysis, anyway. In Chapter 2, we start with simple 1D time series, and from ARMA processes we arrive at the regular and non-singular ones with the Wold decomposition, in an inductive way. We also give a classification of the 1D weakly stationary processes, by distinguishing between different types of singularities. Chapter 3 is devoted to abstract algebraic tools in order to find matrix factorizations that keep the rationality. This is important when we work with rational spectral densities and state space models; however, this chapter can be skipped. Chapter 4 investigates the types of multidimensional stationary processes in a deductive way. First the ones with spectral density matrix of constant rank are introduced that are, in fact, the sliding summations (two-sided moving averages). A special case is the one-sided moving average, that is a (purely) regular, causal process. Types of singularities are also characterized. A subclass of regular processes is constituted by those of rational spectral density, which are nothing else but the VARMA processes, or equivalently, the state space models that also have an MFD; they can be finitely parametrized and predicted by finitely many past observations and shocks. In Chapter 5, we perform predictions in the time domain and introduce the original version of the Wold decomposition. We also establish analogies between the time and frequency domain notions. The chapter is closed with the brief discussion of the Kálmán's filtering and Dynamic Factor Analysis techniques.

We believe that this book fills in the gap between the classical theory and modern techniques, overcoming the curse of dimensionality in multivariate time series. We collect pieces from the very original classical works (H. Cramér, H. Wold, N. Wiener, A.N. Kolmogorov, Yu.A. Rozanov, and R.E. Kálmán) that are "lost in translation" but needed for a deeper understanding of the topic.

Marianna Bolla and Tamás Szabados

Budapest, November 2020
(during the COVID-19 pandemic)

List of Figures

Symbols

Symbol Description (Algebra)

$\mathbb{N}$	Set of natural numbers.
$\mathbb{Z}$	Set of integers.
$\mathbb{R}$	Set of real numbers.
$\mathbb{C}$	Set of complex numbers.
$\mathbb{C}^d$	Set of complex d-dimensional vectors.
$\mathbb{C}^{d\times r}$	Set of complex matrices of size $d \times r$.
T	The unit circle in $\mathbb{C}$.
D	The open unit disc in $\mathbb{C}$.
$a, b, \ldots, x, y$	Real scalars.
$z, \alpha, \beta, \ldots$	Complex scalars.
$\mathbf{a}, \ldots, \mathbf{Z}$	Column vectors (of real or complex) components.
$\boldsymbol{A}, \boldsymbol{B}, \ldots$	Matrices (of real or complex) entries.
$\mathbf{0}$	Zero vector.
$\boldsymbol{O}$	Zero matrix.
$\boldsymbol{I}_d$	$d \times d$ identity matrix.
$\mathbf{a}^T, \boldsymbol{A}^T, \ldots$	Transpose of a (real or complex) vector or matrix.
$\mathbf{a}^*, \boldsymbol{A}^*, \ldots$	Adjoint (conjugate transpose) of a complex vector or matrix.
$\overline{z}, \overline{\mathbf{a}}, \overline{\boldsymbol{A}}, \ldots$	Conjugate of a complex scalar or entrywise conjugate of a complex vector or matrix.
$\mathcal{H}$	Hankel matrix.
$\boldsymbol{A} \otimes \boldsymbol{B}$	Kronecker product of matrices $\boldsymbol{A}$ and $\boldsymbol{B}$.
$\|\boldsymbol{A}\|$	Spectral norm of matrix $\boldsymbol{A}$.
$\|\boldsymbol{A}\|_F$	Frobenius norm of matrix $\boldsymbol{A}$.
$\rho(\boldsymbol{A})$	Spectral radius of matrix $\boldsymbol{A}$.
$\vert\boldsymbol{A}\vert, \det(\boldsymbol{A})$	Determinant of the quadratic matrix $\boldsymbol{A}$.
$\mathrm{tr}(\boldsymbol{A})$	Trace of the quadratic matrix $\boldsymbol{A}$.
$\mathrm{rank}(\boldsymbol{A})$	Rank of the matrix $\boldsymbol{A}$.
$\mathrm{Ker}(\boldsymbol{A})$	Kernel space of the matrix $\boldsymbol{A}$.
$\mathcal{R}(\boldsymbol{A})$, $\mathrm{Range}(\boldsymbol{A})$	Image space (range) of the matrix $\boldsymbol{A}$.
$\mathrm{diag}[\lambda_1, \ldots, \lambda_d]$	Diagonal matrix with $\lambda_1, \ldots, \lambda_d$ in its main diagonal.
$\mathrm{Span}\{\mathbf{a}_j : j \in J\}$	Linear span: all finite linear combinations of a set of vectors.
$\overline{\mathrm{span}}\{\mathbf{a}_j : j \in J\}$	Closure of the linear span of a set of vectors.

δ_{jk}	Kronecker delta, $\begin{cases} 1 & \text{if} \quad j = k; \\ 0 & \text{if} \quad j \neq k. \end{cases}$
$\Sigma = (A, B, C, D)$	Linear system with linear operators A, B, C, D.
$\mathcal{R}$	Reachability matrix.
$\mathcal{O}$	Observability matrix.
ϕ_0	Restricted input/output map.
ϕ	Extended input/output map.
$\mathcal{U}$	Set of input sequences.
$\mathcal{Y}$	Set of output sequences.
$H(z)$	Transfer function.
$\{H_\ell : \ell = 1, 2, \dots\}$	Impulse response function.

Symbol Description (Probability and analysis)

Symbol	Description
$\mathbb{P}$	Probability.
$\mathbb{E}$	Expectation.
Var	Variance.
Cov	Covariance (matrix).
Corr	Correlation.
$X, Y, \xi, \eta, \ldots$	Random variables.
$\mathbf{X}, \mathbf{Y}, \boldsymbol{\xi}, \boldsymbol{\eta}, \ldots$	Random vectors.
$\{\mathbf{X}_t\}_{t \in \mathbb{Z}}$	d-dimensional time series (of complex or real components) with discrete time.
$\boldsymbol{C}(h)$, $h \in \mathbb{Z}$	$d \times d$, hth order autocovariance matrix of a d-dimensional, weakly stationary time series.
$\boldsymbol{C}_n$	$n \times n$ Toeplitz matrix of autocovariances of a 1D, weakly stationary time series.
$\mathfrak{C}_n$	$nd \times nd$ block Toeplitz matrix of autocovariance matrices of a d-dimensional, weakly stationary time series.
$\boldsymbol{f}(\omega) = [f^{ij}(\omega)]$	$d \times d$ spectral density matrix of $\{\mathbf{X}_t\}$, $\omega \in [-\pi, \pi]$ or $[0, 2\pi]$.
$d\boldsymbol{F}(\omega) = [dF^{ij}(\omega)]$	$d \times d$ spectral measure matrix of $\{\mathbf{X}_t\}$, $\omega \in [-\pi, \pi]$ or $[0, 2\pi]$.
S	Right (forward) shift operator (unitary operator).
$L = S^{-1} = S^*$	Left (backward) shift operator.
$\mathrm{WN}(\boldsymbol{\Sigma})$	White noise sequence with covariance matrix $\boldsymbol{\Sigma}$.
ℓ^1	Banach space of absolute summable sequences.
ℓ^2	Hilbert space of square summable sequences.
L^p	Banach space of functions f s.t. $\lvert f\rvert^p$ is integrable, $1 \le p \le \infty$.
$H(\mathbf{X})$	Closure of linear span of the components of a time series $\{\mathbf{X}_t\}$.
$\tilde{X}_T$	Empirical mean.
$\hat{\boldsymbol{C}}(h)$	Empirical covariance matrix function.
$\omega_j = \frac{2\pi j}{n}$	jth Fourier frequency, $j = 0, \ldots, n-1$.
$\rho_j = e^{\omega_j}$	jth primitive root of 1, $j = 0, \ldots, n-1$.
$dF^{Y,X}(\omega)$	Joint spectral measure of $\{(X_t, Y_t)\}$.
$f^{Y\mid X}(\omega)$	Conditional spectral density of $\{Y_t\}$ w.r.t. $\{X_t\}$.
log, ln	Natural logarithm, with e as its base.
$o(f(n))$	'Little o' of $f(n)$, i.e. a function of n such that $\lim_{n\to\infty} \frac{o(f(n))}{f(n)} = 0$.
$O(f(n))$	'Big O' of $f(n)$, i.e. a function of n such that $O(f(n)) \le Cf(n)$ with some constant C, independent of n.

Symbol Description (Verbal abbreviations)

1D	One dimensional (time series).
multi-D	Multidimensional (time series).
w.r.t.	With respect to.
a.e.	Almost everywhere (w.r.t. a measure).
i.i.d.	Independent, identically distributed.
p.d.f.	Probability density function.
c.d.f.	Cumulative distribution function.
SD	Spectral decomposition (of a self-adjoint matrix).
parsimonious SD	Spectral decomposition with minimum number of dyads.
SVD	Singular value decomposition (of a rectangular matrix).
Gram-decomp.	Decomposition $\boldsymbol{G} = \boldsymbol{A}\boldsymbol{A}^*$ of a self-adjoint matrix $\boldsymbol{G}$.
TLF	Time invariant linear filter.
AR(p)	pth order autoregressive process (in 1D).
MA(q)	qth order moving average process (in 1D).
ARMA(p, q)	p-th order autoregressive plus qth order moving average process (in 1D).
VAR(p)	pth order vector autoregressive process (in multi-D).
VMA(q)	qth order vector moving average process (in multi-D).
VARMA(p, q)	p-th order vector autoregressive plus qth order vector moving average process (in multi-D).
RMSE	Root mean square error.
Toeplitz matrix	Quadratic matrix that has the same entries along its main diagonal and along all lines parallel to the main diagonal.
Hankel matrix	Quadratic matrix that has the same entries along its anti-diagonal and along all lines parallel to its anti-diagonal.
block Toeplitz	Block matrix which is Toeplitz in terms of its blocks.
block Hankel	Block matrix which is Hankel in terms of its blocks.
DFT	Discrete Fourier Transform.
IDFT	Inverse Discrete Fourier Transform.

1

Harmonic analysis of stationary time series

1.1 Introduction

A widely applicable notion of random fields is the one of *multidimensional (or multivariate) time series* $\mathbf{X}_t = (X_t^1, \dots, X_t^d)$, where $t \in \mathbb{R}$ is the time, $d \in \mathbb{N}$ is the dimension, and for each t fixed, each coordinate X_t^j $(j = 1, \dots, d)$ is a complex valued random variable on the same probability space $(\Omega, \mathcal{F}, \mathbb{P})$. Sometimes, we investigate the special case when the random variables are real valued. Here we consider time series with discrete time and state space $\mathbb{C}^d$.

We concentrate on *stationary time series*, mainly in the wide sense, the behavior of which is irrespective of time shift. Moreover, the assumption of weak stationarity needs only the first and second moments and cross-moments (second-order processes). This allows us to use a huge machinery of analytical tools and the theory of Hilbert spaces. Fortunately, many time series in practice can be approximated by stationary time series, possibly after some operations, that deprives the time series e.g. from trend and seasonality.

We prove equivalent notions of weak stationarity in terms of autocovariance matrix functions and matrix valued spectral measures. Spectral representation of the process itself is also given by orthogonal increments (Cramér's representation). We also give four different constructions for stationary time series, and discuss ergodicity of the estimates for the mean and covariances. In this way, we are able to make inferences based on a single trajectory, observed for a long enough time. Equivalent forms of a periodogram are also given.

1.2 Covariance function and spectral representation

Definition 1.1. The d-dimensional time series $\mathbf{X}_t$ $(t \in \mathbb{R})$ is strongly stationary (or stationary in the strong sense) if for any $h \in \mathbb{R}$, $n \in \mathbb{N}$, and time instances $t_1 < t_2 < \cdots < t_n$, the joint distribution of $(\mathbf{X}_{t_1+h}, \dots, \mathbf{X}_{t_n+h})$ is the same as the joint distribution of $(\mathbf{X}_{t_1}, \dots, \mathbf{X}_{t_n})$. That is, the joint distributions are invariant for any time shift.

The next conditions of weak stationarity are simpler to check in practice.

Definition 1.2. The d-dimensional time series $\mathbf{X}_t$ $(t \in \mathbb{R})$ is weakly stationary (or stationary in the wide sense) if it has finite expectation and finite covariance function that do not depend on time shift:

$$\mathbb{E}\mathbf{X}_t = \boldsymbol{\mu} = [\mu_1, \dots, \mu_d]^T \in \mathbb{C}^d,$$

$$c_{jk}(h) := \operatorname{Cov}(X^j_{t+h}, X^k_t) = \operatorname{Cov}(X^j_h, X^k_0) = \mathbb{E}\left((X^j_h - \mu_j)\overline{(X^k_0 - \mu_k)}\right),$$

where $t, h \in \mathbb{R}$; $j, k = 1, \dots, d$, and $\mathbf{X}_t = [X^1_t, \dots, X^d_t]^T$ in terms of its components.

Note that strong stationarity implies the weak one if the process has finite second moments. If the time series is Gaussian, then the two notions are equivalent, because Gaussian distributions are uniquely determined by their expectations and covariances. Further, to any weakly stationary process there exists a strongly stationary Gaussian with the same expectation and covariance matrix function. Weakly stationary processes are sometimes called *second-order processes*, as only their first and second moments are used in the theory describing their behavior.

Without loss of generality, from now on we assume that $\boldsymbol{\mu} = \mathbf{0}$, in which case

$$\operatorname{Cov}(X^j_{t+h}, X^k_t) := \mathbb{E}(X^j_{t+h}\overline{X^k_t}), \quad \operatorname{Var}(X^k_t) := \mathbb{E}(|X^k_t|^2),$$

where the complex conjugation on the second factor can be disregarded if the components are real valued.

The main object of the present book is time series *with discrete time.*

Definition 1.3. The d-dimensional time series $\mathbf{X}_t$ $(t \in \mathbb{Z})$ is weakly stationary (or stationary in the wide sense) with discrete time if it has finite expectation and finite covariance function that do not depend on time shift:

$$\mathbb{E}\mathbf{X}_t = \mathbf{0}, \quad c_{jk}(h) := \operatorname{Cov}(X^j_{t+h}, X^k_t) = \operatorname{Cov}(X^j_h, X^k_0) = \mathbb{E}(X^j_h\overline{X^k_0})$$

$(t, h \in \mathbb{Z}; j, k = 1, \dots, d)$, where, as we said above, we may assume that $\mathbb{E}\mathbf{X}_t = \boldsymbol{\mu} = \mathbf{0}$.

From now on, the expression 'stationary time series' will refer to a discrete time weakly stationary process with zero expectation, unless it is explicitly stated otherwise.

Considering complex valued random vectors simplifies the discussion; it is easy to describe the specific case of real valued random vectors whenever it is needed. Since we assume finite second moments, it follows that each component X^j_t, $j = 1, \dots, d$, is square integrable, so belongs to the Hilbert space $L^2(\Omega, \mathcal{F}, \mathbb{P})$ for any time instant $t \in \mathbb{Z}$. Here and below, $\mathbf{X}$ is a column vector and $\mathbf{X}^*$ denotes its adjoint (conjugate transpose), a row vector.

We use the *covariance matrix function* (or *autocovariance matrix function*) $\boldsymbol{C}(h) = [c_{j\ell}(h)] \in \mathbb{C}^{d \times d}$ to describe weakly stationary time series,

$$\boldsymbol{C}(h) := \operatorname{Cov}(\mathbf{X}_{t+h}, \mathbf{X}_t) = \mathbb{E}\left(\mathbf{X}_{t+h}\mathbf{X}^*_t\right), \qquad h \in \mathbb{Z}.$$

The covariance matrix function $\boldsymbol{C}$ does not depend on the time instant $t \in \mathbb{Z}$ because of the assumed weak stationarity. Clearly,

$$c_{j\ell}(-h) = \mathbb{E}\left(X_{t-h}^j \overline{X_t^\ell}\right) = \mathbb{E}\left(X_t^j \overline{X_{t+h}^\ell}\right) = \overline{c_{\ell j}(h)},$$

therefore

$$\boldsymbol{C}(-h) = \boldsymbol{C}^*(h). \tag{1.1}$$

By the Cauchy–Schwartz inequality, for any $j, k = 1, \ldots, d$ and $h \in \mathbb{Z}$,

$$|c_{jk}(h)| = \left|\mathbb{E}\left(X_h^j \overline{X_0^k}\right)\right| \le \left[\mathbb{E}\left|X_h^j\right|^2 \mathbb{E}\left|X_0^k\right|^2\right]^{\frac{1}{2}} = [c_{jj}(0)c_{kk}(0)]^{\frac{1}{2}},$$

and so,

$$|c_{jj}(h)| \le c_{jj}(0). \tag{1.2}$$

While the covariance matrix $\boldsymbol{C}(0)$ is *self-adjoint (Hermitian)* by (1.1), $\boldsymbol{C}(h)$ with a fixed time lag $h \neq 0$ is not self-adjoint in general. However, for an arbitrary $n \geq 1$, let us consider the following $nd \times nd$ matrix, which is a block Toeplitz matrix (see Definition B.9):

$$\mathfrak{C}_n := \begin{bmatrix} \boldsymbol{C}(0) & \boldsymbol{C}(1) & \boldsymbol{C}(2) & \cdots & \boldsymbol{C}(n-1) \\ \boldsymbol{C}^*(1) & \boldsymbol{C}(0) & \boldsymbol{C}(1) & \cdots & \boldsymbol{C}(n-2) \\ \boldsymbol{C}^*(2) & \boldsymbol{C}^*(1) & \boldsymbol{C}(0) & \cdots & \boldsymbol{C}(n-3) \\ \vdots & \vdots & \vdots & \ddots & \vdots \\ \boldsymbol{C}^*(n-1) & \boldsymbol{C}^*(n-2) & \boldsymbol{C}^*(n-3) & \cdots & \boldsymbol{C}(0) \end{bmatrix}. \tag{1.3}$$

It is obviously self-adjoint and positive semidefinite as it is the usual covariance matrix of the compounded nd-dimensional random vector $[\mathbf{X}_1^T, \mathbf{X}_2^T, \ldots, \mathbf{X}_n^T]^T$.

The next two theorems give important characterizations of weakly stationary time series in general.

Theorem 1.1. *The following back-and-forth statements are true.*

(a) If a d-dimensional time series $\{\mathbf{X}_t\}_{t\in\mathbb{Z}}$ is weakly stationary, then its covariance function $\boldsymbol{C}(h)$ $(h \in \mathbb{Z})$ is non-negative definite (positive semidefinite), which means that

$$\sum_{j,k=1}^{n} \mathbf{a}_k^* \boldsymbol{C}(k-j)\, \mathbf{a}_j \geq 0, \qquad \forall n \geq 1, \quad \forall \mathbf{a}_1, \ldots, \mathbf{a}_n \in \mathbb{C}^d. \tag{1.4}$$

Equivalently, the matrix $\mathfrak{C}_n$ in (1.3) is self-adjoint and non-negative definite:

$$\mathbf{a}^* \mathfrak{C}_n \mathbf{a} \geq 0, \qquad \forall n \geq 1, \quad \forall \mathbf{a} \in \mathbb{C}^{nd}.$$

(b) Conversely, to any non-negative definite matrix function $\boldsymbol{C}(h)$ $(h \in \mathbb{Z})$ one can find a weakly stationary time series $\{\mathbf{X}_t\}_{t\in\mathbb{Z}}$, with this covariance function.

Note that the fact that $\boldsymbol{C}(h)$ $(h \in \mathbb{Z})$ is a positive semidefinite matrix function does not mean that the $\boldsymbol{C}(h)$'s are positive semidefinite matrices. However, Equation (1.4) just means that the block Toeplitz matrix $\mathfrak{C}_n$ is positive semidefinite in terms of its blocks.

Theorem 1.2. *The following back-and-forth statements are also true.*

(a) If a d-dimensional time series $\{\mathbf{X}_t\}_{t\in\mathbb{Z}}$ is weakly stationary, then it has a non-negative definite spectral measure matrix $d\boldsymbol{F}$ on $[-\pi,\pi]$ such that the covariance matrix function $\boldsymbol{C}(h)$ $(h \in \mathbb{Z})$ of $\{\mathbf{X}_t\}$ can be represented as the Fourier transform *of $d\boldsymbol{F}$:*

$$\boldsymbol{C}(h) = \int_{-\pi}^{\pi} e^{ih\omega} d\boldsymbol{F}(\omega) \quad (h \in \mathbb{Z}).$$

Note that a measure matrix $d\boldsymbol{F} = [dF^{rs}] \in \mathbb{C}^{d\times d}$ on $[-\pi,\pi]$ is called non-negative definite if for any interval $(\alpha,\beta] \subset [-\pi,\pi]$ and for any $z_1,\dots,z_n \in \mathbb{C}$, it holds that

$$\sum_{r,s=1}^{d} dF^{rs}((\alpha,\beta]) z_r \overline{z_s} \geq 0. \tag{1.5}$$

(b) Conversely, to any non-negative definite measure matrix $d\boldsymbol{F}$ on $[-\pi,\pi]$, one can find a non-negative definite function $\boldsymbol{C}(h)$ $(h \in \mathbb{Z})$, and so one can find a weakly stationary time series $\{\mathbf{X}_t\}_{t\in\mathbb{Z}}$, whose spectral measure matrix is $d\boldsymbol{F}$.

The proof of these two theorems will be given step-by-step in the sequel in this and subsequent sections. First, it is very simple that the matrix $\mathfrak{C}_n$ in formula (1.3) is non-negative definite:

$$\sum_{k,j=1}^{n} \mathbf{a}_k^* \boldsymbol{C}(k-j)\, \mathbf{a}_j = \mathbb{E}\left|\sum_{k=1}^{n} \mathbf{a}_k^* \mathbf{X}_k\right|^2 = \operatorname{Var}\left\{\sum_{k=1}^{n} \mathbf{a}_k^* \mathbf{X}_k\right\} \geq 0 \tag{1.6}$$

for any $n \geq 1$ and $\mathbf{a}_1,\dots,\mathbf{a}_n \in \mathbb{C}^d$, see (1.1) as well. This proves Theorem 1.1(a). Later, a construction in Corollary 1.5 will show that the converse statement in Theorem 1.1(b) is also true.

Theorem 1.3 (Herglotz theorem) below proves Theorem 1.2(a) in the one-dimensional (1D) case. The proof of the general case will be quite circuitous, but relatively simple:

- The 1D case automatically extends to any covariance function $c_{jj}(h)$, $j = 1,\dots,d$, which can be represented by non-negative spectral measures $dF^j(\omega)$, c.f. (1.9).
- Then in Section 1.3 we will use this to give the important spectral representation of the time series $\{\mathbf{X}_t\}$ itself via a stochastic integral w.r.t. a random process $\{\mathbf{Z}_\omega\}$ of orthogonal increments, c.f. (1.20).

- In turn, $\{\mathbf{Z}_\omega\}$ will be used to define the complex spectral measures $dF^{j\ell}(\omega)$, $j, \ell = 1, \ldots, d$, which can represent any covariance function $c_{j\ell}(h)$. This way we are obtaining the spectral measure matrix $d\boldsymbol{F}(\omega)$ that represents the covariance matrix function $\boldsymbol{C}(h)$, see Corollary 1.2.

There exists a shorter way to give first the spectral representation of $\{\mathbf{X}_t\}$ using the spectral theory of normal operators in a Hilbert space. Since this is a quite advanced tool, see e.g. in [49, 12.22 Theorem], this approach will only be briefly sketched in Remark 1.8 later.

Finally, Corollary 1.6 will give a construction to show the converse statement in Theorem 1.2(b).

Naturally, the 1D cases of formulas (1.3) and (1.6) are simpler. If $\{X_t\}_{t\in\mathbb{Z}}$ is a 1D stationary time series and its covariance function is

$$c(h) = \operatorname{Cov}(X_{t+h}, X_t) \quad (t, h \in \mathbb{Z}),$$

then

$$\boldsymbol{C}_n = \begin{bmatrix} c(0) & c(1) & c(2) & \cdots & c(n-1) \\ c(-1) & c(0) & c(1) & \cdots & c(n-2) \\ c(-2) & c(-1) & c(0) & \cdots & c(n-3) \\ \vdots & \vdots & \vdots & \ddots & \vdots \\ c(-n+1) & c(-n+2) & c(-n+3) & \cdots & c(0) \end{bmatrix} \tag{1.7}$$

is an ordinary Toeplitz matrix. Clearly, it is self-adjoint and non-negative definite for any $n \geq 1$:

$$\sum_{j,k=1}^{n} c(k-j)\, z_k \overline{z_j} = \mathbb{E}\left|\sum_{k=1}^{n} z_k X_k\right|^2 = \operatorname{Var}\left\{\sum_{k=1}^{n} z_k X_k\right\} \geq 0 \quad (z_1, \ldots, z_n \in \mathbb{C}).$$

Also, it is the usual covariance matrix of $(X_1, \ldots, X_n)^T$ and it is positive semidefinite for this reason too. This implies that in the d-dimensional case, the Toeplitz matrix of the covariance function of each coordinate X_t^j ($j = 1, \ldots, d$) is also self-adjoint and non-negative definite for any $n \geq 1$.

Theorem 1.3. *(Herglotz theorem) Let $c : \mathbb{Z} \to \mathbb{C}$ be a non-negative definite function. Then c has a spectral representation*

$$c(h) = \int_{-\pi}^{\pi} e^{ih\omega} dF(\omega), \quad h \in \mathbb{Z},$$

where dF is a unique bounded non-negative measure on $[-\pi, \pi]$.

Proof. Since c is non-negative definite, with any $n \geq 1$ and $z_j = e^{-ij\omega}$, $j = 1, \ldots, n$, we get that

$$0 \leq \sum_{j,\ell=1}^{n} c(j-\ell)\, e^{-i(j-\ell)\omega} = \sum_{k=-(n-1)}^{n-1} c(k)\, e^{-ik\omega}\, (n - |k|).$$

Define

$$f_n(\omega) := \frac{1}{2\pi} \sum_{k=-(n-1)}^{n-1} c(k)\, e^{-ik\omega} \left(1 - \frac{|k|}{n}\right). \tag{1.8}$$

Then

$$f_n(\omega) \geq 0, \qquad \int_{-\pi}^{\pi} f_n(\omega) d\omega = c(0) \geq 0 \qquad (n \geq 1).$$

Let dF_n be the measure on $[-\pi, \pi]$ whose non-negative density is f_n for $n \geq 1$. Since these measures are bounded and the interval is compact, by Helly's selection theorem, see e.g. [37, p. 76], there exists a subsequence $\{n'\}$ such that the sequence of measures $\{dF_{n'}\}$ converges weakly to a limit dF; this will be the desired measure of the theorem.

First, for any integer h such that $|h| \leq m$,

$$\int_{-\pi}^{\pi} e^{ih\omega} f_n(\omega) d\omega = c(h) \left(1 - \frac{|h|}{n}\right).$$

Thus for any $h \in \mathbb{Z}$,

$$\lim_{n \to \infty} \int_{-\pi}^{\pi} e^{ih\omega} f_n(\omega) d\omega = c(h).$$

By the definition of weak convergence, for a suitable subsequence $\{n'\}$ one has

$$\lim_{n' \to \infty} \int_{-\pi}^{\pi} e^{ih\omega} dF_{n'}(\omega) = \int_{-\pi}^{\pi} e^{ih\omega} dF(\omega).$$

Comparing these expressions proves the claimed representation.

The uniqueness of the measure dF follows from two facts. First, any continuous function on $[-\pi, \pi]$ can be uniformly approximated by trigonometric polynomials (Weierstrass' theorem), see e.g. [50, 4.25 Theorem]. Second, integrals of continuous functions uniquely determine the measure by the Riesz representation theorem, see e.g. [50, 2.14 Theorem]. □

Later, in Corollary 1.6, we will see that the converse of Herglotz theorem is also true. The Herglotz theorem implies that each covariance function $c_{jj}(h)$ $(h \in \mathbb{Z})$, for $j = 1, \ldots, d$, has a spectral representation

$$c_{jj}(h) = \int_{-\pi}^{\pi} e^{ih\omega} dF^j(\omega), \quad F^j(-\pi) = 0, \quad F^j(\pi) = c_{jj}(0). \tag{1.9}$$

Here $F^j(\omega) := dF^j((-\pi, \omega])$, $\omega \in (-\pi, \pi]$, is a non-decreasing function, a spectral cumulative distribution function (c.d.f.).

Remark 1.1. Here we see two important special cases of Herglotz theorem.

(a) The first case is when the covariance function $c(h)$ is absolutely summable:

$$\sum_{h=-\infty}^{\infty} |c(h)| < \infty,$$

that is, $\{c(h)\}_{h\in\mathbb{Z}}$ is in ℓ^1. Then by (1.8), for $k \leq \ell$ and for any $\omega \in [-\pi, \pi]$,

$$|f_\ell(\omega) - f_k(\omega)| \leq \frac{1}{2\pi} \sum_{|h|\geq k} |c(h)| \to 0$$

as $k \to \infty$. Thus f_k uniformly converges on $[-\pi, \pi]$ to a *continuous spectral density function* f as $k \to \infty$. So dF is absolutely continuous w.r.t. Lebesgue measure, and $c(h)$ becomes the *Fourier coefficient* of the function f:

$$c(h) = \int_{-\pi}^{\pi} e^{ih\omega} f(\omega) d\omega, \quad dF(\omega) = f(\omega) d\omega, \tag{1.10}$$

and, vice versa, f can be expanded into the *Fourier series*

$$f(\omega) = \frac{1}{2\pi} \sum_{h=-\infty}^{\infty} c(h)\, e^{-ih\omega}, \quad f(\omega) \geq 0, \quad \omega \in [-\pi, \pi], \tag{1.11}$$

which is pointwise convergent (in fact, uniformly convergent) in $[-\pi, \pi]$. At this point we mention that analysis textbooks, see e.g. [50, 4.26], typically use different conventions for Fourier series: $c_n = \frac{1}{2\pi}\int_{-\pi}^{\pi} f(t)e^{-int}dt$, $f(t) \sim \sum_{n=-\infty}^{\infty} c_n e^{int}$. This is a matter of convention; several pieces of time series literature use the same convention as we do, since the starting point of Fourier analysis in the theory of time series is Theorem 1.3, the spectral representation of the already defined covariance function.

(b) Another important special case is when the covariance function is square summable:

$$\sum_{h=-\infty}^{\infty} |c(h)|^2 < \infty,$$

that is, $\{c(h)\}_{h\in\mathbb{Z}}$ is in ℓ^2. Though this condition is weaker than the absolute summability, the *Riesz–Fischer theorem* can be applied. This theorem says, see e.g. [50, 4.26], that then there exists a function $f \in L^2[-\pi, \pi]$ such that formula (1.10) holds and the Fourier series (1.11) converges in the $L^2[-\pi, \pi]$ sense.

Remark 1.2. The spectrum of a time series with discrete times $\mathbb{Z}$ has been defined on the interval $[-\pi, \pi]$, more precisely, on $(-\pi, \pi]$. Another usual approach is to define the spectrum on the unit circle $T = \{z \in \mathbb{C} : |z| = 1\}$. Clearly, there is a one-to-one correspondence between the two:

$$(-\pi, \pi] \ni \omega \leftrightarrow e^{-i\omega} \in T.$$

Consequently, there also exists a one-to-one correspondence between functions defined on the two. If $\phi : (-\pi, \pi] \to \mathbb{C}$, then there is a unique function $\Phi : T \to \mathbb{C}$ such that $\phi(\omega) = \Phi(e^{-i\omega})$ for any $\omega \in (-\pi, \pi]$.

Remark 1.3. In this book we need the space $L^2_d(\Omega, \mathcal{F}, \mathbb{P})$ of square integrable d-dimensional complex valued random vectors $\mathbf{X} = (X^1, \dots, X^d)$. We assume that $\mathbb{E}\mathbf{X} = \mathbf{0}$, and define the inner product

$$\langle \mathbf{X}, \mathbf{Y} \rangle := \sum_{j=1}^{d} \mathbb{E}(X^j \overline{Y^j}).$$

Beside the usual linear combinations $a\mathbf{X} + b\mathbf{Y}$ with constants $a, b \in \mathbb{C}$, we also define *linear combinations* $\boldsymbol{A}\mathbf{X} + \boldsymbol{B}\mathbf{Y}$ *with matrices* $\boldsymbol{A}, \boldsymbol{B} \in \mathbb{C}^{d \times d}$ in $L^2_d(\Omega, \mathcal{F}, \mathbb{P})$. A consequence of this is that one can define two types of *linear span* of a set of random vectors $\{\mathbf{X}_\gamma : \gamma \in \Gamma\} \subset L^2_d(\Omega, \mathcal{F}, \mathbb{P})$:

$$M = \operatorname{Span}\{\mathbf{X}_\gamma : \gamma \in \Gamma\} := \left\{ \sum_{j=1}^{n} \sum_{\ell=1}^{d} a_{j\ell} X^\ell_{\gamma_j} : a_{j\ell} \in \mathbb{C}, n \geq 1 \right\}; \tag{1.12}$$

$$M_d = \operatorname{Span}_d\{\mathbf{X}_\gamma : \gamma \in \Gamma\} := \left\{ \sum_{j=1}^{n} \boldsymbol{A}_j \mathbf{X}_{\gamma_j} : \boldsymbol{A}_j \in \mathbb{C}^{d \times d}, n \geq 1 \right\}.$$

Then $M \subset L^2(\Omega, \mathcal{F}, \mathbb{P})$ and $M_d \subset L^2_d(\Omega, \mathcal{F}, \mathbb{P})$. Clearly,

$$M_d = M \times \cdots \times M,$$

a Cartesian product with d factors. One can similarly define two types of *closed linear spans*,

$$\overline{\operatorname{span}}\{\mathbf{X}_\gamma : \gamma \in \Gamma\} \text{ and } \overline{\operatorname{span}}_d\{\mathbf{X}_\gamma : \gamma \in \Gamma\},$$

the closures of the respective linear spans in the respective Hilbert spaces.

The relationship between two second-order random vectors (of not necessarily the same dimension) is described by their *cross-covariance matrix* :

$$\operatorname{Cov}(\mathbf{X}, \mathbf{Y}) = \mathbb{E}(\mathbf{X}\mathbf{Y}^*) = [\mathbb{E}(X^j \overline{Y^k})]_{j,k=1}^{d,d'} \in \mathbb{C}^{d \times d'},$$

where $\mathbf{X}$ is a d-dimensional and $\mathbf{Y}$ is a d'-dimensional complex random vector with $L^2(\Omega, \mathcal{F}, \mathbb{P})$ components with zero expectations. Clearly, $\operatorname{Cov}(\mathbf{Y}, \mathbf{X}) = [\operatorname{Cov}(\mathbf{X}, \mathbf{Y})]^*$, and by Proposition B.1, $\operatorname{Cov}(\mathbf{X}, \mathbf{X})$ is a self-adjoint, non-negative definite (positive semidefinite) $d \times d$ matrix, the usual covariance matrix of $\mathbf{X}$.

We say that $\mathbf{X}$ and $\mathbf{Y}$ are *orthogonal*, denoted $\mathbf{X} \perp \mathbf{Y}$, if $\operatorname{Cov}(\mathbf{X}, \mathbf{Y}) = \boldsymbol{O}$, the zero matrix. This is a more general notion of orthogonality, and applicable to random vectors of different dimensions too. Observe that it is a stronger condition than the standard orthogonality $\langle \mathbf{X}, \mathbf{Y} \rangle = \mathrm{tr}(\mathrm{Cov}(\mathbf{X}, \mathbf{Y})) = 0$ if the two vectors are of the same dimension.

The next lemma describes a slight generalization of the Projection Theorem C.1; see Appendix C. It clarifies that as far as one discusses optimal mean square approximations, orthogonal projections, orthogonality of random vectors, and linear spans of a set of random vectors, one may work with scalar components of random vectors in the scalar Hilbert space $L^2(\Omega, \mathcal{F}, \mathbb{P})$ and its subspaces M defined by (1.12).

Lemma 1.1. *Let* $M_d = M \times \cdots \times M$ *be a closed subspace in* $L^2_d(\Omega, \mathcal{F}, \mathbb{P})$ *and* $\mathbf{Y} \in L^2_d(\Omega, \mathcal{F}, \mathbb{P})$. *Then there exists a unique* $\hat{\mathbf{Y}} \in M_d$ *such that*

$$\|\mathbf{Y} - \hat{\mathbf{Y}}\| \le \|\mathbf{Y} - \mathbf{Z}\| \text{ for any } \mathbf{Z} \in M_d, \tag{1.13}$$

equivalently,

$$\mathrm{Cov}(\mathbf{Y} - \hat{\mathbf{Y}}, \mathbf{Z}) = \boldsymbol{O} \text{ for any } \mathbf{Z} \in M_d, \quad (\mathbf{Y} - \hat{\mathbf{Y}}) \perp M_d. \tag{1.14}$$

For this $\hat{\mathbf{Y}}$, *we have* $\hat{Y}^j = \mathrm{Proj}_M Y^j$ $(j = 1, \dots, d)$, *the standard orthogonal projection of* Y^j *to the closed Hilbert subspace* $M \subset L^2(\Omega, \mathcal{F}, \mathbb{P})$.

Proof. Define $\hat{\mathbf{Y}} = (\hat{Y}^1, \dots, \hat{Y}^d)$ componentwise by the standard projection theorem:

$$\hat{Y}^j := \mathrm{Proj}_M Y^j, \qquad j = 1, \dots, d.$$

Then for arbitrary $\mathbf{Z} \in M_d$ let $\mathrm{Cov}(\mathbf{Y} - \hat{\mathbf{Y}}, \mathbf{Z}) =: [\gamma_{jk}]_{j,k=1}^d$, for which we have

$$\gamma_{jk} = \langle Y^j - \hat{Y}^j, Z^k \rangle = 0 \quad \forall j, k.$$

This proves (1.14).

Also,

$$\|\mathbf{Y} - \mathbf{Z}\|^2 = \sum_{j=1}^d \|Y^j - Z^j\|^2 = \sum_{j=1}^d \left\{ \|Y^j - \hat{Y}^j\|^2 + \|\hat{Y}^j - Z^j\|^2 \right\}, \tag{1.15}$$

which proves (1.13).

Conversely, if we choose any $\mathbf{Z} \in M_d$, $\mathbf{Z} \neq \hat{\mathbf{Y}}$, then (1.15) shows that it cannot have the minimum property (1.13). □

1.3 Spectral representation of multidimensional stationary time series

Let $\mathbf{X}_t = (X_t^1, \dots, X_t^d)$, $t \in \mathbb{Z}$, be a d-dimensional weakly stationary time series. For each $j = 1, \dots, d$ fixed, take the set of all finite linear combinations, with complex coefficients, of the random variables X_t^j ($t \in \mathbb{Z}$), and let $H(X^j)$ denote its closure in the Hilbert space $L^2(\Omega, \mathcal{F}, \mathbb{P})$. Also, let $H(\mathbf{X})$ denote the closure in $L^2(\Omega, \mathcal{F}, \mathbb{P})$ of the linear span of $H(X^1) \cup \cdots \cup H(X^d)$:

$$H(\mathbf{X}) = \overline{\text{span}}\{\mathbf{X}_t : t \in \mathbb{Z}\} := \overline{\text{span}}\{X_t^j : t \in \mathbb{Z},\, j = 1, \ldots, d\}.$$

The right time shift (or forward shift) S is a linear operator given by $SX_t^j = X_{t+1}^j$ for $j = 1, \ldots, d$. The operator S can be extended to $H(\mathbf{X})$ by linearity and continuity. The inverse S^{-1} of S is *the left time shift* L *(or backward shift)*, defined similarly. Thus we may write that $S^t\mathbf{X}_0 = \mathbf{X}_t$ for any $t \in \mathbb{Z}$. Consider the following two equations for any $k, t \in \mathbb{Z}$,

$$\begin{aligned}
&\text{Cov}(\mathbf{X}_{t+k}, \mathbf{X}_t) = \text{Cov}(S^{t+k}\mathbf{X}_0, S^t\mathbf{X}_0) = \text{Cov}(S^k\mathbf{X}_0, S^{*t}S^t\mathbf{X}_0)\\
&\text{Cov}(S^k\mathbf{X}_0, \mathbf{X}_0) = \text{Cov}(\mathbf{X}_k, \mathbf{X}_0).
\end{aligned}$$

They are equal to each other if and only if $S^*S = I$, that is, $S^{-1} = S^*$. Thus the right time shift S is a *unitary operator* in $H(\mathbf{X})$ if and only if the time series $\{\mathbf{X}_t\}_{t\in\mathbb{Z}}$ is weakly stationary.

From now on we assume that $\{\mathbf{X}_t\}_{t\in\mathbb{Z}}$ is a weakly stationary d-dimensional time series. Then, by continuous extension, S is a unitary operator in each $H(X^j)$ ($j = 1, \ldots, d$) and the whole $H(\mathbf{X})$ as well. Also, for any $A \subset \mathbb{Z}$ and for any $k \in \mathbb{Z}$,

$$S^k\left(\overline{\text{span}}\{\mathbf{X}_t : t \in A\}\right) = \overline{\text{span}}\{\mathbf{X}_{t+k} : t \in A\}.$$

The spectral representation of the covariance function can be extended to a *spectral representation of the time series* $\{\mathbf{X}_t\}_{t\in\mathbb{Z}}$ itself. To this end, we introduce a natural isometry. Note that a map ψ from a Hilbert space H onto a Hilbert space G is called an *isometry (isometric isomorphism) between* H *and* G, if we have

$$\langle \psi(X), \psi(Y) \rangle_G = \langle X, Y \rangle_H, \qquad \forall X, Y \in H.$$

First consider a 1D weakly stationary time series $\{X_t\}_{t\in\mathbb{Z}}$ with spectral measure dF. Define the linear map $\psi : H(X) \to L^2([-\pi, \pi], \mathcal{B}, dF)$ for a random variable X_t as

$$\psi(X_t) = \{e^{it\omega} : \omega \in (-\pi, \pi]\}.$$

Here $\mathcal{B}$ denotes the σ-field of Borel sets in $[-\pi, \pi]$. We emphasize that the image of the random variable X_t is the *function* $\omega \mapsto e^{it\omega}$ from $(-\pi, \pi]$ onto the unit circle T. Then ψ is indeed an isometry:

$$\begin{aligned}
\langle X_t, X_s \rangle_{H(X)} &= \mathbb{E}\left(X_t\overline{X_s}\right) = c(t-s) = \int_{-\pi}^{\pi} e^{i(t-s)\omega}dF(\omega)\\
&= \int_{-\pi}^{\pi} e^{it\omega}e^{-is\omega}dF(\omega) = \langle e^{it\omega}, e^{is\omega} \rangle_{dF}.
\end{aligned}$$

Next, this isometry ψ can be extended to finite linear combinations as

$$\psi\left(\sum_{k=1}^{m} a_k X_{t_k}\right) = \sum_{k=1}^{m} a_k e^{it_k\omega}, \quad \omega \in (-\pi, \pi],$$

and finally to the whole Hilbert space $H(X)$ by continuity. The image of $H(X)$ is the closure of the set of trigonometric polynomials.

Proposition 1.1. *The closure of the set of trigonometric polynomials in $L^2([-\pi,\pi],\mathcal{B},dF)$ is the whole space $L^2([-\pi,\pi],\mathcal{B},dF)$.*

Proof. First, the space of continuous functions on $[-\pi,\pi]$ is dense in $L^2([-\pi,\pi],\mathcal{B},dF)$, see e.g. [50, 3.14 Theorem]. Second, the set of trigonometric polynomials is dense in the space of continuous functions on $[-\pi,\pi]$, see e.g. [50, 4.25 Theorem]. □

Returning to the case of d-dimensional weakly stationary time series $\{\mathbf{X}_t\}$, for any index j and component $\{X_t^j\}$, $j = 1,\dots,d$, we can define an isometry $\psi^j : H(X^j) \to L^2([-\pi,\pi],\mathcal{B},dF^j)$ as described above. In this isometric isomorphism, the application of S^k in $H(X^j)$ corresponds to a multiplication with the function $e^{ik\omega}$ in $L^2([-\pi,\pi],\mathcal{B},dF^j)$. It means that for any square-integrable periodic function on $[-\pi,\pi]$, there exists a unique random variable in $H(X^j)$ by $(\psi^j)^{-1}$. In particular, for any $\omega \in (-\pi,\pi]$ and indicator $\mathbf{1}_{(-\pi,\omega]}$, there exists a unique complex valued random variable of 0 expectation:

$$Z_\omega^j := (\psi^j)^{-1}(\mathbf{1}_{(-\pi,\omega]}) \in H(X^j),$$

and we define $Z_{-\pi}^j := 0$. Moreover, for any $B \in \mathcal{B}$, there exists a unique random variable $Z_B^j = (\psi^j)^{-1}(\mathbf{1}_B) \in H(X^j)$.

Then the process $(Z_\omega^j)_{\omega\in(-\pi,\pi]}$ has orthogonal increments. Indeed, if $-\pi \le a < b \le c < d \le \pi$, then

$$\mathbb{E}\left((Z_b^j - Z_a^j)\overline{(Z_d^j - Z_c^j)}\right) = \langle Z_b^j - Z_a^j, Z_d^j - Z_c^j\rangle_{H(X^j)}$$

$$= \langle \mathbf{1}_{(a,b]}, \mathbf{1}_{(c,d]}\rangle_{dF^j} = \int_{-\pi}^{\pi} \mathbf{1}_{(a,b]}(\omega)\mathbf{1}_{(c,d]}(\omega)dF^j(\omega) = 0, \tag{1.16}$$

likewise,

$$\mathbf{E}(|Z_b^j - Z_a^j|^2) = \int_{-\pi}^{\pi} \mathbf{1}_{(a,b]}(\omega)dF^j(\omega) = F^j(b) - F^j(a). \tag{1.17}$$

In order to introduce *stochastic integration* of non-random functions w.r.t. $\{Z_\omega^j\}$, let us start with *step functions* (or *simple functions*). With an arbitrary positive integer N, let

$$\phi_{\text{step}}(\omega) := \sum_{r=1}^{N-1} a_r \mathbf{1}_{(\omega_r,\omega_{r+1}]}(\omega) \quad (a_r \in \mathbb{C})$$

in $L^2([-\pi,\pi],\mathcal{B},dF^j)$, where each ω_r is a continuity point of F^j and $-\pi = \omega_1 < \omega_2 < \cdots < \omega_N = \pi$. Define the stochastic integral of a step function by

$$\int_{-\pi}^{\pi} \phi_{\text{step}}(\omega) dZ_\omega := \sum_{r=1}^{N-1} a_r (Z_{\omega_{r+1}} - Z_{\omega_r}). \tag{1.18}$$

This stochastic integration also establishes an isometry between step functions and random variables of the form (1.18). Since any $\phi \in L^2([-\pi,\pi],\mathcal{B},dF^j)$ can be approximated by step functions, this isometry extends to a Hilbert space isometry. So we get the stochastic integral of a non-random L^2-function:

$$(\psi^j)^{-1}(\phi) = \int_{-\pi}^{\pi} \phi(\omega) dZ_\omega^j, \qquad \phi \in L^2([-\pi,\pi],\mathcal{B},dF^j).$$

Apply this for $\phi(\omega) = e^{it\omega}$:

$$X_t^j = \int_{-\pi}^{\pi} e^{it\omega} dZ_\omega^j \qquad (j = 1,\dots,d). \tag{1.19}$$

In this way the following important fact is proved.

Theorem 1.4. *With the notation* $\mathbf{X}_t := [X_t^1,\dots,X_t^d]^T$ $(t \in \mathbb{Z})$ *and* $\mathbf{Z}_\omega := [Z_\omega^1,\dots,Z_\omega^d]^T$, $\omega \in [-\pi,\pi]$, *one can write that*

$$\mathbf{X}_t = \int_{-\pi}^{\pi} e^{it\omega} d\mathbf{Z}_\omega \qquad (t \in \mathbb{Z}). \tag{1.20}$$

This is the spectral representation (sometimes called Cramér's representation) of the stationary time series $\{\mathbf{X}_t\}_{t\in\mathbb{Z}}$.

Now we are ready to define the complex measures corresponding to entries $c_{j\ell}(h)$ of the autocovariance matrix. Let us recall again that an isometric isomorphism between two Hilbert spaces is an isomorphism between the inner products as well. For any $B \in \mathcal{B}$, define the complex valued set function

$$dF^{j\ell}(B) := \mathbb{E}\left(Z_B^j \, \overline{Z_B^\ell}\right) \quad (j,\ell = 1,\dots,d).$$

In particular, for $\omega \in (-\pi,\pi]$, define spectral cumulative distribution functions (spectral c.d.f.):

$$F^{j\ell}(\omega) := dF^{j\ell}((-\pi,\omega]) = \mathbb{E}\left(Z_\omega^j \, \overline{Z_\omega^\ell}\right) \quad (j,\ell = 1,\dots,d). \tag{1.21}$$

By isometry, it is consistent with our previous definition of the function $F^j(\omega)$:

$$F^{jj}(\omega) = \mathbb{E}(|Z_\omega^j|^2) = \int_{-\pi}^{\pi} |\mathbf{1}_{(-\pi,\omega]}(t)|^2 \, dF^j(t) = \int_{-\pi}^{\omega} dF^j(t) = F^j(\omega). \tag{1.22}$$

The Cauchy–Schwarz inequality, (1.9), and (1.22) show that for any $B \in \mathcal{B}$ we have

$$|dF^{j\ell}(B)| = \mathbb{E}\left(Z_B^j \, \overline{Z_B^\ell}\right) \le \left\{\mathbb{E}(|Z_B^j|^2)\mathbb{E}(|Z_B^\ell|^2)\right\}^{\frac{1}{2}} \\ \le \left\{F^j(\pi)F^\ell(\pi)\right\}^{\frac{1}{2}} = \left\{c_{jj}(0)c_{\ell\ell}(0)\right\}^{\frac{1}{2}}. \tag{1.23}$$

This inequality and the next lemma imply that $dF^{j\ell}$ has bounded variation.

Lemma 1.2. *Let μ be a complex valued set function on a σ-field $\mathcal{F}$ and suppose that $|\mu(F)| \le c < \infty$ for any $F \in \mathcal{F}$. Then μ has bounded variation.*

Proof. Let $\mu_{re} := \mathrm{Re}(\mu)$ and $\mu_{im} := \mathrm{Im}(\mu)$. By our assumption, $|\mu_{re}(F)| \le c$ and $|\mu_{im}(F)| \le c$ for any $F \in \mathcal{F}$. Indirectly, suppose that at least one of the signed measures μ_{re} and μ_{im} has infinite variation. Then there exists a sequence $(F_j)_{j=1}^\infty$ of pairwise disjoint sets in $\mathcal{F}$ such that $\sum_{j=1}^\infty |\mu_{re}(F_j)| > 2c$, say.

Define $J_+ := \{j : \mu_{re}(F_j) > 0\}$ and $J_- := \{j : \mu_{re}(F_j) < 0\}$. Then

$$|\mu_{re}(\bigcup_{j \in J_+} F_j)| > c \quad \text{or} \quad |\mu_{re}(\bigcup_{j \in J_-} F_j)| > c \quad \text{or both,}$$

which is a contradiction. □

Thus $F^{j\ell}(\omega)$ can be decomposed into four bounded, non-decreasing functions:

$$F^{j\ell}(\omega) = F^{j\ell}_{re+}(\omega) - F^{j\ell}_{re-}(\omega) + i(F^{j\ell}_{im+}(\omega) - F^{j\ell}_{im-}(\omega)) \quad (-\pi < \omega \le \pi).$$

It means that $F^{j\ell}(\omega)$ defines a complex measure on $([-\pi, \pi], \mathcal{B})$. Also, we have the equality of bilinear functions:

$$\mathbb{E}(Z_B^j \, \overline{Z_{B'}^\ell}) = \int_{-\pi}^{\pi} \mathbf{1}_B(\omega)\mathbf{1}_{B'}(\omega) dF^{j\ell}(\omega) \quad (j, \ell = 1, \dots, d).$$

In particular, if $B \cap B' = \emptyset$, then

$$\mathbb{E}(Z_B^j \, \overline{Z_{B'}^\ell}) = 0, \quad dF^{j\ell}(B \cup B') = dF^{j\ell}(B) + dF^{j\ell}(B'),$$

for any $j, \ell \in \{1, \dots, d\}$. (Of course, the second equality is consistent with the fact that $dF^{j\ell}$ is a complex measure.)

Similarly to the Herglotz theorem (Theorem 1.3), one can exhibit and prove spectral representation of the non-diagonal entries of the covariance matrix:

$$c_{j\ell}(h) = \int_{-\pi}^{\pi} e^{ih\omega} dF^{j\ell}(\omega), \quad F^{j\ell}(-\pi) = 0, \quad F^{j\ell}(\pi) = c_{j\ell}(0). \tag{1.24}$$

The only difference is that here the spectral measure is complex valued in general, even if the time series has real valued components.

Let us introduce the *spectral measure matrix* $\boldsymbol{dF} := [dF^{j\ell}]_{d\times d}$. First, it is self-adjoint:

$$c_{j\ell}(h) = \int_{-\pi}^{\pi} e^{ih\omega} dF^{j\ell}(\omega) = \overline{c_{\ell j}(-h)} = \int_{-\pi}^{\pi} e^{ih\omega} \overline{dF^{\ell j}(\omega)} \quad \Rightarrow \quad dF^{j\ell} = \overline{dF^{\ell j}}, \tag{1.25}$$

since by Weierstrass' theorem the trigonometric polynomials are dense in $C[-\pi, \pi]$, the space of continuous functions over $[-\pi, \pi]$, and thus determine the measure by the Riesz representation theorem. Also, $\boldsymbol{dF}$ is non-negative definite:

$$\sum_{r,s=1}^{m} \mathbf{a}_r^* \, \boldsymbol{dF}(B_r \cap B_s) \, \mathbf{a}_s = \mathbb{E} \left| \sum_{r=1}^{m} \mathbf{a}_r^* \, \mathbf{Z}_{B_r} \right|^2 \geq 0 \tag{1.26}$$

for any $m \geq 1$ and $\mathbf{a}_1, \dots, \mathbf{a}_m \in \mathbb{C}^n$; $B_1, \dots, B_m \in \mathcal{B}$. In particular, definition (1.5) of non-negative definiteness of a spectral measure matrix holds too.

Corollary 1.1. *In the special case when each $dF^{j\ell}$ are absolutely continuous w.r.t. Lebesgue measure in $[-\pi, \pi]$, that is, $dF^{j\ell}(\omega) = f^{j\ell}(\omega)\, d\omega$ for $j, \ell = 1, \dots, d$, it follows that the time series has a* spectral density matrix $\boldsymbol{f} := [f^{j\ell}]_{d\times d}$, *which is self-adjoint and non-negative definite.*

Proof. The self-adjointness of $\boldsymbol{f}$ follows from (1.25), while the non-negative definiteness of $\boldsymbol{f}$ follows from (1.26). □

Corollary 1.2. *We can write* (1.24) *in matrix form:*

$$\boldsymbol{C}(h) = \int_{-\pi}^{\pi} e^{ih\omega} \boldsymbol{dF}(\omega), \qquad h \in \mathbb{Z},$$

or in the case of an absolutely continuous spectral measure:

$$\boldsymbol{C}(h) = \int_{-\pi}^{\pi} e^{ih\omega} \boldsymbol{f}(\omega) d\omega, \qquad h \in \mathbb{Z}. \tag{1.27}$$

This proves Theorem 1.2(a).

Corollary 1.3. *By formula* (1.21), *we have the following relationship between the orthogonal increment process $\mathbf{Z}_\omega$ and the spectral c.d.f. $\mathbf{F}(\omega)$:*

$$\mathbf{F}(\omega) = \mathbb{E}(\mathbf{Z}_\omega \mathbf{Z}_\omega^*) \quad (\omega \in [-\pi, \pi]).$$

Corollary 1.4. *Let $\psi_q^j \in L^2([-\pi, \pi], \mathcal{B}, dF^j)$ for $j = 1, \dots, d$ and $q = 1, 2$. Define the random variables*

$$Y_q^j := \int_{-\pi}^{\pi} \psi_q^j(\omega) dZ_\omega^j \in H(\mathbf{X}) \subset L^2(\Omega, \mathcal{F}, \mathbb{P}),$$

and the random vectors $\mathbf{Y}_q := (Y_q^1, \dots, Y_q^d)$ *for* $q = 1, 2$. *Then by isometry, we may write the cross-covariance matrix of* $\mathbf{Y}_1$ *and* $\mathbf{Y}_2$ *as*

$$\begin{aligned}\mathbb{E}(\mathbf{Y}_1\mathbf{Y}_2^*) &= \left[\langle Y_1^j, Y_2^\ell\rangle_{H(\mathbf{X})}\right]_{d\times d} = \left[\mathbb{E}\left\{\int_{-\pi}^{\pi}\psi_1^j(\omega)dZ_\omega^j \int_{-\pi}^{\pi}\overline{\psi_2^\ell(\omega)dZ_\omega^\ell}\right\}\right]_{d\times d} \\ &= \left[\langle\psi_1^j, \psi_2^\ell\rangle_{dF^{j\ell}}\right]_{d\times d} = \left[\int_{-\pi}^{\pi}\psi_1^j(\omega)\overline{\psi_2^\ell(\omega)}dF^{j\ell}(\omega)\right]_{d\times d}. \qquad (1.28)\end{aligned}$$

If $d\boldsymbol{F}$ *is absolutely continuous with respect to Lebesgue measure with spectral density* $\boldsymbol{f}$, *then*

$$\mathbb{E}(\mathbf{Y}_1\mathbf{Y}_2^*) = \left[\int_{-\pi}^{\pi}\psi_1^j(\omega)f^{j\ell}(\omega)\overline{\psi_2^\ell(\omega)}d\omega\right]_{d\times d}.$$

Definition 1.4. An important special case of weakly stationary time series is a so-called **white noise process**. Given a self-adjoint non-negative definite matrix $\mathbf{\Sigma} = [\sigma_{jk}] \in \mathbb{C}^{d\times d}$, the stationary sequence $\{\boldsymbol{\xi}_t\}_{t\in\mathbb{Z}}$ is called a white noise process with covariance matrix $\mathbf{\Sigma}$, denoted $\mathrm{WN}(\mathbf{\Sigma})$, if $\mathbb{E}(\boldsymbol{\xi}_t) = 0$ for all $t \in \mathbb{Z}$ and its autocovariance function is given by

$$\boldsymbol{C}_{\boldsymbol{\xi}}(h) = \mathbb{E}(\boldsymbol{\xi}_{t+h}\boldsymbol{\xi}_t^*) = \delta_{h0}\mathbf{\Sigma} \qquad (h \in \mathbb{Z}), \qquad (1.29)$$

where $\delta_{jk} = 1$ if $j = k$, and $\delta_{jk} = 0$ if $j \neq k$ (*Kronecker delta*). It means that the values of $\{\boldsymbol{\xi}_t\}$ are orthogonal (uncorrelated) longitudinally for different time instants, while its coordinates may be correlated cross-sectionally, that is, at the same time instant.

The special case $\mathrm{WN}(\boldsymbol{I}_d)$ is an **orthonormal sequence**, whose coordinates are uncorrelated cross-sectionally as well.

By Remark 1.4 and (1.29), $\{\boldsymbol{\xi}_t\} \sim \mathrm{WN}(\mathbf{\Sigma})$ has spectral density $\boldsymbol{f}_{\boldsymbol{\xi}}(\omega) = [f_{\boldsymbol{\xi}}^{jk}(\omega)]_{d\times d}$:

$$f_{\boldsymbol{\xi}}^{jk}(\omega) = \frac{1}{2\pi}\sigma_{jk}, \quad \boldsymbol{f}_{\boldsymbol{\xi}}(\omega) = \frac{1}{2\pi}\mathbf{\Sigma}. \qquad (1.30)$$

This explains the name ‘white noise’: the spectral density is the same constant at every frequency ω, like the spectrum of ideal white light.

Remark 1.4. Remark 1.1 can be extended to the present case as well.

(a) If for each $j, \ell = 1, \dots, d$ we have

$$\sum_{h=-\infty}^{\infty} |c_{j\ell}(h)| < \infty, \qquad (1.31)$$

that is, $\{\boldsymbol{C}(h)\}_{h\in\mathbb{Z}} \in \ell^1$, then the time series $\{\mathbf{X}_t\}$ has a *continuous* spectral density matrix $\boldsymbol{f}(\omega) = [f^{j\ell}(\omega)]_{d\times d}$ such that (1.27) holds and

$$\boldsymbol{f}(\omega) = \frac{1}{2\pi}\sum_{h=-\infty}^{\infty}\boldsymbol{C}(h)\,e^{-ih\omega}, \qquad \omega \in [-\pi, \pi], \qquad (1.32)$$

where the series converges pointwise.

(b) By the Riesz–Fischer theorem, condition (1.31) can be weakened as

$$\sum_{h=-\infty}^{\infty} |c_{j\ell}(h)|^2 < \infty, \quad \forall j, \ell = 1, \dots, d,$$

that is, $\{\boldsymbol{C}(h)\}_{h\in\mathbb{Z}} \in \ell^2$. Then the series (1.32) converges in $L^2([-\pi,\pi],\mathcal{B},d\omega)$ entrywise, where $d\omega$ is Lebesgue measure, so the spectral density $\boldsymbol{f}(\omega)$ exists and (1.27) holds.

Remark 1.5. Let $\{\mathbf{X}_t\}$ be a d-dimensional stationary time series with *real components* and with absolute continuous spectral measure with density matrix $\boldsymbol{f}$. By (1.1) we have $\boldsymbol{C}(-h) = \boldsymbol{C}^*(h)$ for the covariance matrix of the process. However, now $\boldsymbol{C}(h)$ is a matrix with real entries, hence $\boldsymbol{C}(-h) = \boldsymbol{C}^T(h)$ and also $\overline{\boldsymbol{C}(h)} = \boldsymbol{C}(h)$ (entrywise complex conjugation) for any $h \in \mathbb{Z}$. Therefore from the equation (1.27), we obtain that

$$\overline{\boldsymbol{C}(-h)} = \int_{-\pi}^{\pi} e^{ih\omega}\overline{\boldsymbol{f}(\omega)}d\omega = \boldsymbol{C}(-h) = \int_{-\pi}^{\pi} e^{ih\omega}\boldsymbol{f}(-\omega)d\omega.$$

Since the Fourier transform is a.e. uniquely determines $\boldsymbol{f}$, it follows that for real valued processes we have

$$\boldsymbol{f}(-\omega) = \overline{\boldsymbol{f}(\omega)}, \quad \omega \in [-\pi, \pi]. \tag{1.33}$$

(We do not distinguish two density functions which are equal almost everywhere.) It implies that in this case it is enough to consider only the half frequency interval $[0, \pi]$.

Remark 1.6. Here we discuss some properties of time series which have autocovariance matrix function with real entries, which are absolute summable.

(a) Let $\{\mathbf{X}_t\}$ be a d-dimensional weakly stationary time series of complex components. Denoting by $\boldsymbol{C}(h) = [c_{pq}(h)]$ the $d \times d$ autocovariance matrix function, $\boldsymbol{C}(-h) = \boldsymbol{C}^*(h)$, $h \in \mathbb{Z}$, in the time domain, assume that their entries are absolutely summable, i.e. $\sum_{h=0}^{\infty} |c_{pq}(h)| < \infty$ for $p, q = 1, \dots, d$. By Remark 1.4(a), then the spectral density matrix $\boldsymbol{f}(\omega)$ exists in the frequency domain, and it can be computed as (1.32). In view of Corollary 1.1 it is always self-adjoint, non-negative definite (positive semidefinite). Further, if $\boldsymbol{C}(h)$ is a matrix of real entries $\forall h \in \mathbb{Z}$, then the relation

$$\boldsymbol{f}(-\omega) = \boldsymbol{f}(2\pi - \omega) = \overline{\boldsymbol{f}(\omega)}$$

(with entrywise conjugation) holds $\forall \omega \in [0, 2\pi]$. The last statement follows from (1.33) and also from (1.32) by substituting $-\omega$ or $2\pi - \omega$ for ω.

(b) We have the following equivalent forms for $2\pi\boldsymbol{f}(\omega)$:

$$2\pi\boldsymbol{f}(\omega) = \sum_{h=-\infty}^{\infty} \boldsymbol{C}(h)e^{-ih\omega} = \boldsymbol{C}(0) + \sum_{h=1}^{\infty}[\boldsymbol{C}(h)e^{-ih\omega} + \boldsymbol{C}^*(h)e^{ih\omega}]$$

$$= \boldsymbol{C}(0) + \sum_{h=1}^{\infty}[(\boldsymbol{C}(h) + \boldsymbol{C}^*(h))\cos(h\omega) + i(\boldsymbol{C}^*(h) - \boldsymbol{C}(h))\sin(h\omega)].$$

The first line shows again that $\boldsymbol{f}(\omega)$ is self-adjoint. The second line shows that whenever $\boldsymbol{C}(h)$ is a real matrix and so, $\boldsymbol{C}^*(h) = \boldsymbol{C}^T(h)$, then $\boldsymbol{C}(h) + \boldsymbol{C}^T(h)$ is symmetric and $\boldsymbol{C}^T(h) - \boldsymbol{C}(h)$ is anti-symmetric with 0 diagonal, but $i(\boldsymbol{C}^T(h) - \boldsymbol{C}(h))$ is self-adjoint. Actually, $\sum_{h=1}^{\infty}(\boldsymbol{C}(h) + \boldsymbol{C}^T(h))\cos(h\omega)$ is the real and $\sum_{h=1}^{\infty}(\boldsymbol{C}^T(h) - \boldsymbol{C}(h))\sin(h\omega)$ is the imaginary part of $2\pi\boldsymbol{f}(\omega)$, the (entrywise) conjugate of which is $2\pi\boldsymbol{f}(-\omega) = 2\pi\boldsymbol{f}(2\pi - \omega)$.

Remark 1.7. If $\{X_t\}$ is a 1D weakly stationary, real time series, then $f(\omega) \geq 0$ is real, and $f(-\omega) = f(\omega)$, $\forall\omega \in [0, 2\pi]$. If $\{\mathbf{X}_t\}$ is a d-dimensional weakly stationary time series of real components, then $\boldsymbol{C}(h)$ is a matrix of real entries ($\forall h \in \mathbb{Z}$), so $\boldsymbol{f}(-\omega) = \overline{\boldsymbol{f}(\omega)}$ holds, $\forall\omega \in [0, 2\pi]$. But it is not necessary for a time series to have real components so that $\boldsymbol{C}(h)$ be a real matrix. For example, let $\{\mathbf{Y}_t\}$ be a d-dimensional weakly stationary time series of real components with expectation $\mathbf{0}$. Let $\boldsymbol{\mu} \in \mathbb{C}^d$ be a vector of at least one coordinate containing a nonzero imaginary part. Then the time series $\mathbf{X}_t = \mathbf{Y}_t + \boldsymbol{\mu}$ has at least one complex (not purely real) coordinate, still its autocovariance matrix sequence is the same as that of $\mathbf{Y}_t$, so $\boldsymbol{C}(h)$s have real entries.

When $\boldsymbol{C}(h)$ has complex entries too, then there is no such simple relationship between $\boldsymbol{f}(-\omega)$ and $\boldsymbol{f}(\omega)$ in general. Indeed, $\boldsymbol{A} := \sum_{h=1}^{\infty}(\boldsymbol{C}(h) + \boldsymbol{C}^*(h))\cos(h\omega)$ and $\boldsymbol{B} := \sum_{h=1}^{\infty}(\boldsymbol{C}^*(h) - \boldsymbol{C}(h))\sin(h\omega)$ are complex matrices, say $\boldsymbol{A} = \boldsymbol{A}_1 + i\boldsymbol{A}_2$, $\boldsymbol{B} = \boldsymbol{B}_1 + i\boldsymbol{B}_2$. Then

$$\boldsymbol{f}(\omega) = \boldsymbol{C}_0 + \boldsymbol{A} + i\boldsymbol{B} = \boldsymbol{C}_0 + (\boldsymbol{A}_1 - \boldsymbol{B}_2) + i(\boldsymbol{A}_2 + \boldsymbol{B}_1),$$

whereas

$$\boldsymbol{f}(-\omega) = \boldsymbol{C}_0 + \boldsymbol{A} - i\boldsymbol{B} = \boldsymbol{C}_0 + (\boldsymbol{A}_1 + \boldsymbol{B}_2) + i(\boldsymbol{A}_2 - \boldsymbol{B}_1),$$

and the latter is neither the same, nor the conjugate of the former.

Summarizing, there is no extension from the real to the complex case. Consequently, in the real case we can confine ourselves to $[0, \pi]$, while in the complex case the whole $[0, 2\pi]$ or $[-\pi, \pi]$ should be used.

Remark 1.8. It should be noted that an alternative approach for the spectral representation of the process $\{\mathbf{X}_t\}_{t\in\mathbb{Z}}$ could start from the spectral representation of the unitary operator S in the Hilbert subspace $H(\mathbf{X})$, and then there follows the spectral representation of the covariance matrix. See this approach e.g. in [34] and [47], and see the underlying spectral theorem of normal operators e.g. in [49]. Here we just briefly summarize the main points.

If T is a bounded operator in a Hilbert space H and T is normal: $TT^* = T^*T$, then there exists an orthogonal projection measure E on the Borel subsets of the spectrum $\sigma(T)$ of T which satisfies

$$T = \int_{\sigma(T)} \lambda \, dE(\lambda).$$

The spectrum $\sigma(T)$ is the subset of $\mathbb{C}$ such that $\lambda \in \sigma(T)$ if and only if $T - \lambda I$ is not invertible:

- $T - \lambda I$ is not one-to-one (that is, λ is an eigenvalue of T), or
- $T - \lambda I$ is not onto H.

If the operator S we consider is a unitary operator: $SS^* = I = S^*S$, then its spectrum $\sigma(S)$ is a subset of the unit circle:

$$S = \int_{-\pi}^{\pi} e^{i\omega} dE(\omega).$$

Since $\mathbf{X}_t = S^t \mathbf{X}_0$, we get that

$$\mathbf{X}_t = \int_{-\pi}^{\pi} e^{it\omega} d\mathbf{\Phi}(\omega),$$

where $d\mathbf{\Phi}(\omega) := dE(\omega)\mathbf{X}_0$. This formula corresponds to (1.20).

1.4 Constructions of stationary time series

There are some standard constructions of a stationary time series with a given covariance function or with a given spectral measure.

1.4.1 Construction 1

Suppose that $\boldsymbol{C} : \mathbb{Z} \to \mathbb{C}^{d\times d}$ is a non-negative definite function:

$$\sum_{j,k=1}^{n} \mathbf{a}_k^* \boldsymbol{C}(k-j)\, \mathbf{a}_j \geq 0, \qquad \forall n \geq 1, \quad \forall \mathbf{a}_1, \dots, \mathbf{a}_n \in \mathbb{C}^d.$$

Equivalently, the block Toeplitz matrix $\mathfrak{C}_n$ defined by (1.3) is self-adjoint and non-negative definite for any $n \geq 1$.

We are going to construct a d-dimensional time series $\{\mathbf{X}_t\}_{t\in\mathbb{Z}}$ with the given covariance function

$$\mathbb{E}(\mathbf{X}_t \mathbf{X}_s^*) = \boldsymbol{C}(t-s), \qquad t, s \in \mathbb{Z}. \tag{1.34}$$

The construction goes by induction, defining the value of the time series at $t = 0$, then at $t = 1$, then at $t = -1$, then at $t = 2$, then at $t = -2$, and so on. However, the very first thing is to choose an orthonormal sequence of random variables $\{\xi_j\}_{j=0}^{\infty}$ with expectation 0 on a probability space $(\Omega, \mathcal{F}, \mathbb{P})$. For example, one can choose a sequence of independent tossing of a fair coin

$\mathbb{P}(\xi_j = \pm 1) = \frac{1}{2}$ $(j = 0, 1, 2, \dots)$; or, one can choose a sequence $\{\xi_j\}_{j=0}^{\infty}$ of independent standard normal variables with the probability density function

$$\phi(x) = \frac{1}{\sqrt{2\pi}} e^{-\frac{x^2}{2}} \qquad (x \in \mathbb{R}).$$

We define the Hilbert space H that is going to be a basis of the construction as the closed linear span $H := \overline{\text{span}}\{\xi_j : j = 0, 1, 2, \dots\} \subset L^2(\Omega, \mathcal{F}, \mathbb{P})$.

At the beginning we set

$$\boldsymbol{C}(0) = \boldsymbol{A}_0 \boldsymbol{A}_0^*, \quad r := \operatorname{rank} \boldsymbol{C}(0), \quad \mathbf{X}_0 := \boldsymbol{A}_0 \boldsymbol{\xi}_r,$$

where $\boldsymbol{\xi}_r := [\xi_0, \dots, \xi_{r-1}]^T$, using the parsimonious Gram-decomposition (B.5) of the Appendix.

Then each step of the induction will consist of two sub-steps. Assuming for example that a sequence $(\mathbf{X}_{-n}, \mathbf{X}_{-n+1}, \dots, \mathbf{X}_0, \dots, \mathbf{X}_{n-1}, \mathbf{X}_n)$ has already been defined for some $n \geq 0$, *the first sub-step* will take a preliminary sequence

$$\mathfrak{X}_{-n,n+1} := \left[\tilde{\mathbf{X}}_{-n}, \tilde{\mathbf{X}}_{-n+1}, \dots, \tilde{\mathbf{X}}_0, \dots, \tilde{\mathbf{X}}_n, \tilde{\mathbf{X}}_{n+1}\right]^T. \tag{1.35}$$

After that a second sub-step will result the new vector $\mathbf{X}_{n+1}$ in the constructed time series. Then the construction of a new vector $\mathbf{X}_{-n-1}$ goes similarly, so not detailed.

For $n \geq 0$ fixed, we would like to define the sequence $\mathfrak{X}_{-n,n+1}$ so that its covariance function be $\boldsymbol{C}(h)$, $h \in \mathbb{Z}$:

$$\begin{aligned}
&\mathfrak{C}_{-n,n+1} \\
&:= \mathbb{E}(\mathfrak{X}_{-n,n+1}\mathfrak{X}^*_{-n,n+1}) = \begin{bmatrix} \boldsymbol{C}(0) & \boldsymbol{C}(-1) & \cdots & \boldsymbol{C}(-2n-1) \\ \boldsymbol{C}(1) & \boldsymbol{C}(0) & \cdots & \boldsymbol{C}(-2n) \\ \vdots & \vdots & \ddots & \vdots \\ \boldsymbol{C}(2n+1) & \boldsymbol{C}(2n) & \cdots & \boldsymbol{C}(0) \end{bmatrix}.
\end{aligned}$$

Clearly, $\mathfrak{C}_{-n,n+1}$ is a self-adjoint, non-negative definite block Toeplitz matrix of rank $r \leq 2n + 2$. (The value of r can be different at each step of the construction!) Thus by equation (B.5) in the Appendix, it has a parsimonious Gram-decomposition:

$$\mathfrak{C}_{-n,n+1} = \boldsymbol{A}_{-n,n+1} \cdot \boldsymbol{A}^*_{-n,n+1}, \qquad \boldsymbol{A}_{-n,n+1} \in \mathbb{C}^{(2n+2)\times r}.$$

We set

$$\mathfrak{X}_{-n,n+1} := \boldsymbol{A}_{-n,n+1}\boldsymbol{\xi}_r, \qquad \boldsymbol{\xi}_r := [\xi_0, \dots, \xi_{r-1}]^T. \tag{1.36}$$

Then, really,

$$\mathbb{E}(\mathfrak{X}_{-n,n+1}\mathfrak{X}^*_{-n,n+1}) = \boldsymbol{A}_{-n,n+1}\mathbb{E}(\boldsymbol{\xi}_r\boldsymbol{\xi}_r^*)\boldsymbol{A}^*_{-n,n+1} = \mathfrak{C}_{-n,n+1}.$$

Now comes *the second sub-step* of the induction. Assume that the sequence $(\mathbf{X}_{-n}, \mathbf{X}_{-n+1}, \dots, \mathbf{X}_0, \dots, \mathbf{X}_{n-1}, \mathbf{X}_n)$ has been already defined for some $n \geq 0$

and has the covariance function (1.34) for $s, t \in \{-n, \dots, n\}$. We have also defined the preliminary sequence $\mathfrak{X}_{-n,n+1}$ by (1.35) and (1.36). Then define the operator T_{2n+1} by

$$T_{2n+1}\tilde{\mathbf{X}}_t = \mathbf{X}_t \qquad (t = -n, \dots, n).$$

By the construction it follows that T_{2n+1} has the following important property:

$$\mathbb{E}\left((T_{2n+1}\tilde{\mathbf{X}}_t)(T_{2n+1}\tilde{\mathbf{X}}_s)^*\right) = \mathbb{E}(\tilde{\mathbf{X}}_t\tilde{\mathbf{X}}_s^*) = \boldsymbol{C}(t-s), \quad s, t \in \{-n, \dots, n\}. \tag{1.37}$$

By linearity, T_{2n+1} can be extended to an isomorphy between the finite dimensional spaces

$$\tilde{H}_{2n+1} := \mathrm{Span}_d\{\tilde{\mathbf{X}}_t : t = -n, \dots, n\},$$
$$H_{2n+1} := \mathrm{Span}_d\{\mathbf{X}_t : t = -n, \dots, n\},$$

so that we still have

$$\mathbb{E}\left((T_{2n+1}\tilde{\mathbf{X}})(T_{2n+1}\tilde{\mathbf{Y}})^*\right) = \mathbb{E}(\tilde{\mathbf{X}}\tilde{\mathbf{Y}}^*), \quad \tilde{\mathbf{X}}, \tilde{\mathbf{Y}} \in \tilde{H}_{2n+1}.$$

By Lemma 1.1, we may write that

$$\tilde{\mathbf{X}}_{n+1} = \tilde{\mathbf{X}}_{n+1}^- + \tilde{\mathbf{X}}_{n+1}^+, \quad \tilde{\mathbf{X}}_{n+1}^- \in \tilde{H}_{2n+1}, \quad \tilde{\mathbf{X}}_{n+1}^+ \perp \tilde{H}_{2n+1}. \tag{1.38}$$

If $\tilde{\mathbf{X}}_{n+1}^+ = \mathbf{0}$, then we are ready: $\mathbf{X}_{n+1} := T_{2n+1}\tilde{\mathbf{X}}_{n+1}$ belongs to the already defined subspace H_{2n+1}. Otherwise, set $\mathbf{U}_{n+1} := \mathbf{X}_{n+1}^+/\|\mathbf{X}_{n+1}^+\|$ and define $\tilde{H}_{2n+1}^+ := \mathrm{Span}_d\{\tilde{H}_{2n+1}, \mathbf{U}_{n+1}\}$.

Let ξ_k be the first random variable in the sequence $\{\xi_j\}_{j=0}^\infty$ that has not been used so far in the construction of the sequence $(\mathbf{X}_{-n}, \dots, \mathbf{X}_n)$. Define $T_{2n+1}\mathbf{U}_{n+1} = \mathbf{V}_{n+1} := [\xi_k, \xi_{k+1}, \dots, \xi_{k+d-1}]^T/\sqrt{d}$ and $H_{2n+1}^+ := \mathrm{Span}_d\{H_{2n+1}, \mathbf{V}_{n+1}\}$. Extend T_{2n+1} between $\tilde{H}_{2n+1}^+$ and H_{2n+1}^+ by linearity, and define $\mathbf{X}_{n+1} := T_{2n+1}\tilde{\mathbf{X}}_{n+1}$. Then by (1.37) and (1.38),

$$\mathbb{E}(\mathbf{X}_{n+1}\mathbf{X}_t^*) = \mathbb{E}\left((T_{2n+1}\tilde{\mathbf{X}}_{n+1})(T_{2n+1}\tilde{\mathbf{X}}_t)^*\right) = \mathbb{E}(\tilde{\mathbf{X}}_{n+1}\tilde{\mathbf{X}}_t^*) = \boldsymbol{C}(n+1-t)$$

for any $t = -n, \dots, n$. This completes the induction.

Corollary 1.5. *Formula* (1.6) *shows that the covariance function of any stationary time series is non-negative definite. Conversely, Construction 1 proves that for any non-negative definite function $\boldsymbol{C}(h)$, $h \in \mathbb{Z}$, one can construct a stationary time series with this covariance function.*

1.4.2 Construction 2

Assume that we are given a $d \times d$ matrix $d\boldsymbol{F} = [dF^{rs}]_{r,s=1}^d$ whose entries are finite complex measures on $([-\pi, \pi], \mathcal{B})$ and which is non-negative definite:

$$\sum_{r,s=1}^d dF^{rs}((\alpha, \beta])z_r\bar{z}_s = \sum_{r,s=1}^d \int_{-\pi}^{\pi} z_r\bar{z}_s\, \mathbf{1}_{(\alpha,\beta]}(\omega)\, dF^{rs}(\omega) \geq 0, \tag{1.39}$$

for any interval $(\alpha,\beta]\subset[-\pi,\pi]$ and $z_1,\dots,z_d\in\mathbb{C}$. Equivalently, the matrix

$$\Delta_{\alpha\beta}\boldsymbol{F}:=[\Delta_{\alpha\beta}F^{rs}]_{r,s=1}^{d}:=[F^{rs}(\beta)-F^{rs}(\alpha)]_{r,s=1}^{d}$$

is non-negative definite for any $(\alpha,\beta]\subset[-\pi,\pi]$, that is,

$$\sum_{r,s=1}^{d}\Delta_{\alpha\beta}F^{rs}z_r\bar{z}_s\geq 0. \tag{1.40}$$

We would like to construct a time series whose spectral measure matrix is the given $d\boldsymbol{F}$. One way to do it is to show that $d\boldsymbol{F}$ defines a non-negative definite function $\boldsymbol{C}(h)$, $h\in\mathbb{Z}$, and then using Construction 1 to complete the construction.

It should be noted that inequality (1.39) can be extended to sums of the

$$\sum_{r,s=1}^{d}\int_{-\pi}^{\pi}\sum_{j=1}^{n}z_r(j)\bar{z}_s(j)\,\mathbf{1}_{(\alpha_j,\beta_j]}(\omega)\,dF^{rs}(\omega)\geq 0,\quad n\geq 1, \tag{1.41}$$

where $(\alpha_j,\beta_j]\subset[-\pi,\pi]$ and $z_1(j),\dots,z_d(j)\in\mathbb{C}$ for any $n\geq 1$ and $j=1,\dots,n$. It is clear that the class of step-functions

$$g(\omega):=\sum_{j=1}^{n}z(j)\,\mathbf{1}_{(\alpha_j,\beta_j]}(\omega),\quad n\geq 1,\quad \omega\in[-\pi,\pi],$$

is dense in $L^2([-\pi,\pi],\mathcal{B},\,\mathrm{tr}(d\boldsymbol{F}(\omega)))$, where $\mathrm{tr}(d\boldsymbol{F}(\omega))$ denotes the trace of $d\boldsymbol{F}$, which dominates any measure entry in $d\boldsymbol{F}$. Thus one can extend inequality (1.41) to the case

$$\sum_{r,s=1}^{d}\int_{-\pi}^{\pi}g_r(\omega)\bar{g}_s(\omega)\,dF^{rs}(\omega)\geq 0, \tag{1.42}$$

where $g_1,\dots,g_d\in L^2([-\pi,\pi],\mathcal{B},\,\mathrm{tr}(d\boldsymbol{F}(\omega)))$.

Define

$$\boldsymbol{C}(h):=\int_{-\pi}^{\pi}e^{ih\omega}d\boldsymbol{F}(\omega)\in\mathbb{C}^{d\times d},\qquad h\in\mathbb{Z}.$$

Take an arbitrary integer $n\geq 1$ and arbitrary vectors $\mathbf{a}_k=(a_k^1,\dots,a_k^d)\in\mathbb{C}^d$ for $k=1,\dots,n$. Define the trigonometric polynomials

$$\zeta_r(\omega):=\sum_{k=1}^{n}a_k^r\,e^{-ik\omega},\qquad r=1,\dots,d.$$

Then by inequality (1.42) we have

$$\sum_{j,k=1}^{n}\mathbf{a}_k^*\boldsymbol{C}(k-j)\,\mathbf{a}_j=\sum_{j,k=1}^{n}\mathbf{a}_k^*\int_{-\pi}^{\pi}e^{i(k-j)\omega}d\boldsymbol{F}(\omega)\,\mathbf{a}_j$$

$$=\sum_{r,s=1}^{d}\int_{-\pi}^{\pi}\zeta_r(\omega)\bar{\zeta}_s(\omega)dF^{rs}(\omega)\geq 0.$$

Thus the matrix function $\boldsymbol{C}(h)$, $h \in \mathbb{Z}$, is non-negative definite, so by Construction 1 a stationary time series can be constructed with this covariance matrix function.

Corollary 1.6. *Equation* (1.26) *shows that the spectral measure matrix of any stationary time series is non-negative definite. Conversely, Construction 2 proves that for any non-negative definite measure matrix, one can construct a stationary time series with this spectral measure.*

If $d = 1$, then (1.40) *holds if and only if F is a right continuous, non-decreasing function on $[-\pi, \pi]$ such that $F(-\pi) = 0$, $F(\pi) < \infty$. It implies the converse of Herglotz theorem, Theorem 1.3. For any such distribution function F, its Fourier transform*

$$c(r) = \int_{-\pi}^{\pi} e^{ir\omega} dF(\omega) \quad (r \in \mathbb{Z})$$

defines a non-negative definite function c.

If $d = 2$, then (1.40) *holds if and only if for any $-\pi \le \alpha \le \beta \le \pi$,*

$$\Delta_{\alpha\beta} F^{rr} \ge 0 \quad (r = 1, 2) \text{ and } \begin{vmatrix} \Delta_{\alpha\beta} F^{11} & \Delta_{\alpha\beta} F^{12} \\ \Delta_{\alpha\beta} F^{21} & \Delta_{\alpha\beta} F^{22} \end{vmatrix} \ge 0.$$

1.4.3 Construction 3

Let us assume that the d-dimensional stationary time series $\{\mathbf{X}_t\}_{t\in\mathbb{Z}}$ we would like to construct has an absolutely continuous spectral measure with given density matrix $\boldsymbol{f}$ (which is a self-adjoint, non-negative definite matrix valued function) and suppose that $\boldsymbol{f}(\omega)$ has constant rank $r \le d$ for a.e. $\omega \in [-\pi, \pi]$. Then we can take the parsimonious Gram-decomposition (B.5) of $2\pi\boldsymbol{f}$:

$$\boldsymbol{f}(\omega) = \frac{1}{2\pi} \boldsymbol{\phi}(\omega) \boldsymbol{\phi}^*(\omega), \qquad \boldsymbol{\phi}(\omega) \in \mathbb{C}^{d\times r},$$

for a.e. $\omega \in [-\pi, \pi]$. (Compare with Theorem 4.1.)

Then we may define a d-dimensional stationary Gaussian time series $\{\mathbf{X}_t\}$ with spectral density $\boldsymbol{f}$ using Itô's stochastic integration. Let $\mathbf{B}(\omega)$ be a standard r-dimensional Brownian motion (Wiener process) on the interval $[-\pi, \pi]$. Define a time series as

$$\mathbf{X}_t := \frac{1}{\sqrt{2\pi}} \int_{-\pi}^{\pi} e^{it\omega} \boldsymbol{\phi}(\omega) d\mathbf{B}(\omega), \qquad t \in \mathbb{Z}. \tag{1.43}$$

It is well-known that then $\{\mathbf{X}_t\}$ is a Gaussian process, $\mathbb{E}\mathbf{X}_t = \mathbf{0}$ for any t, and by Itô isometry, the covariance function is

$$\boldsymbol{C}(h) = \mathbb{E}(\mathbf{X}_{t+h}\mathbf{X}_t^*) = \frac{1}{2\pi} \int_{-\pi}^{\pi} e^{i(t+h)\omega} \boldsymbol{\phi}(\omega) e^{-it\omega} \boldsymbol{\phi}^*(\omega) d\omega = \int_{-\pi}^{\pi} e^{ih\omega} \boldsymbol{f}(\omega) d\omega$$

for any $h \in \mathbb{Z}$. Thus this time series is stationary with spectral density $\boldsymbol{f}$: this proves the correctness of the construction. In practice one would approximate the stochastic integral in (1.43) by a stochastic sum; for the approximation there are several approaches. For example, one may use Lévy's construction of Brownian motion, see e.g. [42, p. 7]; or an approximation of Brownian motion by simple, symmetric random walks [52].

1.4.4 Construction 4

1.4.4.1 Discrete Fourier Transform

First let us review the *Discrete Fourier Transform (DFT)* in a way that is consistent with our previous setting. For simplicity, choose a positive *odd* integer $2N+1$ and define $\Delta\omega := \frac{2\pi}{2N+1}$. Suppose that the spectral measure of the investigated d-dimensional stationary time series $\{\mathbf{X}_t\}$ is absolutely continuous with density matrix function $\boldsymbol{f}$. We assume that $\boldsymbol{f}$, or an estimate of it, is given at the discrete points $\omega_j := j\Delta\omega \in [-\pi, \pi]$, $j = -N, \dots, N$, called *Fourier frequencies.* Then the DFT of $\boldsymbol{f}$ is defined as

$$\hat{\boldsymbol{C}}(k) = \Delta\omega \sum_{j=-N}^{N} \boldsymbol{f}(\omega_j) e^{ik\omega_j}, \quad k = -N, \dots, N. \tag{1.44}$$

This finite sequence is a natural estimate of the covariance matrix function, see (1.27),

$$\boldsymbol{C}(k) = \int_{-\pi}^{\pi} e^{ik\omega} \boldsymbol{f}(\omega) d\omega \qquad (k \in \mathbb{Z}),$$

if $\boldsymbol{f}$ is Riemann integrable and N is large enough. A property of DFT is that it is periodic with period $2N+1$: $\hat{\boldsymbol{C}}(k + 2N + 1) = \hat{\boldsymbol{C}}(k)$ for any k.

Conversely, assume that the covariance matrix function $\boldsymbol{C}(k)$, $k \in \mathbb{Z}$, or an estimate of it, is given and we would like to find an estimate of the spectral density $\boldsymbol{f}$. Then the *inverse DFT (IDFT)* is defined by

$$\hat{\boldsymbol{f}}(\omega_j) = \frac{1}{2\pi} \sum_{k=-N}^{N} \boldsymbol{C}(k)\, e^{-ik\omega_j}, \quad j = -N, \dots, N. \tag{1.45}$$

It is a natural estimate of the spectral density matrix, see (1.32):

$$\boldsymbol{f}(\omega) = \frac{1}{2\pi} \sum_{k=-\infty}^{\infty} \boldsymbol{C}(k) e^{-ik\omega}, \quad \omega \in [-\pi, \pi],$$

if the entries of $\boldsymbol{C}(k)$ are negligible for $|k| > N$. If the entries of $\boldsymbol{C}$ are absolute summable: $\boldsymbol{C} \in \ell^1$, and N is large enough, then this condition holds. A property of IDFT is that it is also periodic with period $2N+1$: $\hat{\boldsymbol{f}}(\omega_{j+2N+1}) = \hat{\boldsymbol{f}}(\omega_j)$ for any j.

If the chosen positive integer is *even*: $2N$, then everything goes similarly as above, except that the indices run from $-N+1$ to N.

It is well-known that $\{e^{ik\omega}\}_{k\in\mathbb{Z}}$ is an orthonormal sequence of functions in $L^2([0,2\pi],\mathcal{B},d\omega)$:

$$\langle e^{ik\omega}, e^{i\ell\omega}\rangle := \frac{1}{2\pi}\int_{-\pi}^{\pi} e^{ik\omega}\overline{e^{i\ell\omega}}\,d\omega = \delta_{k\ell} \quad (k,\ell\in\mathbb{Z}).$$

Similarly, $\{j \mapsto e^{ik\omega_j}\}_{\{k=-N,\dots,N\}}$ is an orthonormal sequence of functions on the discrete set of points $\{\omega_j : j=-N,\dots,N\}$ in the following sense:

$$\begin{aligned}\langle j\mapsto e^{ik\omega_j}, j\mapsto e^{i\ell\omega_j}\rangle &:= \frac{1}{2N+1}\sum_{j=-N}^{N} e^{ik\omega_j}\,\overline{e^{i\ell\omega_j}}\\ &= \frac{1}{2N+1}\sum_{j=-N}^{N} e^{i(k-\ell)\omega_j} = \delta_{k\ell},\end{aligned}$$

for any $k,\ell\in\mathbb{Z}$. The next proposition shows that this property implies that the IDFT is really the inverse transformation of the DFT.

Proposition 1.2. *Assume that* $\hat{\boldsymbol{C}}(k)$, $k=-N,\dots,N$, *is the DFT of an approximate spectral density* $\hat{\boldsymbol{f}}$ *as defined by* (1.44). *Then the IDFT defined by* (1.45) *gives*

$$\begin{aligned}\frac{1}{2\pi}\sum_{k=-N}^{N}\hat{\boldsymbol{C}}(k)e^{-ik\omega_j} &= \frac{1}{2\pi}\sum_{k=-N}^{N} e^{-ik\omega_j}\Delta\omega\sum_{\ell=-N}^{N}\hat{\boldsymbol{f}}(\omega_\ell)e^{ik\omega_\ell}\\ &= \sum_{\ell=-N}^{N}\hat{\boldsymbol{f}}(\omega_\ell)\frac{1}{2N+1}\sum_{k=-N}^{N} e^{i(\ell-j)k\Delta\omega} = \hat{\boldsymbol{f}}(\omega_j),\end{aligned}$$

for $j=-N,\dots,N$.

Similarly, assume that $\hat{\boldsymbol{f}}(\omega_j)$, $j=-N,\dots,N$, *is the IDFT of an approximate covariance matrix function* $\hat{\boldsymbol{C}}$ *as defined by* (1.45). *Then the DFT defined by* (1.44) *gives*

$$\begin{aligned}\Delta\omega\sum_{j=-N}^{N}\hat{\boldsymbol{f}}(\omega_j)e^{ik\omega_j} &= \Delta\omega\sum_{j=-N}^{N} e^{ik\omega_j}\frac{1}{2\pi}\sum_{\ell=-N}^{N}\hat{\boldsymbol{C}}(\ell)e^{-i\ell\omega_j}\\ &= \sum_{\ell=-N}^{N}\hat{\boldsymbol{C}}(\ell)\frac{1}{2N+1}\sum_{j=-N}^{N} e^{i(k-\ell)\omega_j} = \hat{\boldsymbol{C}}(k),\end{aligned}$$

for $k=-N,\dots,N$.

The DFT and IDFT are efficient from an algorithmic point of view, because when $N=2^n$, they can be evaluated in $O(N\log N)$ steps using Fast Fourier Transform (FFT).

1.4.4.2 The construction

Like in Construction 3, let us assume that the d-dimensional stationary time series $\{\mathbf{X}_t\}_{t\in\mathbb{Z}}$ we would like to construct has an absolutely continuous spectral measure with given density matrix $\boldsymbol{f}$ (which is a self-adjoint, non-negative definite matrix valued function) and suppose that $\boldsymbol{f}(\omega)$ has constant rank $r \le d$ for a.e. $\omega \in [-\pi, \pi]$ and is Riemann integrable on $[-\pi, \pi]$. Take the parsimonious Gram-decomposition (B.5) of $2\pi\boldsymbol{f}$:

$$\boldsymbol{f}(\omega) = \frac{1}{2\pi}\boldsymbol{\phi}(\omega)\boldsymbol{\phi}^*(\omega), \qquad \boldsymbol{\phi}(\omega) \in \mathbb{C}^{d\times r},$$

for a.e. $\omega \in [-\pi, \pi]$. (Compare with Theorem 4.1.)

The basis of the construction is the spectral representation (1.20) of $\{\mathbf{X}_t\}$:

$$\mathbf{X}_t = \int_{-\pi}^{\pi} e^{it\omega} d\mathbf{Z}_\omega \quad (t \in \mathbb{Z}), \tag{1.46}$$

where $\{\mathbf{Z}_\omega\}_{\omega\in[-\pi,\pi]}$ is a d-dimensional process with orthogonal increments. Like above, assume that we have chosen an odd positive integer $2N+1$, $\Delta\omega = \frac{2\pi}{2N+1}$, and the Fourier frequencies $\omega_j = j\Delta\omega$, $j = -N, \dots, N$. Define

$$\Delta\mathbf{Z}(\omega_j) := (2N+1)^{-1/2}\boldsymbol{\phi}(\omega_j)\mathbf{V}_j, \qquad j = -N, \dots, N,$$

where

$$\mathbf{V}_j := [e^{iU_j^1}, \dots, e^{iU_j^r}]^T,$$

and $\{U_j^k : k = 1, \dots, r; j = -N, \dots, N\}$ are independent random variables, uniformly distributed on $[-\pi, \pi]$. Here $\Delta\mathbf{Z}(\omega_j)$ gives a random vector measure of the interval $[\omega_j, \omega_{j+1}]$. It is an increment of a process with orthogonal increments, see (1.16) and (1.17):

$$\begin{aligned}\mathbb{E}\left(\Delta\mathbf{Z}(\omega_j)\Delta\mathbf{Z}^*(\omega_\ell)\right) &= \Delta\omega\frac{1}{2\pi}\boldsymbol{\phi}(\omega_j)\mathbb{E}(\mathbf{V}_j\mathbf{V}_\ell^*)\boldsymbol{\phi}^*(\omega_\ell)\\ &= \delta_{j\ell}\Delta\omega\boldsymbol{f}(\omega_j),\end{aligned} \tag{1.47}$$

since

$$\mathbb{E}\left(e^{iU_j^k}e^{-iU_\ell^m}\right) = \delta_{j\ell}\delta_{km}, \qquad \mathbb{E}(\mathbf{V}_j\mathbf{V}_\ell^*) = \delta_{j\ell}\boldsymbol{I}_r.$$

As an approximation of (1.46), for $t = 0, \dots, 2N$ define

$$\hat{\mathbf{X}}_t := \sum_{j=-N}^{N} e^{it\omega_j}\,\Delta\mathbf{Z}(\omega_j) = \frac{1}{\sqrt{2\pi}}\sum_{j=-N}^{N} e^{it\omega_j}\,\boldsymbol{\phi}(\omega_j)\mathbf{V}_j\sqrt{\Delta\omega}, \tag{1.48}$$

which is a periodic sequence with period $2N+1$. It is of the form of DFT (1.44) with coefficients $(\Delta\omega)^{-1}\Delta\mathbf{Z}(\omega_j)$. Compare also the last expression in (1.48) with construction (1.43).

By (1.47) and (1.48), the covariance matrix function of $\{\hat{\mathbf{X}}_t\}$ is

$$\text{Cov}(\hat{\mathbf{X}}_{t+h}, \hat{\mathbf{X}}_t) = \mathbb{E}(\hat{\mathbf{X}}_{t+h}\hat{\mathbf{X}}_t^*) = \Delta\omega \sum_{j=-N}^{N} \boldsymbol{f}(\omega_j)\, e^{ih\omega_j} = \hat{\boldsymbol{C}}(h).$$

By (1.44) and Proposition 1.2 it follows that the spectral density of the sequence $\{\hat{\mathbf{X}}_t\}$ is exactly the given $\boldsymbol{f}(\omega_j)$ at the Fourier frequencies $\{\omega_j : j = -N, \dots, N\}$.

1.5 Estimating parameters of stationary time series

1.5.1 Estimation of the mean

It is important in practice whether one can estimate parameters of a stationary time series by observing a single trajectory of the process for a long enough time. The first thing to estimate is the mean $\mu \in \mathbb{C}$ of a process, which could differ from 0 now. It is enough to consider a one-dimensional time series $\{X_t\}_{t\in\mathbb{Z}}$, because expectation can be taken componentwise. If

$$X_t^{\mu} := X_t + \mu, \quad \mathbb{E}X_t = 0, \quad \mu \in \mathbb{C} \quad (t \in \mathbb{Z}),$$

then one gets a natural approximation of μ by taking a positive integer T and computing the *empirical mean*, that is, the average of a single trajectory for $t = 0, 1, \dots, T-1$:

$$\tilde{X}_T^{\mu} := \frac{1}{T}\sum_{t=0}^{T-1} X_t^{\mu} = \mu + \frac{1}{T}\sum_{t=0}^{T-1} X_t = \mu + \tilde{X}_T.$$

If we have convergence of the time average $\tilde{X}_T^{\mu}$ to the theoretical expectation μ in mean square then it is called *ergodicity for the mean*; it is a 'law of large numbers'. Obviously, for this it is necessary and sufficient that $\tilde{X}_T \to 0$ in mean square; this is the case that we are going to investigate in the sequel.

Let $H(\mathbf{X}) \subset L^2(\Omega, \mathcal{F}, \mathbb{P})$ be the Hilbert space defined in Section 1.3. In this section each random variable is considered as a vector in $H(\mathbf{X})$. So equality of two random variables means that they are $\mathbb{P}$-a.s. equal. Also, convergence of random variables is always understood in this Hilbert space, that is, as convergence in mean square.

Now we introduce two subspaces of $H(\mathbf{X})$. First, let S denote the unitary operator of forward shift in $H(\mathbf{X})$ and

$$\mathcal{I} := \{\xi \in H(\mathbf{X}) : S\xi = \xi\}$$

the subspace of translation invariant random variables in $H(\mathbf{X})$. It is clear that $\mathcal{I}$ is a closed subspace in $H(\mathbf{X})$. Second, define

$$\mathcal{N} := \overline{\text{span}}\{\eta \in H(\mathbf{X}) : \eta = S\zeta - \zeta \text{ for some } \zeta \in H(\mathbf{X})\}.$$

Let $\mathcal{N}^{\perp}$ denote the orthogonal complement of $\mathcal{N}$ in $H(\mathbf{X})$, so $H(\mathbf{X}) = \mathcal{N} \oplus \mathcal{N}^{\perp}$, where $\oplus$ denotes orthogonal direct sum.

Theorem 1.5. *Let $\{X_t\}_{t\in\mathbb{Z}}$ be a stationary time series (with mean 0). Then the time average $\tilde{X}_T$ converges to a random variable Y in mean square as $T \to \infty$:*

$$\lim_{T\to\infty} \mathbb{E}|\tilde{X}_T - Y|^2 = 0,$$

where Y is the orthogonal projection of X_0 to the subspace $\mathcal{I}$, denoted as $Y = P_{\mathcal{I}} X_0$; moreover, $\mathbb{E}Y = 0$.

Proof. Since $X_t = S^t X_0$ $(t \in \mathbb{Z})$, let us introduce the following linear operators in $H(\mathbf{X})$:

$$V_T := \frac{1}{T} \sum_{t=0}^{T-1} S^t \quad (T \geq 1).$$

Obviously, if $\xi \in \mathcal{I}$, then

$$V_T\,\xi = \xi \text{ for any } T \geq 1, \text{ so } \lim_{T\to\infty} V_T\,\xi = \xi.$$

On the other hand, if $\eta = S\zeta - \zeta$ for some $\zeta \in H(\mathbf{X})$, then

$$V_T\,\eta = V_T(S\zeta - \zeta) = \frac{1}{T} \sum_{t=0}^{T-1} (S^{t+1} - S^t)\zeta = \frac{1}{T}(S^T - S^0)\zeta.$$

Since $\|S\| = 1$, and also $\|S^T\| = 1$, we get that

$$\lim_{T\to\infty} V_T\,\eta = 0 \quad \forall \eta \in \mathcal{N}.$$

Next we want to show that $\mathcal{I} = \mathcal{N}^{\perp}$. First let $\xi \in \mathcal{I}$. Then

$$\langle \xi, S\zeta - \zeta \rangle = \langle \xi, S\zeta \rangle - \langle \xi, \zeta \rangle = \langle S^*\xi, \zeta \rangle - \langle \xi, \zeta \rangle = \langle \xi, \zeta \rangle - \langle \xi, \zeta \rangle = 0,$$

where $\zeta \in H(\mathbf{X})$ is arbitrary, since ξ is invariant under $S^* = S^{-1}$ as well. Hence

$$\xi \perp (S\zeta - \zeta) \text{ for any } \zeta \in H(\mathbf{X}).$$

This implies that $\xi \in \mathcal{N}^{\perp}$, that is, $\mathcal{I} \subset \mathcal{N}^{\perp}$.

Conversely, assume that $\xi \in \mathcal{N}^{\perp}$. Then for any $\zeta \in H(\mathbf{X})$,

$$0 = \langle \xi, S\zeta - \zeta \rangle = \langle \xi, S\zeta \rangle - \langle \xi, \zeta \rangle = \langle S^*\xi - \xi, \zeta \rangle.$$

This means that $(S^*\xi - \xi) \perp \zeta$ for any $\zeta \in H(\mathbf{X})$. Thus $S^{-1}\xi = \xi$, that is, $\xi \in \mathcal{I}$, which implies that $\mathcal{N}^{\perp} \subset \mathcal{I}$. Consequently, $\mathcal{I} = \mathcal{N}^{\perp}$.

Finally, it follows that any $\xi \in H(\mathbf{X})$ can be written as $\xi = \xi_{\mathcal{I}} + \xi_{\mathcal{N}}$, where $\xi_{\mathcal{I}} = P_{\mathcal{I}}\xi \in \mathcal{I}$ and $\xi_{\mathcal{N}} \in \mathcal{N}$. Then by the first part of the proof,

$$\lim_{T\to\infty} V_T\left(\xi_{\mathcal{I}} + \xi_{\mathcal{N}}\right) = \xi_{\mathcal{I}}.$$

This shows that

$$\lim_{T\to\infty} \tilde{X}_T = \lim_{T\to\infty} V_T X_0 = P_{\mathcal{I}} X_0 =: Y \in \mathcal{I}.$$

Also, $\mathbb{E}Y = 0$, as the expectation of any random variable in $H(\mathbf{X})$ is zero. These prove the theorem. □

The next theorem gives a necessary and sufficient condition of ergodicity for the mean.

Theorem 1.6. *Let $\{X_t\}_{t\in\mathbb{Z}}$ be a stationary time series with mean 0 and with covariance function $c(j)$ $(j \in \mathbb{Z})$. Then the time average $\tilde{X}_T$ converges to $Y = 0$ in mean square as $T \to \infty$ if and only if*

$$\lim_{T\to\infty} \frac{1}{T} \sum_{j=0}^{T-1} c(j) = 0. \tag{1.49}$$

Proof. By Theorem 1.5 there exists a random variable $Y \in H(\mathbf{X})$ such that $\tilde{X}_T \to Y$ in mean square as $T \to \infty$. Thus for any $t \in \mathbb{Z}$ we have

$$\langle Y, X_t \rangle = \lim_{T\to\infty} \frac{1}{T} \sum_{j=t}^{t+T-1} \langle X_j, X_t \rangle = \lim_{T\to\infty} \frac{1}{T} \sum_{j=0}^{T-1} \langle X_j, X_0 \rangle = \lim_{T\to\infty} \frac{1}{T} \sum_{j=0}^{T-1} c(j).$$

In the case of $Y = 0$, this implies that (1.49) holds. Conversely, if (1.49) holds, then $\langle Y, X_t \rangle = 0$ for any $t \in \mathbb{Z}$. Since Y belongs to the space spanned by $\{X_t\}$, $Y = 0$ follows. □

Corollary 1.7. *It is an elementary analysis fact that $\lim_{j\to\infty} c(j) = 0$ implies (1.49), thus it is a sufficient (but not necessary) condition of ergodicity for the mean. An even stronger sufficient condition is that $\sum_{j=-\infty}^{\infty} |c(j)| < \infty$ holds.*

Another approach to ergodicity for the mean is to use spectral representation of the time series.

Theorem 1.7. *Let $\{X_t\}_{t\in\mathbb{Z}}$ be a stationary time series with mean 0, with random spectral measure dZ_ω, and spectral measure $dF(\omega)$, for $\omega \in [-\pi, \pi]$.*

(a) Then the time average $\tilde{X}_T$ converges in mean square to $Y = Z_{\{0\}}$, the atom (point mass) of the random spectral measure at $\{0\}$, as $T \to \infty$.

(b) Ergodicity for the mean holds if and only if $dF(\{0\}) = 0$, that is, the spectral measure has no atom at $\{0\}$.

Proof. Take the spectral representation

$$X_t = \int_{-\pi}^{\pi} e^{it\omega} dZ_\omega \quad (t \in \mathbb{Z}).$$

Then

$$\tilde{X}_T = \int_{-\pi}^{\pi} \frac{1}{T} \sum_{t=0}^{T-1} e^{it\omega} dZ(\omega) = \int_{-\pi}^{\pi} g_T(\omega) dZ_\omega,$$

where

$$g_T(\omega) = \begin{cases} \frac{e^{iT\omega}-1}{T(e^{i\omega}-1)} & (\omega \neq 0) \\ 1 & (\omega = 0) \end{cases}.$$

Fix an arbitrary $\epsilon > 0$. By Corollary 1.4, we can write that

$$\mathbb{E}\left|\tilde{X}_T - Z_{\{0\}}\right|^2 = \int_{0<|\omega|\leq\delta} |g_T(\omega)|^2 dF(\omega) + \int_{\delta<|\omega|\leq\pi} |g_T(\omega)|^2 dF(\omega), \quad (1.50)$$

where $\delta > 0$ will be suitably chosen.

The first integral on the right of (1.50) can be estimated as

$$\int_{0<|\omega|\leq\delta} \left|\frac{e^{iT\omega}-1}{T(e^{i\omega}-1)}\right|^2 dF(\omega) \leq \int_{0<|\omega|\leq\delta} 2\, dF(\omega) < \frac{\epsilon}{2},$$

if δ is small enough. Then the second integral on the right of (1.50) can be estimated as

$$\int_{\delta<|\omega|\leq\pi} \left|\frac{e^{iT\omega}-1}{T(e^{i\omega}-1)}\right|^2 dF(\omega) \leq \int_{\delta<|\omega|\leq\pi} \left(\frac{2}{T\sqrt{2(1-\cos\delta)}}\right)^2 dF(\omega) < \frac{\epsilon}{2},$$

if T is large enough. This proves (a).

It follows from (a) that the time series $\{X_t\}$ is ergodic for the mean if and only if $Y = Z_{\{0\}} = 0$. By Corollary 1.3,

$$dF(\{0\}) = \mathbb{E}|Z_{\{0\}}|^2,$$

thus ergodicity for the mean is equivalent to the fact that dF has no atom at $\{0\}$. This proves (b). □

Remark 1.9. By Theorem 1.7(b), any stationary time series whose spectral measure dF is absolutely continuous w.r.t. Lebesgue measure (that is, has a spectral density f) is ergodic for the mean, since then the spectral measure dF does not have atoms.

Example 1.1. A simple example for a time series which is *not* ergodic for the mean:

$$X_k = \sum_{j=0}^{n} A_j e^{ik\omega_j} \quad (k \in \mathbb{Z}),$$

where $0 = \omega_0 < \cdots < \omega_n < 2\pi$; $A_0, \dots, A_n$ are uncorrelated random variables with mean 0 and variance $\sigma_j^2 > 0$ $(j = 0, \dots, n)$. (The A_j's can be e.g. Gaussian random variables.)

This process is weakly stationary with

$$c(k) = \mathbb{E}(X_{m+k}\overline{X_m}) = \sum_{j=0}^{n} \mathbb{E}(|A_j|^2)e^{ik\omega_j} = \sum_{j=0}^{n} \sigma_j^2 e^{ik\omega_j} = \int_0^{2\pi} e^{ik\omega}\, dF(\omega),$$

where

$$F(\omega) = \sum_{\omega_j \le \omega} \sigma_j^2 \quad (\omega \in [0, 2\pi]).$$

(See Subsection 2.10.1(a) later as well.) By Theorem 1.7(b), this process is not ergodic for the mean, because dF has an atom $\sigma_0^2 > 0$ at $\{0\}$.

Theorem 1.7(b) shows that a time series is not ergodic for the mean if and only if it contains a component which is a nonzero time-constant random variable, like A_0 in the present example.

1.5.2 Estimation of the covariances

Let $\{\mathbf{X}_t\}_{t\in\mathbb{Z}}$ be a d-dimensional stationary time series with mean $\mathbf{0}$ and with covariance matrix function $\boldsymbol{C}(h) = [c_{jk}(h)]_{d\times d}$ $(h \in \mathbb{Z})$. For any positive integer T and for any $j, k = 1, \dots, d$, a natural estimator of a covariance $c_{jk}(h)$ is

$$\tilde{c}_{jk}(h) := \frac{1}{T-h} \sum_{t=0}^{T-1-h} X_{t+h}^j \overline{X_t^k} \quad (0 \le h \le T-1).$$

Clearly, this is an unbiased estimator:

$$\mathbb{E}(\tilde{c}_{jk}(h)) = \frac{1}{T-h} \sum_{t=0}^{T-h-1} \mathbb{E}(X_{t+h}^j \overline{X_t^k}) = c_{jk}(h) \quad (0 \le h \le T-1).$$

Another useful estimator is the *empirical covariance*

$$\hat{c}_{jk}(h) = \hat{c}_{jk}^{(T)}(h) := \frac{1}{T} \sum_{t=0}^{T-1-h} X_{t+h}^j \overline{X_t^k} \quad (0 \le h \le T-1).$$

This is only asymptotically unbiased if $0 \le h \le h_T = o(T)$ as $T \to \infty$, that is, $h_T/T \to 0$. However, the *empirical covariance matrix function*

$$\hat{\boldsymbol{C}}(h) := [\hat{c}_{jk}(h)]_{d\times d} = \frac{1}{T} \sum_{t=0}^{T-1-h} \mathbf{X}_{t+h}\mathbf{X}_t^* \quad (0 \le h \le T-1) \tag{1.51}$$

has the desirable property that the following $Td \times Td$ block Toeplitz matrix

$$\hat{\mathfrak{C}}_T := \begin{bmatrix} \hat{C}(0) & \hat{C}(1) & \cdots & \hat{C}(T-1) \\ \hat{C}(1)^* & \hat{C}(0) & \cdots & \hat{C}(T-2) \\ \vdots & \vdots & \ddots & \vdots \\ \hat{C}(T-1)^* & \hat{C}(T-2)^* & \cdots & \hat{C}(0) \end{bmatrix}$$

is non-negative definite for arbitrary $T \geq 1$. This follows from a factorization of $\hat{\mathfrak{C}}_T$. Consider the $Td \times 2Td$ matrix

$$\mathfrak{X}_T := \begin{bmatrix} \mathbf{X}_0 & \mathbf{X}_1 & \cdots & \mathbf{X}_{T-1} & 0 & 0 & \cdots & 0 & 0 \\ 0 & \mathbf{X}_0 & \cdots & \mathbf{X}_{T-2} & \mathbf{X}_{T-1} & 0 & \cdots & 0 & 0 \\ \vdots & \vdots & \ddots & \vdots & \vdots & \vdots & \ddots & \vdots & \vdots \\ 0 & 0 & \cdots & \mathbf{X}_0 & \mathbf{X}_1 & \mathbf{X}_2 & \cdots & \mathbf{X}_{T-1} & 0 \end{bmatrix}.$$

Then $\hat{\mathfrak{C}}_T = T^{-1}\mathfrak{X}_T\mathfrak{X}_T^*$. Hence, for any $\mathbf{a}_1, \dots, \mathbf{a}_T \in \mathbb{C}^d$, we have

$$\sum_{j,k=1}^{T} \mathbf{a}_j^* \hat{C}(j-k)\mathbf{a}_k = \mathbf{a}^* \frac{\hat{\mathfrak{C}}_T}{T}\mathbf{a} = \frac{1}{T}(\mathbf{a}^*\mathfrak{X}_T)(\mathbf{a}^*\mathfrak{X}_T)^* \geq 0,$$

where

$$\mathbf{a} := \begin{bmatrix} \mathbf{a}_1 \\ \vdots \\ \mathbf{a}_T \end{bmatrix}.$$

This is another proof for the positive semidefiniteness of the large block Toeplitz matrix.

We mention that in the case when the mean $\boldsymbol{\mu}$ of the time series is not $\mathbf{0}$, the empirical covariances are defined as

$$\hat{c}_{jk}(h) = \hat{c}_{jk}^{(T)}(h) := \frac{1}{T}\sum_{t=0}^{T-1-h} (X_{t+h}^j - \tilde{X}_T^j)(\overline{X_t^k} - \overline{\tilde{X}_T^k}) \quad (0 \leq h \leq T-1),$$

where $\tilde{X}_T^j$ and $\tilde{X}_T^k$ are the empirical means of $\{X_t^j\}$ and $\{X_t^k\}$, respectively.

The weakly stationary time series $\{\mathbf{X}_t\}_{t\in\mathbb{Z}}$ is called *ergodic for the covariance* if it is ergodic for the mean, plus for each $j, k = 1, \dots, d$, the empirical covariance $\hat{c}_{jk}(h)$ converges in mean square to the covariance $c_{jk}(h)$ for any $0 \leq h \leq h_T$ as $T \to \infty$, where $h_T = o(T)$.

Thus fix $j, k \in \{1, \dots, d\}$ and an $h \in \mathbb{Z}$ from now on. Define the following closed subspace of $L^2(\Omega, \mathcal{F}, \mathbb{P})$:

$$H(X(j,k,h)) := \overline{\text{span}}\{X_{t+h}^j \bar{X}_t^k - c_{jk}(h) : t \in \mathbb{Z}\}.$$

So by defining the process

$$Y_t := X_{t+h}^j \bar{X}_t^k - c_{jk}(h), \qquad t \in \mathbb{Z}, \tag{1.52}$$

it follows that $\mathbb{E}(Y_t) = 0$ for any $t \in \mathbb{Z}$. If we also assume that $\mathbb{E}(Y_{t+s}\bar{Y}_t) = \mathbb{E}(Y_s\bar{Y}_0)$, equivalently,

$$\mathbb{E}\left(X^j_{t+s+h}\bar{X}^k_{t+s}\overline{X^j_{t+h}\bar{X}^k_t}\right) = \mathbb{E}\left(X^j_{s+h}\bar{X}^k_s\overline{X^j_h\bar{X}^k_0}\right) \quad \forall t, s \in \mathbb{Z}, \tag{1.53}$$

then $\{Y_t\}_{t\in\mathbb{Z}}$ becomes a weakly stationary time series with expectation 0. The right shift (forward shift) operator S is defined and unitary in the Hilbert subspace $H(X(j,k,h)) = H(Y) = \overline{\text{span}}\{Y_t : t \in \mathbb{Z}\}$ as well:

$$SY_0 = Y_t, \quad \text{that is,} \quad S(X^j_h\bar{X}^k_0) = X^j_{t+h}\bar{X}^k_t \quad \text{for all} \quad t \in \mathbb{Z}.$$

Similarly as in Subsection 1.5.1, we may define

$$\begin{aligned}&\mathcal{I}(j,k,h) := \{\xi \in H(X(j,k,h)) : S\xi = \xi\},\\ &\mathcal{N}(j,k,h)\\ &:= \overline{\text{span}}\{\eta \in H(X(j,k,h)) : \eta = S\zeta - \zeta \text{ for some } \zeta \in H(X(j,k,h))\}.\end{aligned}$$

As in the proof of Theorem 1.5, $H(X(j,k,h)) = \mathcal{I}(j,k,h) \oplus \mathcal{N}(j,k,h)$. We arrive at the following theorem, see [16, Chapter X, Theorem 7.1].

Theorem 1.8. *Assume that $\{\mathbf{X}_t\}$ is a d-dimensional complex weakly stationary time series with expectation $\mathbf{0}$ and property* (1.53) *holds for each $j, k = 1, \dots, d$ and $h \geq 0$. Take the empirical covariance matrix function $\hat{\boldsymbol{C}}(h) = [\hat{c}^{(T)}_{jk}(h)]_{d\times d}$ defined by* (1.51) *for $0 \leq h \leq h_T$, where $h_T = o(T)$.*

(a) Then for each $j, k = 1, \dots, d$ and $0 \leq h \leq h_T$,

$$\operatorname*{lim}_{T\to\infty} \hat{c}^{(T)}_{jk}(h) - c_{jk}(h) = \operatorname*{lim}_{T\to\infty} \frac{1}{T}\sum_{t=0}^{T-1-h} X^j_{t+h}\bar{X}^k_t - c_{jk}(h) = Y(j,k,h),$$

where lim *denotes here limit in mean square and the random variable $Y(j,k,h)$ is the orthogonal projection of $X^j_h\bar{X}^k_0 - c_{jk}(h)$ to the subspace $\mathcal{I}(j,k,h)$; $\mathbb{E}Y(j,k,h) = 0$.*

(b) The time series $\{\mathbf{X}_t\}$ is ergodic for the covariance if and only if

$$\lim_{T\to\infty} \frac{1}{T}\sum_{s=0}^{T-1-h} \mathbb{E}\left(X^j_{s+h}\bar{X}^k_s\overline{X^j_h\bar{X}^k_0}\right) = |c_{jk}(h)|^2, \tag{1.54}$$

equivalently, $Y(j,k,h) = 0$, for any $j, k = 1, \dots, d$ and $h \geq 0$.

Proof. Statement (a) follows from Theorem 1.5, applied to the stationary time series $\{Y_t\}$ defined in (1.52). The only difference in the proof is that now

$$V_T := \frac{1}{T}\sum_{t=0}^{T-1-h} S^t \quad (T \geq 1, 0 \leq h \leq h_T).$$

However, this difference asymptotically vanishes as $T \to \infty$.

The proof of statement (b) follows the lines of the proof of Theorem 1.6. Fix $j, k = 1, \dots, d$ and $h \geq 0$. By statement (a), there exists a random variable $Y(j,k,h) \in H(X(j,k,h))$ such that $\hat{c}_{jk}^{(T)}(h) - c_{jk}(h)$ converges to $Y(j,k,h)$ in mean square as $T \to \infty$. Thus for any $t \in \mathbb{Z}$ we have

$$\begin{aligned}
&\left\langle Y(j,k,h), X_{t+h}^j \bar{X}_t^k - c_{jk}(h) \right\rangle \\
&= \lim_{T\to\infty} \frac{1}{T} \sum_{s=t}^{t+T-1-h} \left\langle X_{s+h}^j \bar{X}_s^k - c_{jk}(h), X_{t+h}^j \bar{X}_t^k - c_{jk}(h) \right\rangle \\
&= \lim_{T\to\infty} \frac{1}{T} \sum_{s=0}^{T-1-h} \left\langle X_{s+h}^j \bar{X}_s^k - c_{jk}(h), X_h^j \bar{X}_0^k - c_{jk}(h) \right\rangle \\
&= \lim_{T\to\infty} \frac{1}{T} \sum_{s=0}^{T-1-h} \mathbb{E}\left(X_{s+h}^j \bar{X}_s^k \overline{X_h^j \bar{X}_0^k} \right) - |c_{jk}(h)|^2.
\end{aligned}$$

In the case of $Y(j,k,h) = 0$, this implies that (1.54) holds. Conversely, if (1.54) holds, then

$$\left\langle Y(j,k,h), X_{t+h}^j \bar{X}_t^k - c_{jk}(h) \right\rangle = 0 \quad \forall t \in \mathbb{Z}.$$

Since $Y(j,k,h)$ belongs to the space spanned by $X_{t+h}^j \bar{X}_t^k - c_{jk}(h)$, $t \in \mathbb{Z}$, the statement $Y(j,k,h) = 0$ follows. □

1.5.3 Periodograms

First we extend the Discrete Fourier Transform (DFT) and its inverse IDFT discussed in Subsection 1.4.4 to a random sample of a d-dimensional complex stationary time series $\{\mathbf{X}_t\}$ with $\mathbf{0}$ expectation. Assume that for a positive integer T, a *random sample* $\{\mathbf{X}_0, \mathbf{X}_1, \dots, \mathbf{X}_{T-1}\}$ is given. Then we define $\Delta\omega := \frac{2\pi}{T}$ and the Fourier frequencies $\omega_j = j\Delta\omega \in [0, 2\pi]$ $(j = 0, \dots, T-1)$. (In this subsection it is simpler to work with the frequency interval $[0, 2\pi]$ than with $[-\pi, \pi]$.)

By the IDFT (1.45), one may assign d-dimensional *spectral amplitudes* $\hat{\mathbf{Z}}_{\omega_j}$ to the sample:

$$\hat{\mathbf{Z}}_{\omega_j} := T^{-\frac{1}{2}} \sum_{t=0}^{T-1} \mathbf{X}_t e^{-it\omega_j}, \quad j = 0, \dots, T-1.$$

Then similarly as in Proposition 1.2, one gets that $\{\mathbf{X}_t\}$ can be obtained by DFT from the spectral amplitudes:

$$\mathbf{X}_t = T^{-\frac{1}{2}} \sum_{j=0}^{T-1} \hat{\mathbf{Z}}_{\omega_j} e^{it\omega_j}, \quad t = 0, 1, \dots, T-1.$$

This is a discrete version of the spectral representation of the time series $\{\mathbf{X}_t\}$, see (1.20).

By definition, the *periodogram* of the sample is the sequence

$$\{\boldsymbol{I}_T(\omega_0), \boldsymbol{I}_T(\omega_1), \ldots, \boldsymbol{I}_T(\omega_{T-1})\}$$

of $d \times d$ *intensity matrices*:

$$\boldsymbol{I}_T(\omega_j) := \hat{\mathbf{Z}}_{\omega_j} \hat{\mathbf{Z}}^*_{\omega_j} = \frac{1}{T} \sum_{t,s=0}^{T-1} \mathbf{X}_t \mathbf{X}_s^* e^{-i(t-s)\omega_j}. \tag{1.55}$$

Substituting $t = s + h$ into (1.55) and rearranging the terms, it follows that

$$\boldsymbol{I}_T(\omega_j) = \sum_{h=-T+1}^{T-1} \frac{1}{T} \sum_{s=0}^{T-1-|h|} \mathbf{X}_{s+h} \mathbf{X}_s^* \, e^{-ih\omega_j} = \sum_{h=-T+1}^{T-1} \hat{\boldsymbol{C}}(h) \, e^{-ih\omega_j}, \tag{1.56}$$

where

$$\hat{\boldsymbol{C}}(h) = \frac{1}{T} \sum_{s=0}^{T-1-|h|} \mathbf{X}_{s+h} \mathbf{X}_s^* \qquad (h = -T+1, \ldots, T-1), \tag{1.57}$$

is the empirical covariance matrix function of the sample, see (1.51).

Assume that the spectral measure of the time series $\{\mathbf{X}_t\}$ is absolutely continuous with density matrix $\boldsymbol{f}$. Then comparing the IDFT (1.45) with (1.56) shows that

$$\hat{\boldsymbol{f}}_T(\omega_j) := \frac{1}{2\pi} \boldsymbol{I}_T(\omega_j), \quad \omega_j = j\frac{2\pi}{T}, \quad j = 0, 1, \ldots, T-1, \tag{1.58}$$

is a discrete estimate of the spectral density $\boldsymbol{f}(\omega)$, $\omega \in [0, 2\pi)$. The next proposition shows that estimate (1.58) is *asymptotically unbiased*.

Proposition 1.3. *Suppose that $\boldsymbol{C}(h)$, $h \in \mathbb{Z}$, is absolutely summable. For $\omega \in [0, 2\pi)$ set $\omega^{(T)} := \lfloor \omega/\Delta\omega \rfloor \Delta\omega$, where $\Delta\omega := \frac{2\pi}{T}$. Then for the estimate* (1.58) *we have*

$$\lim_{T\to\infty} \frac{1}{2\pi} \mathbb{E} \boldsymbol{I}_T(\omega^{(T)}) = \boldsymbol{f}(\omega),$$

uniformly in $\omega \in [0, 2\pi)$.

Proof. By (1.56) and (1.57),

$$\mathbb{E}\hat{\boldsymbol{C}}(h) = \frac{T - |h|}{T} \boldsymbol{C}(h), \quad \frac{1}{2\pi} \mathbb{E} \boldsymbol{I}_T(\omega_j) = \frac{1}{2\pi} \sum_{h=-T+1}^{T-1} \left(1 - \frac{|h|}{T}\right) \boldsymbol{C}(h) e^{-ih\omega_j}.$$

By (1.32),

$$\boldsymbol{f}(\omega) = \frac{1}{2\pi} \sum_{h=-\infty}^{\infty} \boldsymbol{C}(h) e^{-ih\omega}, \quad \omega \in [0, 2\pi).$$

Since $\boldsymbol{C}$ is absolutely summable, for any $\epsilon > 0$ there exists T_0 such that

$$\frac{1}{2\pi} \sum_{|h|>T_0} |c_{jk}(h)e^{-ih\omega}| < \epsilon$$

for any $j, k = 1, \dots, d$ and $\omega \in [0, 2\pi)$. This implies that for any $T \geq T_0$,

$$\frac{1}{2\pi} \sum_{T_0<|h|<T} \left| \left(1 - \frac{|h|}{T}\right) c_{jk}(h)e^{-ih\omega^{(T)}} \right| \leq \frac{1}{2\pi} \sum_{T_0<|h|<T} |c_{jk}(h)| < \epsilon$$

for any $j, k = 1, \dots, d$ and $\omega \in [0, 2\pi)$.

On the other hand, there exists $T_1 > T_0$ such that for any $T \geq T_1$, we have

$$\begin{aligned} &\frac{1}{2\pi} \sum_{h=-T_0}^{T_0} \left| c_{jk}(h)e^{-ih\omega} - \left(1 - \frac{|h|}{T}\right) c_{jk}(h)e^{-ih\omega^{(T)}} \right| \\ &\leq \frac{1}{2\pi} \sum_{h=-T_0}^{T_0} |c_{jk}(h)| \left| e^{-ih\omega} - \left(1 - \frac{|h|}{T}\right) e^{-ih(\omega+\Delta\omega)} \right| \\ &\leq \frac{1}{2\pi} \sum_{h=-T_0}^{T_0} |c_{jk}(h)| \left| 1 - \left(1 - \frac{T_0}{T}\right) e^{-iT_0 2\pi/T} \right| < \epsilon \end{aligned}$$

for any $j, k = 1, \dots, d$ and $\omega \in [0, 2\pi)$. The above obtained inequalities together prove the proposition. □

Unfortunately, the estimate $\hat{\boldsymbol{f}}_T$ defined in (1.58) is not consistent and does not converge to the theoretical spectral density function $\boldsymbol{f}$ as $T \to \infty$, see e.g. Brillinger's [9, Chapter 5]. Definition (1.55) shows that $\hat{\boldsymbol{f}}_T$ is a rank 1 matrix, which can be very far from the theoretical matrix $\boldsymbol{f}$. There exists a considerable amount of literature on methods that treat this problem; several approaches are mentioned in [9] and [56].

1.6 Summary

The d-dimensional, complex-valued, weakly stationary time series $\{\mathbf{X}_t\}_{t\in\mathbb{Z}}$ can be uniquely characterized by its first and second moments that do not depend on time shift:

$$\boldsymbol{\mu} = \mathbb{E}\mathbf{X}_t, \quad \boldsymbol{C}(h) = \mathbb{E}[(\mathbf{X}_{t+h} - \boldsymbol{\mu})(\mathbf{X}_t - \boldsymbol{\mu})^*], \quad h \in \mathbb{Z},$$

where $\boldsymbol{C}(h)$ is called (auto)covariance matrix function. We usually assume that $\boldsymbol{\mu} = \mathbf{0}$. Note that $\boldsymbol{C}(-h) = \boldsymbol{C}^*(h)$, $h \in \mathbb{Z}$. More generally, we speak of second-order processes whenever assume that the above first and second moments determine the process. The following are equivalent to the fact that $\{\mathbf{X}_t\}$ is weakly stationary:

- $\boldsymbol{C}(h)$ is non-negative definite, i.e. $\sum_{k,r=1}^{n} \mathbf{a}_k^* \boldsymbol{C}(k-r)\, \mathbf{a}_r \geq 0$ for $n \geq 1$ and $\mathbf{a}_1, \dots, \mathbf{a}_n \in \mathbb{C}^d$. Equivalently, the self-adjoint matrix

$$\mathfrak{C}_n := \begin{bmatrix} \boldsymbol{C}(0) & \boldsymbol{C}(1) & \boldsymbol{C}(2) & \cdots & \boldsymbol{C}(n-1) \\ \boldsymbol{C}^*(1) & \boldsymbol{C}(0) & \boldsymbol{C}(1) & \cdots & \boldsymbol{C}(n-2) \\ \boldsymbol{C}^*(2) & \boldsymbol{C}^*(1) & \boldsymbol{C}(0) & \cdots & \boldsymbol{C}(n-3) \\ \vdots & \vdots & \vdots & \ddots & \vdots \\ \boldsymbol{C}^*(n-1) & \boldsymbol{C}^*(n-2) & \boldsymbol{C}^*(n-3) & \cdots & \boldsymbol{C}(0) \end{bmatrix}$$

  is positive semidefinite. Note that $\mathfrak{C}_n$ is the covariance matrix of the nd-dimensional random vector $(\mathbf{X}_1^T, \dots, \mathbf{X}_n^T)^T$ and it is a block Toeplitz matrix.

- It has a non-negative definite spectral measure matrix $d\boldsymbol{F}$ on $[-\pi, \pi]$ such that its autocovariance matrix function can be represented as the Fourier transform of $d\boldsymbol{F}$: $\boldsymbol{C}(h) = \int_{-\pi}^{\pi} e^{ih\omega} d\boldsymbol{F}(\omega)$, $h \in \mathbb{Z}$.

When each dF^{jk} is absolutely continuous w.r.t. the Lebesgue measure on $[-\pi, \pi]$, that is, $dF^{jk}(\omega) = f^{jk}(\omega)\, d\omega$ for $j, k = 1, \dots, d$, then there exists the $d \times d$ spectral density matrix $\boldsymbol{f}(\omega) = [f^{jk}(\omega)]$. A sufficient condition for this is that the entries of $\boldsymbol{C}(h)$ are absolutely summable (w.r.t. h). Under this condition, the functions $f^{jk}(\omega)$ are also continuous ($j, k = 1, \dots, d$).

The matrix $\boldsymbol{f}(\omega)$ is self-adjoint and positive semidefinite for $\omega \in [-\pi, \pi]$. In particular, when the state space is $\mathbb{R}^d$, then $\boldsymbol{f}(-\omega) = \overline{\boldsymbol{f}(\omega)}$, so we can confine ourselves to the $[0, \pi]$ interval. Also, if the entries of $\boldsymbol{C}(h)$ are absolutely summable, then

$$\boldsymbol{C}(h) = \int_{-\pi}^{\pi} e^{ih\omega} \boldsymbol{f}(\omega)\, d\omega \quad \Longleftrightarrow \quad \boldsymbol{f}(\omega) = \frac{1}{2\pi} \sum_{h=-\infty}^{\infty} \boldsymbol{C}(h) e^{-ih\omega},$$

where $-\pi \leq \omega \leq \pi$. $\boldsymbol{F}$ is sometimes called *spectral distribution matrix*, while $\boldsymbol{f}$ *spectral density matrix*. The diagonal entries of $\boldsymbol{f}$ are real functions, whereas the off-diagonal ones are usually complex. If we write them in polar form, then we get the so-called amplitude and phase spectrum, respectively.

Cramér's representation: the weakly stationary time series itself can be represented (with probability 1) as

$$\mathbf{X}_t = \int_{-\pi}^{\pi} e^{it\omega} d\mathbf{Z}_\omega, \quad t \in \mathbb{Z}$$

with the orthogonal increment process $\mathbf{Z}_\omega$, where $\boldsymbol{F}(\omega) = \mathbb{E}(\mathbf{Z}_\omega \mathbf{Z}_\omega^*)$, $\omega \in [-\pi, \pi]$. Note that this generalizes the case of the superposition of sinusoids, where the process has point spectrum. More generally, if the time is continuous, then a spectral measure matrix $d\boldsymbol{F}$ can also be defined on the whole $\mathbb{R}$, see [12].

Based on the finite set $\mathbf{X}_1, \dots, \mathbf{X}_T$ of observations, the parameters of the process are estimated as follows:

$$\hat{\boldsymbol{C}}(h) = \begin{cases} \frac{1}{T}\sum_{t=1}^{T-h}(\mathbf{X}_{t+h} - \tilde{\mathbf{X}}_T)(\mathbf{X}_t - \tilde{\mathbf{X}}_T)^*, & 0 \le h \le T-1 \\ \hat{\boldsymbol{C}}^*(-h), & -T+1 \le h < 0, \end{cases}$$

while $\hat{\boldsymbol{\mu}} = \tilde{\mathbf{X}}_T = \frac{1}{T}\sum_{t=1}^{T}\mathbf{X}_t$. In practice, we usually estimate from a single trajectory ($T \to \infty$), so ergodicity is of distinguished importance. It is proved that $\tilde{\mathbf{X}}_T$ is ergodic for the mean if and only if $d\boldsymbol{F}(\{0\}) = 0$, that is, the spectral measure has no atom at $\{0\}$. It surely holds if $d\boldsymbol{F}$ is absolutely continuous w.r.t. Lebesgue measure (that is, has a spectral density matrix $\boldsymbol{f}$). The time series $\{\mathbf{X}_t\}$ is called *ergodic for the covariance* if it is ergodic for the mean, and in addition, for each $j,k = 1,\dots,d$, the empirical covariance $\hat{c}_{jk}(h)$ converges in mean square to the true covariance $c_{jk}(h)$ for any $0 \le h \le h_T$ as $T \to \infty$, where $h_T = o(T)$. A necessary and sufficient condition for this is described in Theorem 1.8.

By the inverse DFT, we can assign d-dimensional *spectral amplitudes* $\hat{\mathbf{Z}}_{\omega_j}$ to the sample:

$$\hat{\mathbf{Z}}_{\omega_j} := T^{-\frac{1}{2}}\sum_{t=1}^{T}\mathbf{X}_t e^{-it\omega_j}, \quad j = 0,1,\dots,T-1,$$

where $\omega_j = j\frac{2\pi}{T}$ is the jth Fourier frequency ($j = 0,1,\dots,T-1$). Conversely, $\{\mathbf{X}_t\}$ can be obtained by DFT from the spectral amplitudes:

$$\mathbf{X}_t = T^{-\frac{1}{2}}\sum_{j=0}^{T-1}\hat{\mathbf{Z}}_{\omega_j}e^{it\omega_j}, \quad t = 1,\dots,T.$$

This is a discrete version of the spectral representation of the time series $\{\mathbf{X}_t\}$.

The *periodogram* of the sample is the sequence

$$\{\boldsymbol{I}_T(\omega_0), \boldsymbol{I}_T(\omega_1), \dots, \boldsymbol{I}_T(\omega_{T-1})\}$$

of the $d \times d$ *intensity matrices*

$$\boldsymbol{I}_T(\omega_j) := \hat{\mathbf{Z}}_{\omega_j}\hat{\mathbf{Z}}^*_{\omega_j} = \frac{1}{T}\sum_{t,s=1}^{T}\mathbf{X}_t\mathbf{X}_s^* e^{-i(t-s)\omega_j}.$$

Also,

$$\boldsymbol{I}_T(\omega_j) = \sum_{h=-T+1}^{T-1}\hat{\boldsymbol{C}}(h)\, e^{-ih\omega_j}.$$

So $\hat{\boldsymbol{f}}(\omega_j) := \frac{1}{2\pi}\boldsymbol{I}_T(\omega_j)$ is a discrete estimate of the spectral density $\boldsymbol{f}(\omega)$, $\omega \in [0,2\pi)$. This estimate is asymptotically unbiased, but usually not consistent.

This is explained with the fact that the matrix $\boldsymbol{I}_T(\omega_j)$ is a dyad, so has rank 1, and it uses the estimates $\hat{\boldsymbol{C}}(h)$ for $h = 0, 1, \dots, T-1$, where only those with $h = o(T)$ support the ergodicity for the covariances. There is a wide literature of overcoming this difficulty with smoothing the periodogram with weight functions and windows.

Note that the spectrum of the block Toeplitz matrix $\mathfrak{C}_T$ asymptotically comprises the union of the spectra of the matrices $\boldsymbol{f}(\omega_j)$, $j = 0, 1, \dots, T-1$, as $T \to \infty$. We will further discuss this relationship between the time and frequency domain calculations in Chapter 5.

2

ARMA, regular, and singular time series in 1D

2.1 Introduction

In this chapter, we collect some basic facts about important classes of 1D (one-dimensional) stationary time series as a motivation for the multidimensional studies. We do it inductively, while proceeding from the simplest 1D processes to more and more general ones. There is a huge literature on the analysis of 1D stationary time series; to mention just the most significant ones, we refer to [6, 9, 10, 11, 34]. In particular, in this chapter we borrowed some ideas that we very much liked in Lamperti's book [38].

Through the technique of linear filtering (i.e. applying a time-invariant linear filter TLF), parametric families, such as MA (moving average), AR (autoregressive), and ARMA (both AR and MA) processes are defined. For any 1D real, weakly stationary process with continuous spectral density f, it is possible to find both a causal AR and an invertible MA process with spectral density arbitrarily close to f. This is because the ARMA processes have rational spectral densities. Therefore, ARMA processes are vital in modelling 1D time series. Also, the linear structure of them is in close relation to the prediction theory of Chapter 5.

Under certain conditions, a TLF, applied to a white noise process, results in a sliding summation (two-sided MA). We will see in Chapter 4 that in the multidimensional case, these are the processes with spectral density matrix of constant rank. The special class of them, the MA(∞) processes (one-sided MA) are the regular ones. In contrast, singular (in other words, deterministic) processes are completely determined by their remote past. Regular (in other words, purely non-deterministic) processes have no remote past at all. As a mixture of them, non-singular processes cannot be completely predicted based on their past values, but there are added values (innovations) of the newcoming observations.

We discuss the Wold decomposition of a non-singular weakly stationary time series into a regular and singular part. Again, the regular part is MA(∞), i.e. a causal (future-independent) TLF. We also consider the spectral form of the Wold decomposition and the types of singularities. It is important, that in the frame of non-singular processes, 1D regular processes can coexist only with Type (0) singular ones (their spectral measure is singular w.r.t. Lebesgue measure); while adding a regular part to Type (1) or Type (2) singularities (they do have an absolutely continuous spectral measure, though, their spectral density do not obey certain conditions), makes them regular.

2.2 Time invariant linear filtering

Time invariant linear filtering (TLF) of a stationary time series $\{X_t\}_{t\in\mathbb{Z}}$ is an important tool. Suppose that $\{X_t\}$ has spectral measure $dF = dF^X$ and spectral representation

$$X_t = \int_{-\pi}^{\pi} e^{it\omega} dZ_\omega.$$

Assume that a finite sequence of complex 'weights' $(b_{-N}, b_{-N+1}, \dots, b_N)$ is given, where N is an arbitrary positive integer. Then the filtered process

$$Y_t := \sum_{j=-N}^{N} b_j X_{t-j} \quad (t \in \mathbb{Z})$$

is represented by

$$Y_t = \int_{-\pi}^{\pi} \sum_{j=-N}^{N} b_j e^{i(t-j)\omega} dZ_\omega = \int_{-\pi}^{\pi} e^{it\omega}\phi(\omega) dZ_\omega, \tag{2.1}$$

where

$$\phi(\omega) := \sum_{j=-N}^{N} b_j e^{-ij\omega}.$$

Thus by the isometry discussed in Section 1.3,

$$\begin{aligned} c_Y(h) &= \operatorname{Cov}(Y_{t+h}, Y_t) = \mathbb{E}(Y_{t+h}\bar{Y}_t) = \langle Y_{t+h}, Y_t \rangle_{H_X} \\ &= \langle e^{i(t+h)\omega}\phi(\omega), e^{it\omega}\phi(\omega)\rangle_{dF} = \int_{-\pi}^{\pi} e^{i(t+h)\omega}\phi(\omega)\overline{e^{it\omega}\phi(\omega)}\, dF(\omega) \\ &= \int_{-\pi}^{\pi} e^{ih\omega}|\phi(\omega)|^2\, dF(\omega) \qquad (t, h \in \mathbb{Z}). \end{aligned}$$

Similarly,

$$\begin{aligned} &\operatorname{Cov}(Y_{t+h}, X_t) = \mathbb{E}(Y_{t+h}\bar{X}_t) = \langle Y_{t+h}, X_t \rangle_{H_X} = \langle e^{i(t+h)\omega}\phi(\omega), e^{it\omega}\rangle_{dF} \\ &= \int_{-\pi}^{\pi} e^{i(t+h)\omega}\phi(\omega)\overline{e^{it\omega}}\, dF(\omega) = \int_{-\pi}^{\pi} e^{ih\omega}\phi(\omega)\, dF(\omega) \qquad (t, h \in \mathbb{Z}). \end{aligned}$$

This implies that $\{Y_t\}$ is also a weakly stationary sequence, $\{Y_t\} \subset H(X)$, with spectral measure

$$dF^Y(\omega) := |\phi(\omega)|^2 \, dF^X(\omega) \tag{2.2}$$

and the pair (X_t, Y_t) $(t \in \mathbb{Z})$ has a *joint spectral measure* (a complex measure in general)

$$dF^{Y,X}(\omega) := \phi(\omega) \, dF^X(\omega). \tag{2.3}$$

The covariance function of $\{Y_t\}$ is

$$c_Y(h) = \mathbb{E}\left(\sum_{j=-N}^{N} b_j X_{t+h-j} \sum_{k=-N}^{N} \bar{b}_k \bar{X}_{t-k} \right) = \sum_{j,k=-N}^{N} b_j \bar{b}_k c_X(h+k-j),$$

for $h \in \mathbb{Z}$.

It is important that the above formulas are still valid when $N = \infty$ if and only if the condition

$$\phi(\omega) := \sum_{j=-\infty}^{\infty} b_j e^{-ij\omega} \in L^2([-\pi, \pi], \mathcal{B}, dF) \tag{2.4}$$

holds. This fact follows from the isometry between the Hilbert spaces $H(X)$ and $L^2([-\pi, \pi], \mathcal{B}, dF)$, discussed in Section 1.3. Then the series

$$Y_t := \sum_{j=-\infty}^{\infty} b_j X_{t-j} \qquad (t \in \mathbb{Z}) \tag{2.5}$$

converges in mean square and we can say that the stationary time series $\{Y_t\}_{t \in \mathbb{Z}}$ is a *filtration of* or *subordinated to* the process $\{X_t\}_{t \in \mathbb{Z}}$. Clearly, it is equivalent to the assumption that $Y_t \in H(X)$ for all $t \in \mathbb{Z}$.

A sufficient condition of (2.4) and so the mean square convergence of the filtration (2.5) is as follows.

Proposition 2.1. *If $\{X_t\}$ is a weakly stationary time series and $\sum_{j=-\infty}^{\infty} |b_j| < \infty$, then the filtered series* (2.5) *converges almost surely and in mean square.*

Proof. The monotone convergence theorem and the weak stationarity of $\{X_t\}$ imply that

$$\mathbb{E}\left\{ \sum_{j=-\infty}^{\infty} |b_j X_{t-j}| \right\} = \lim_{N\to\infty} \mathbb{E}\left\{ \sum_{j=-N}^{N} |b_j X_{t-j}| \right\} \le \sup_{t \in \mathbb{Z}} \mathbb{E}|X_t| \lim_{N\to\infty} \sum_{j=-N}^{N} |b_j|$$

$$\le \left\{ \mathbb{E}(|X_0|^2) \right\}^{1/2} \sum_{j=-\infty}^{\infty} |b_j| < \infty.$$

Then it follows that (2.5) converges almost surely.

Choose arbitrary positive integers $N > M$. Then by inequality (1.2),

$$\mathbb{E}\left|\sum_{M<|j|\leq N} b_j X_{t-j}\right|^2 = \sum_{M<|j|,|k|\leq N} b_j \bar{b}_k \mathbb{E}(X_{t-j}\bar{X}_{t-k})$$
$$\leq c(0)\left\{\sum_{M<|j|\leq N} |b_j|\right\}^2 \to 0 \quad \text{as} \quad M, N \to \infty.$$

Thus the series (2.5) converges in mean square as well. □

Observe the role of ϕ in the representations (2.1), (2.2), and (2.3). Because of these we use the notation

$$f^{Y|X}(\omega) := \phi(\omega) = \frac{dF^{Y,X}(\omega)}{dF^X(\omega)} \tag{2.6}$$

for the Radon–Nikodym derivative and call it the *conditional spectral density* of Y w.r.t. X.

Assume that $\{Y_t^j\}_{t\in\mathbb{Z}}$ ($j = 1, 2$) are two stationary time series subordinated to $\{X_t\}_{t\in\mathbb{Z}}$:

$$Y_t^j = \int_{-\pi}^{\pi} e^{it\omega} f^{Y^j|X}(\omega) dZ_\omega \quad (t \in \mathbb{Z}),$$

where $f^{Y^j|X}(\omega) \in L^2([-\pi, \pi], \mathcal{B}, dF^X)$ ($j = 1, 2$). Then using formula (1.28), we get that

$$\operatorname{Cov}(Y_{t+h}^1, Y_t^2) = \int_{-\pi}^{\pi} e^{ih\omega} dF^{Y^1,Y^2}(\omega) = \int_{-\pi}^{\pi} e^{ih\omega} f^{Y^1|X}(\omega)\, \overline{f^{Y^2|X}(\omega)}\, dF^X(\omega).$$

This implies that

$$dF^{Y^1,Y^2}(\omega) = f^{Y^1|X}(\omega)\, \overline{f^{Y^2|X}(\omega)}\, dF^X(\omega). \tag{2.7}$$

2.3 Moving Average processes

Let $\{\xi_t : t \in \mathbb{Z}\}$ be a WN(1) white noise sequence, see Definition 1.4, that is, an *orthonormal sequence* of complex valued random variables: $\mathbb{E}\xi_t = 0$ and $\mathbb{E}(\xi_t \bar{\xi}_s) = \delta_{ts}$. By (1.30), its spectral density function is

$$f^\xi(\omega) = \frac{1}{2\pi}, \qquad \omega \in [-\pi, \pi].$$

Define a two-sided infinite *moving average (MA) process* (a so-called *sliding summation*) by

$$X_t = \sum_{k=-\infty}^{\infty} b_k \xi_{t-k} = \sum_{j=-\infty}^{\infty} b_{t-j} \xi_j, \qquad t \in \mathbb{Z}, \tag{2.8}$$

where we assume that the sequence of the non-random complex coefficients $\{b_k\}$ is in ℓ^2, that is, $\sum_k |b_k|^2 < \infty$. Then by the Riesz–Fischer theorem, (2.8) is convergent in $L^2(\Omega, \mathcal{F}, \mathbb{P})$, i.e., in mean square.

If $\xi_j = 0$ for any $j \neq j_0$, then (2.8) shows that $X_t = b_{t-j_0} \xi_{j_0}$, $t \in \mathbb{Z}$. That is why the sequence $\{b_k\}_{k \in \mathbb{Z}}$ is called the *impulse response function.* Since the single impulse ξ_{j_0} at time j_0 can create nonzero responses not only for times $t \geq j_0$, but also for times $t < j_0$, a sliding summation process $\{X_t\}$ is *non-causal* in general.

It is customary to introduce the *operator of left (backward) shift* L, and the formal power series $H(L) := \sum_k b_k L^k$, so we may write (2.8) as $X_t = H(L)\xi_t$, $t \in \mathbb{Z}$. Further, it is also customary to denote the operator L by the indeterminate z as well and to call the power series

$$H(z) = \sum_{k=-\infty}^{\infty} b_k z^k$$

the *transfer function* of the sliding summation $\{X_t\}$ or the *z-transform* of $H(L)$. Since by our assumption $\sum_k |b_k|^2 < \infty$, we have $H(z) \in L^2(T)$, where T denotes the unit circle of the complex plane $\mathbb{C}$.

By the formulas in Section 2.2, the covariance function of $\{X_t\}$ is

$$c(h) = \mathbb{E}(X_{t+h} \bar{X}_t) = \sum_{j=-\infty}^{\infty} b_j \bar{b}_{j-h}, \qquad h \in \mathbb{Z},$$

so that $\{X_t\}$ is a stationary process. It is not difficult to show that $c(h) \to 0$ as $|h| \to \infty$.

In the special case when $b_k = 0$ whenever $k < 0$, one obtains a *causal (one-sided, future-independent) MA(∞) process:*

$$X_t = \sum_{k=0}^{\infty} b_k \xi_{t-k} = \sum_{j=-\infty}^{t} b_{t-j} \xi_j, \qquad t \in \mathbb{Z},$$

with the covariance function $c(h) = \sum_{k=0}^{\infty} b_{k+h} \bar{b}_k \quad (h \geq 0), \quad c(-h) = \bar{c}(h)$.

Definition 2.1. If q is a nonnegative integer, a *qth order moving average process*, denoted MA(q), is

$$X_t = \sum_{k=0}^{q} \beta_k \xi_{t-k}, \qquad t \in \mathbb{Z},$$

where each $\beta_k \in \mathbb{C}$. Introducing the *MA polynomial*

$$\beta(z) = H(z) := \sum_{k=0}^{q} \beta_k z^k, \qquad z \in \mathbb{C}, \tag{2.9}$$

and using the left (backward) shift operator L, we may write that

$$X_t = \beta(L)\xi_t, \quad X_t = \beta(z)\xi_t \quad t \in \mathbb{Z}.$$

The *covariance function of a MA(q) process* is

$$c(h) = \sum_{k=0}^{q} b_{k+h}\bar{b}_k \quad (0 \le h \le q), \quad c(-h) = \bar{c}(h), \quad c(h) = 0 \text{ if } |h| > q.$$

Conversely, the following is true.

Remark 2.1. ([11, Proposition 3.2.1]). If the covariance function of a stationary process is such that $c(h) = 0$ for $|h| > q$ and $c(q) \neq 0$, then it is a MA(q) process.

The *spectral density of an MA(q) process* by (2.2) is

$$f^X(\omega) = \frac{1}{2\pi}|\phi(\omega)|^2, \qquad \phi(\omega) = f^{X|\xi}(\omega) := \sum_{k=0}^{q} \beta_k e^{-ik\omega} = \beta(e^{-i\omega}), \tag{2.10}$$

where $\beta(z) = H(z)$ is the MA polynomial or transfer function of the MA(q) process. Similar statement is valid for $q = \infty$ too when $\sum_{k=0}^{\infty} |b_k|^2 < \infty$:

$$f^X(\omega) = \frac{1}{2\pi}|\phi(\omega)|^2, \qquad \phi(\omega) = \sum_{k=0}^{\infty} b_k e^{-ik\omega} = H(e^{-i\omega}), \tag{2.11}$$

where $H(z)$ is the transfer function of the process. Likewise, the spectral density of a sliding summation process (2.8) is

$$f^X(\omega) = \frac{1}{2\pi}|\phi(\omega)|^2, \quad \phi(\omega) := \sum_{k=-\infty}^{\infty} b_k e^{-ik\omega} = H(e^{-i\omega}), \quad \sum_{k=-\infty}^{\infty} |b_k|^2 < \infty.$$

Example 2.1. Figure 2.1 shows a typical trajectory of a MA(4) process and its prediction based on the finite past $X_0, \dots, X_{n-1}$, see Subsection 5.2.1. The second and third panel show its covariance function and spectral density. The MA polynomial is $\beta(z) = 1 - \frac{1}{2}z + \frac{1}{2}z^2 + \frac{1}{3}z^3 - \frac{1}{3}z^4$. Observe that the covariance is 0 if $h > q = 4$.

The first panel on Figure 2.2 shows the mean square prediction error as a function of n as X_n is predicted. The prediction error goes to 1, which is the non-zero impulse response b_0, since it is a regular process, see Section 2.6. The last panel shows $\det(\boldsymbol{C}_n)$.

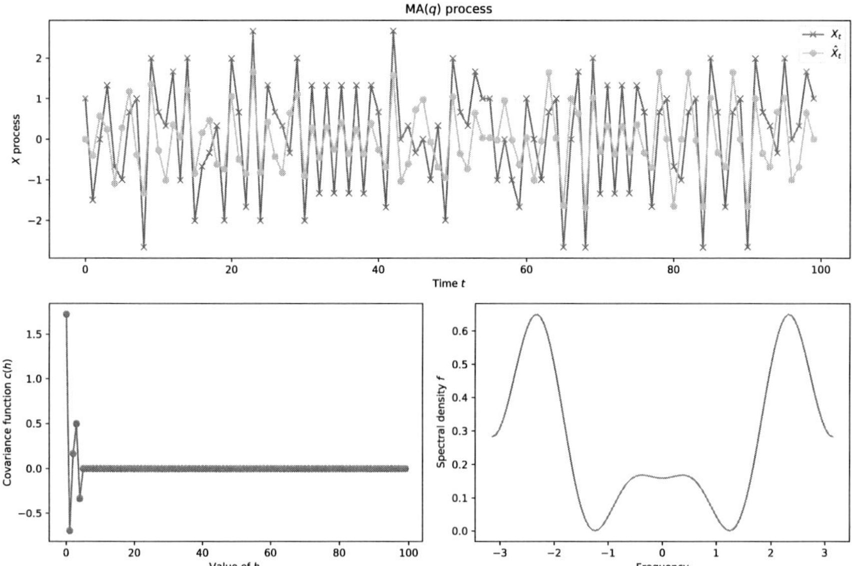

FIGURE 2.1
A typical trajectory and its prediction of an MA(4) process, with its covariance function and spectral density in Example 2.1.

Example 2.2. Figure 2.3 shows a typical trajectory of a MA(∞) process and its prediction based on the finite past $X_0, \ldots, X_{n-1}$, see Subsection 5.2.1. The second and third panel show its covariance function and spectral density. The impulse response function is $b_k = 1/(k+1)$ for $k \geq 0$.

The first panel on Figure 2.4 shows the mean square prediction error as a function of n as X_n is predicted . The prediction error goes to 1, which is the non-zero impulse response b_0, since it is a regular process, see Section 2.6. The last panel shows $\det(\boldsymbol{C}_n)$.

Example 2.3. Figure 2.5 shows a typical trajectory of a sliding summation process and its prediction based on the finite past $X_0, \ldots, X_{n-1}$, see Subsection 5.2.1. The second and third panel show its covariance function and spectral density. The parameters are

$$b_k = \begin{cases} 1 & (k=0) \\ 1/|k| & (k \neq 0) \end{cases} .$$

Observe that $|\log f|$ is significantly larger here than in the case of MA(∞).

The first panel on Figure 2.6 shows the mean square prediction error as a function of n as X_n is predicted. The prediction error decreases. The last panel shows $\det(\boldsymbol{C}_n)$.

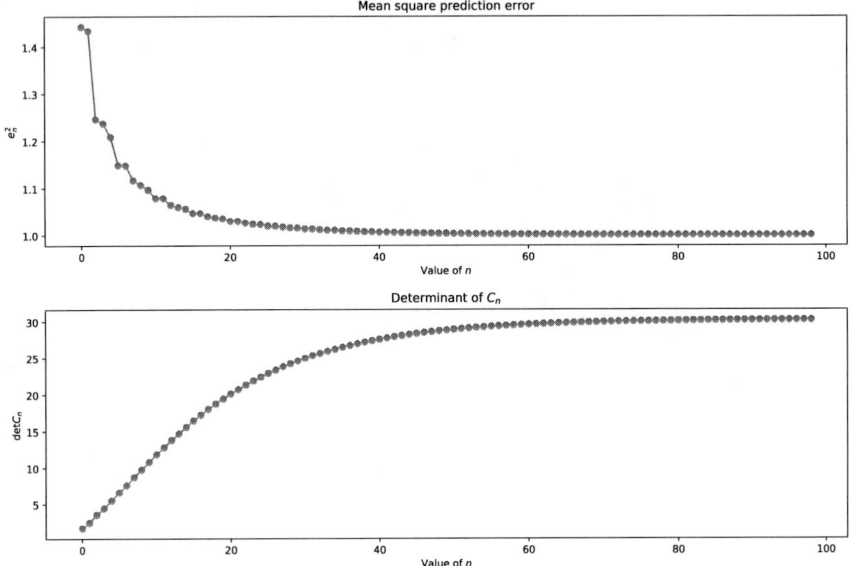

FIGURE 2.2
The mean square prediction error and $\det(\boldsymbol{C}_n)$ of an MA(4) process in Example 2.1.

2.4 Autoregressive processes

Next let us consider a *first order autoregressive process*, denoted AR(1), in fact, a first order stochastic linear difference equation:

$$X_t = \alpha X_{t-1} + \beta \xi_t, \quad t \in \mathbb{Z},$$

where α and $\beta \neq 0$ are complex constants. We assume that $\{\xi_t\}$ is a WN(1) (orthonormal) sequence and each X_t depends only on the present and past values $(\xi_t, \xi_{t-1}, \dots)$ of the driving white noise process. Iterating the equation k times we get

$$X_t = \beta \xi_t + \alpha \beta \xi_{t-1} + \alpha^2 \beta \xi_{t-2} + \cdots + \alpha^k \beta \xi_{t-k} + \alpha^{k+1} X_{t-k-1}.$$

In order that the right side have bounded norm as $k \to \infty$, it is necessary that $|\alpha| < 1$. If that is so the series $\sum_{k=0}^{\infty} \alpha^k \xi_{t-k}$ converges and $\alpha^{k+1} X_{t-k-1} \to 0$ as $k \to \infty$, since we are looking for a stationary process, where $\|X_t\| := \left(\mathbb{E}(|X_t|^2)\right)^{1/2}$ is constant. Then

$$X_t = \beta \sum_{k=0}^{\infty} \alpha^k \xi_{t-k}, \quad t \in \mathbb{Z},$$

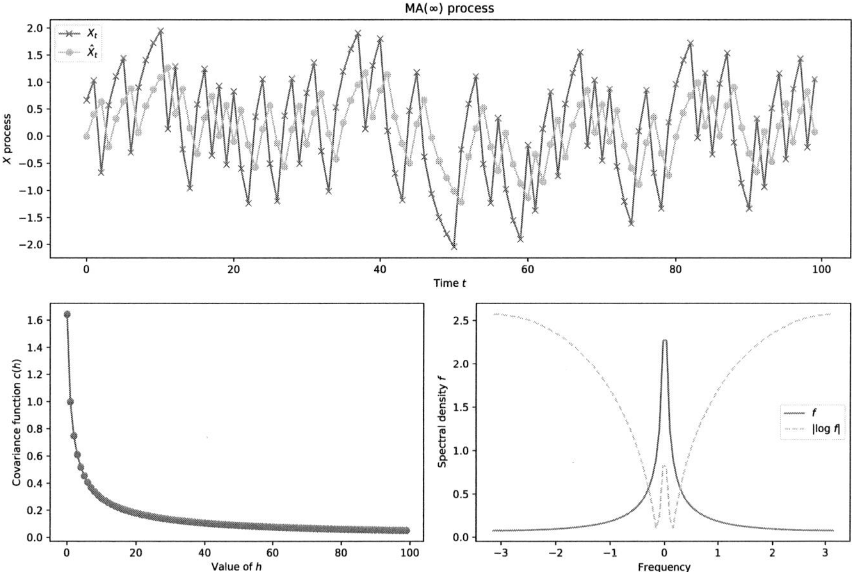

FIGURE 2.3
A typical trajectory and its prediction of an MA(∞) process, with its covariance function and spectral density in Example 2.2.

converges in mean square, it is the only stationary solution of the AR(1) process and it is a causal MA(∞) process. Its covariance function is

$$c(h) = |\beta|^2 \frac{\alpha^h}{1 - |\alpha|^2} \quad (h \in \mathbb{Z}),$$

converging to 0 exponentially fast.

Definition 2.2. A *pth order autoregressive process*, denoted AR(p), in fact a pth order stochastic linear difference equation, is a natural generalization:

$$\sum_{j=0}^{p} \alpha_j X_{t-j} = \beta \xi_t, \quad t \in \mathbb{Z}, \quad \alpha_0 = 1, \quad \beta \neq 0. \tag{2.12}$$

Using the *AR polynomial*

$$\alpha(z) := \sum_{j=0}^{p} \alpha_j z^j, \qquad z \in \mathbb{C}, \tag{2.13}$$

we may concisely write that

$$\alpha(L) X_t = \beta \xi_t, \quad \text{or} \quad \alpha(z) X_t = \beta \xi_t, \quad t \in \mathbb{Z}.$$

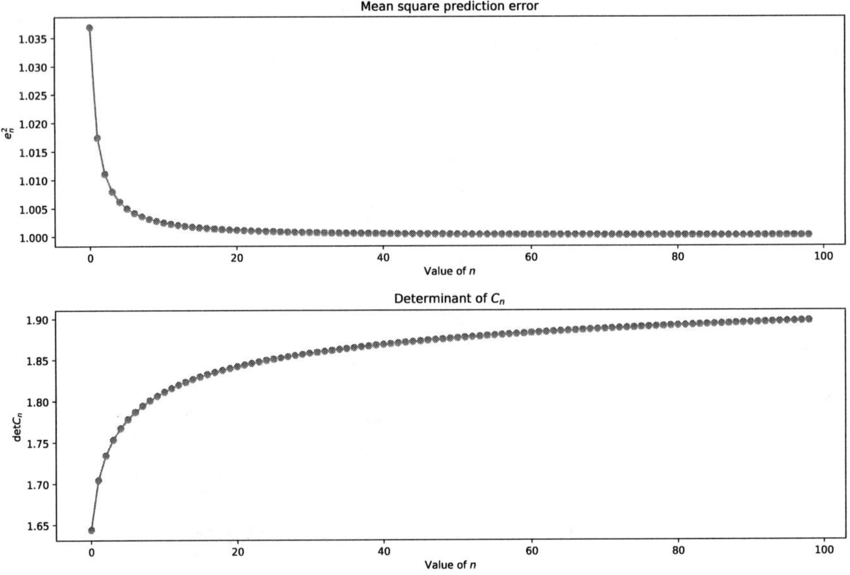

FIGURE 2.4
The mean square prediction error and $\det(\boldsymbol{C}_n)$ of an MA(∞) process in Example 2.2.

We have changed the notation used in AR(1) a little bit, though we may equivalently write that

$$X_t = -\sum_{j=1}^{p} \alpha_j X_{t-j} + \beta \xi_t, \qquad t \in \mathbb{Z}.$$

Quite often in the literature, the negative sign on the right hand side is replaced by positive sign.

Here we always assume that each X_t depends only on the present and past values $(\xi_t, \xi_{t-1}, \dots)$ of the driving white noise process, so there is no remote past, see Section 2.6.

To find such a stationary solution of equation (2.12), let us look for a causal MA(∞) solution of form

$$X_t = \sum_{k=0}^{\infty} b_k \xi_{t-k}, \quad t \in \mathbb{Z}, \quad b_k \in \mathbb{C}. \tag{2.14}$$

Substitute this in the equation (2.12) and equate the coefficients of ξ_ℓ on both sides, starting with ξ_t and working toward the past:

$$\sum_{k=0}^{\infty} \sum_{j=0}^{p} \alpha_j b_k \xi_{t-j-k} = \beta \xi_t. \tag{2.15}$$

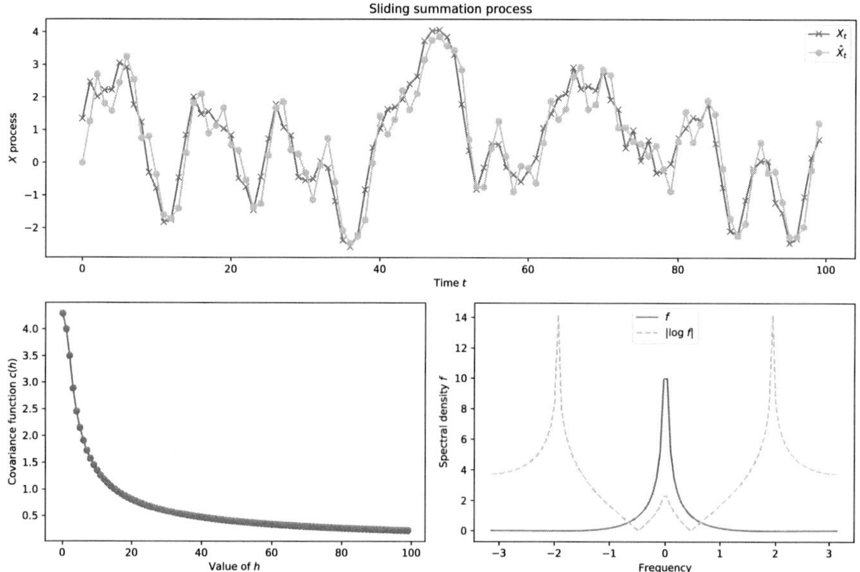

FIGURE 2.5
A typical trajectory and its prediction of a sliding summation process, with its covariance function and spectral density in Example 2.3.

The result will be ($\alpha_0 = 1$):

$$\begin{aligned}
b_0 &= \beta, \\
b_1 + \alpha_1 b_0 &= 0, \\
b_2 + \alpha_1 b_1 + \alpha_2 b_0 &= 0, \\
&\vdots \\
b_{p-1} + \alpha_1 b_{p-2} + \cdots + \alpha_{p-1} b_0 &= 0, \\
b_{p+k} + \alpha_1 b_{p+k-1} + \cdots + \alpha_p b_k &= 0 \qquad (k \geq 0).
\end{aligned} \tag{2.16}$$

If $\{\alpha_j : j = 1, \ldots, p\}$ and β are known, these equations uniquely determine the coefficients $\{b_k : k \geq 0\}$ by recursion. The question is whether or not this sequence will be square summable, so the proposed solution will converge in mean square.

Proposition 2.2. *An AR(p) process has a unique causal MA(∞) solution* (2.14)*, convergent in mean square, if and only if the corresponding AR polynomial $\alpha(z)$ has no zeros in the closed unit disc $\{z : |z| \leq 1\}$. In this case the AR(p) process is called* stable.

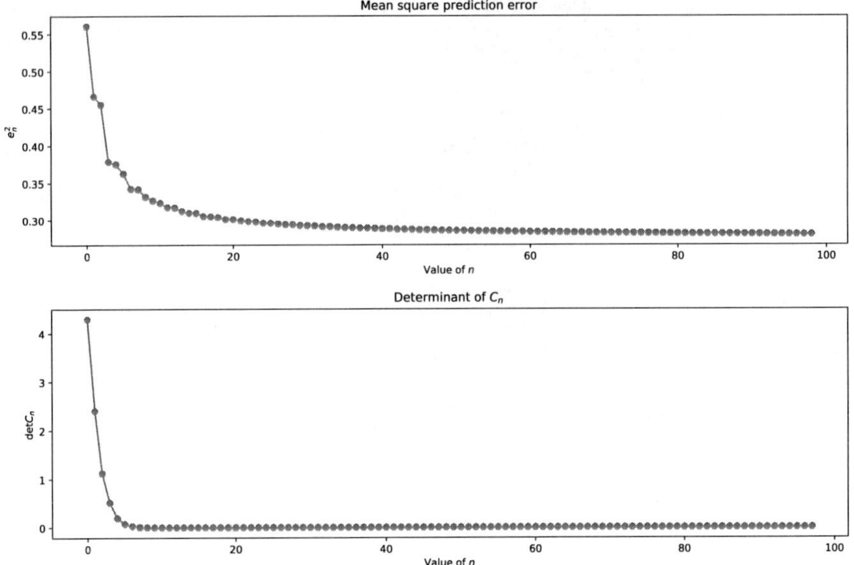

FIGURE 2.6
The mean square prediction error and $\det(\boldsymbol{C}_n)$ of a sliding summation process in Example 2.3.

Proof. Here we sketch an elementary proof. The numerical (not random) homogeneous linear difference equation

$$\sum_{j=0}^{p} \alpha_j b_{k-j} = 0 \quad (k \geq p) \tag{2.17}$$

in (2.16) has a nonzero solution $b_k = w^k \in \mathbb{C}$ if and only if $w \neq 0$ is a root of the polynomial

$$\sum_{j=0}^{p} \alpha_j w^{p-j} = w^p \sum_{j=0}^{p} \alpha_j (1/w)^j = w^p \alpha(1/w).$$

If the p complex roots of the AR polynomial $\alpha(z)$ are distinct, say $z_1, z_2, \ldots, z_p$, and none of them is 0, then the general solution of equation (2.17) is

$$b_k = \sum_{j=1}^{p} B_j w_j^k = \sum_{j=1}^{p} B_j (1/z_j)^k, \qquad k \geq 0, \tag{2.18}$$

where B_j $(j = 1, \ldots, p)$ are complex constants. It is easy to see that the solution space of equation (2.17) is p-dimensional. The constants $B_1, \ldots, B_p$ can be uniquely determined from the first p equations of (2.16), which is a system of linear equations for them.

If each $|z_j| > 1$, then the sequence $\{b_j\}_{j\geq 0}$ will be square summable. It is not difficult to show that this conclusion still holds even if some of the roots coincide. Thus if the AR polynomial $\alpha(z)$ has no roots in the closed unit disc, then (2.14) and (2.18) give a causal MA(∞) stationary solution of the AR(p) process (2.12), which is convergent in mean square.

Conversely, if for one or more roots z_j of the AR polynomial we have $|z_j| \leq 1$, then it is clear from (2.18) that the sequence $\{b_k\}_{k\geq 0}$ cannot be square summable. □

The covariance function of a stable AR(p) can be obtained in terms of the coefficients $\{b_k\}$ by (2.14):

$$c(h) = \mathbb{E}(X_{t+h}\bar{X}_t) = \sum_{k=0}^{\infty} b_{k+h}\bar{b}_k \quad (h \geq 0), \quad c(-h) = \bar{c}(h), \tag{2.19}$$

and $c(h)$ converges to 0 exponentially fast as $|h| \to \infty$.

However, it is more important to describe the covariance function from the defining formula (2.12). Multiply the complex conjugate of (2.12) by X_{t-k}, $k \geq 0$, and take expectation:

$$\sum_{j=0}^{p} \bar{\alpha}_j \mathbb{E}(X_{t-k}\bar{X}_{t-j}) = \mathbb{E}(X_{t-k}\bar{\beta}\bar{\xi}_t).$$

This way we obtain the important *Yule–Walker equations*:

$$\sum_{j=0}^{p} c(j-k)\bar{\alpha}_j = \delta_{0k}|\beta|^2, \qquad k \geq 0, \tag{2.20}$$

see (2.14) and the first equation $b_0 = \beta$ in (2.16).

The significance of the Yule–Walker equations is that taking them for $0 \leq k \leq p$, one obtains a system of linear equations for the unknowns $\alpha_1, \dots, \alpha_p$ ($\alpha_0 = 1$) and $|\beta|^2$ if the covariances $c(0), c(1), \dots, c(p)$ are known:

$$\begin{bmatrix} c(0) & c(1) & c(2) & \dots & c(p) \\ \bar{c}(1) & c(0) & c(1) & \dots & c(p-1) \\ \bar{c}(2) & \bar{c}(1) & c(0) & \dots & c(p-2) \\ \vdots & \vdots & \vdots & \ddots & \vdots \\ \bar{c}(p) & \bar{c}(p-1) & \bar{c}(p-2) & \dots & c(0) \end{bmatrix} \begin{bmatrix} 1 \\ \bar{\alpha}_1 \\ \bar{\alpha}_2 \\ \vdots \\ \bar{\alpha}_p \end{bmatrix} = \begin{bmatrix} |\beta|^2 \\ 0 \\ 0 \\ \vdots \\ 0 \end{bmatrix}.$$

Rearrange this system into a standard form:

$$\begin{bmatrix} c(0) & c(1) & \dots & c(p-1) \\ \bar{c}(1) & c(0) & \dots & c(p-2) \\ \vdots & \vdots & \ddots & \vdots \\ \bar{c}(p-1) & \bar{c}(p-2) & \dots & c(0) \end{bmatrix} \begin{bmatrix} \bar{\alpha}_1 \\ \bar{\alpha}_2 \\ \vdots \\ \bar{\alpha}_p \end{bmatrix} = - \begin{bmatrix} \bar{c}(1) \\ \bar{c}(2) \\ \vdots \\ \bar{c}(p) \end{bmatrix},$$

$$|\beta|^2 = c(0) + c(1)\bar{\alpha}_1 + c(2)\bar{\alpha}_2 + \dots + c(p)\bar{\alpha}_p. \tag{2.21}$$

The matrix of the system is a Toeplitz matrix $\boldsymbol{C}_p$, defined in (1.7), self-adjoint and non-negative definite. Introducing $\boldsymbol{\alpha} := [\alpha_1, \dots, \alpha_p]^T$, $\mathbf{c} := [c(1), \dots, c(p)]^T$, we may write the above system of linear equations as

$$\boldsymbol{C}_p \bar{\boldsymbol{\alpha}} = -\bar{\mathbf{c}}. \tag{2.22}$$

If $\boldsymbol{C}_p$ is positive definite, then there exists a unique solution: $\bar{\boldsymbol{\alpha}} = -\boldsymbol{C}_p^{-1}\bar{\mathbf{c}}$. Otherwise, one may use the Moore–Penrose inverse $\boldsymbol{C}_p^{+}$ instead, see Definition B.4 in Appendix B. The system (2.22) is always consistent, always has a solution, because it is a *Gauss normal equation*, see Appendix C. For, taking $\mathbf{X} := [X_{p-1}X_{p-2}\dots X_0]^T$, we have

$$\boldsymbol{C}_p = \mathbb{E}(\mathbf{X}\mathbf{X}^*), \quad \bar{\mathbf{c}} = \mathbb{E}(\mathbf{X}\bar{X}_p),$$

and this shows that the right hand side of (2.22) is in the column space of the left hand side.

Assume that the AR(p) process is stable. Introduce the $\infty \times p$ lower trapezoidal Toeplitz matrix

$$\boldsymbol{B} := \begin{bmatrix} b_0 & 0 & \cdots & 0 \\ b_1 & b_0 & \cdots & 0 \\ \vdots & \vdots & \ddots & \vdots \\ b_{p-1} & b_{p-2} & \cdots & b_0 \\ b_p & b_{p-1} & \cdots & b_1 \\ b_{p+1} & b_p & \cdots & b_2 \\ \vdots & \vdots & & \vdots \end{bmatrix} \tag{2.23}$$

and the infinite column matrix $\mathbf{b} := [b_1 b_2 b_3 \cdots]^T$ using the coefficients $\{b_0, b_1, b_2, \dots\}$ introduced in (2.14). Then the infinite system of linear equations (2.16) can be written as

$$\boldsymbol{B}\boldsymbol{\alpha} = -\mathbf{b},$$

and in the light of (2.19), equation (2.22) is equivalent to

$$\boldsymbol{B}^*\boldsymbol{B}\boldsymbol{\alpha} = -\boldsymbol{B}^*\mathbf{b}, \quad \boldsymbol{C}_p^T = \boldsymbol{B}^*\boldsymbol{B}, \quad \mathbf{c} = \boldsymbol{B}^*\mathbf{b}.$$

It is a Cholesky-type decomposition of $\boldsymbol{C}_p^T$. By (2.16), $b_0 = \beta \neq 0$, so (2.23) shows that the range of $\boldsymbol{B}$ is p-dimensional and this p-dimensional subspace of ℓ^2 is mapped by $\boldsymbol{B}^*$ onto $\mathbb{C}^p$.

Corollary 2.1. *If the AR(p) process is stable, then* $\operatorname{rank}(\boldsymbol{C}_p) = p$, *so the system* (2.22) *has a unique solution* $\boldsymbol{\alpha} = -(\boldsymbol{C}_p^T)^{-1}\mathbf{c}$.

Since the covariances can be easily estimated from a random sample, see Subsection 1.5.2, the resulting estimated version of (2.21) gives a practical method for estimating the coefficients $\alpha_1, \dots, \alpha_p$ and $|\beta|$ of an AR(p) process.

On the other hand, if the coefficients $\alpha_1, \dots, \alpha_p$ and β are known, then one can determine the covariances $c(0), c(1), \dots, c(p)$ from a $(p+1) \times (p+1)$ system of linear equations based on the Yule–Walker equations (2.20), and then one can determine $c(p+1), c(p+2), \dots$ recursively from (2.20).

Let us determine the *spectral density of an AR(p) process* $\{X_t\}$, assuming that the AR polynomial $\alpha(z)$ has no roots in the closed unit disc. By (2.2), we get the following equation for the spectral densities of the left and right hand side of (2.12):

$$|\psi(\omega)|^2 f^X(\omega) = \beta \frac{1}{2\pi}, \qquad \psi(\omega) := \sum_{j=0}^{p} \alpha_j e^{-ij\omega} = \alpha(e^{-i\omega}),$$

where $f^X(\omega)$ denotes the spectral density of the AR(p) process $\{X_t\}$ and $\alpha(z)$ is its AR polynomial. Thus

$$f^X(\omega) = \frac{1}{2\pi} \frac{\beta}{|\psi(\omega)|^2} = \frac{1}{2\pi} \frac{\beta}{|\alpha(e^{-i\omega})|^2}, \qquad \omega \in [-\pi, \pi]. \tag{2.24}$$

Example 2.4. Figure 2.7 shows a typical trajectory of a stable AR(4) process and its prediction based on the finite past $X_0, \dots, X_{n-1}$, see Subsection 5.2.1. The second, third, and fourth panel show its impulse response b_k coefficients, covariance function, and spectral density. The AR polynomial is $\alpha(z) = 1 - \frac{7}{6}z + \frac{1}{2}z^2 + \frac{1}{12}z^3 - \frac{1}{12}z^4$ and $\beta = \frac{1}{2}$. The last panel shows the exponential decrease of the covariance function.

The first panel on Figure 2.8 shows the mean square prediction error of this process as a function of n as X_n is predicted. The prediction error goes to 1, which is the non-zero impulse response b_0, since it is a regular process, see Section 2.6. The last panel shows $\det(\boldsymbol{C}_n)$.

2.5 Autoregressive moving average processes

Definition 2.3. The *autoregressive, moving average processes*, denoted ARMA(p, q) ($p \geq 0$, $q \geq 0$), are generalizations of both AR(p) and MA(q) processes:

$$\sum_{j=0}^{p} \alpha_j X_{t-j} = \sum_{\ell=0}^{q} \beta_\ell \xi_{t-\ell}, \quad t \in \mathbb{Z}, \quad \alpha_0 = 1, \quad \beta_0 \neq 0. \tag{2.25}$$

Concisely,

$$\alpha(L) X_t = \beta(L) \xi_t, \quad \text{or} \quad \alpha(z) X_t = \beta(z) \xi_t, \quad t \in \mathbb{Z}, \tag{2.26}$$

where L is the left shift operator, $\alpha(z)$ is the AR polynomial (2.13), and $\beta(z)$ is the MA polynomial (2.9).

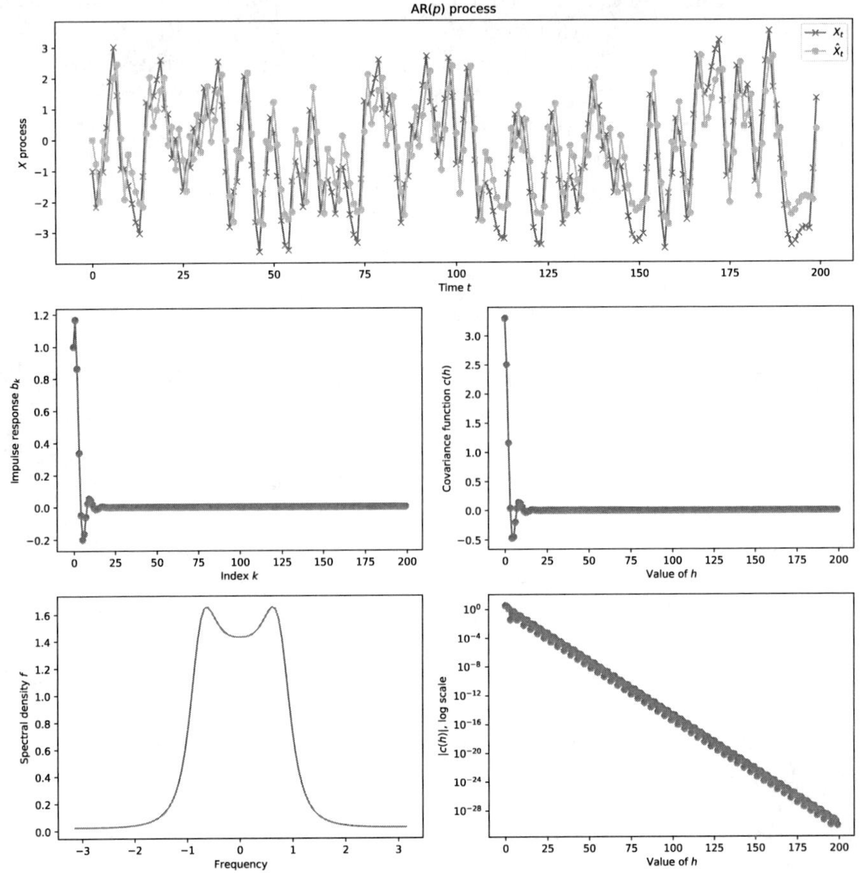

FIGURE 2.7
A typical trajectory and its prediction of a stable AR(4) process, with its b_k coefficients, covariance function, and spectral density in Example 2.4.

Again, we want to find a causal stationary solution $\{X_t\}$ of this equation, that is, a MA(∞) solution of form (2.14). Substituting this into the equation, now we get

$$\sum_{k=0}^{\infty}\sum_{j=0}^{p} \alpha_j b_k \xi_{t-j-k} = \sum_{\ell=0}^{q} \beta_\ell \xi_{t-\ell},$$

similarly to (2.15). Here too, we can equate the coefficients of ξ_ℓ on both sides, starting with ξ_t and working toward the past. Thus we get a similar system of equations as (2.16), the only difference being that on the right hand side we may have now nonzero numbers in the first $q+1$ equations instead of only

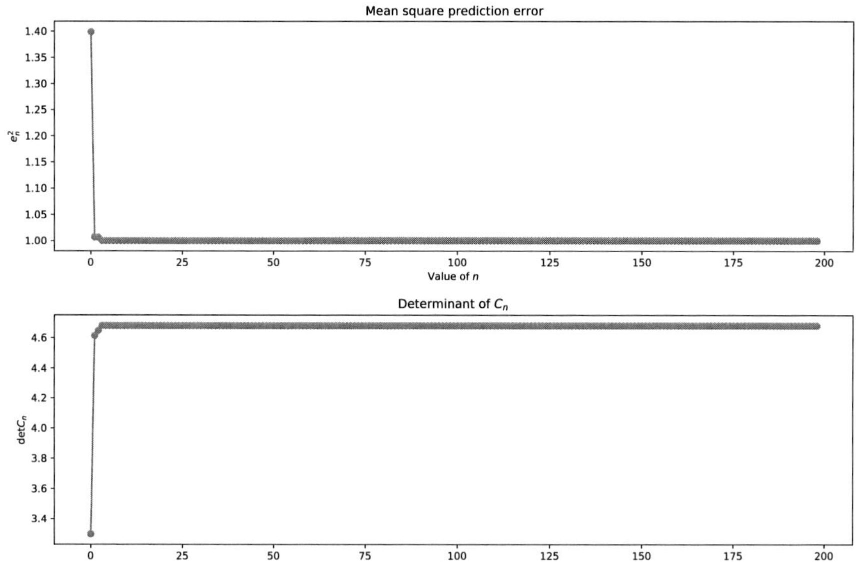

FIGURE 2.8
The mean square prediction error and $\det(\boldsymbol{C}_n)$ of a stable AR(4) process in Example 2.4.

in the first equation:

$$
\begin{aligned}
b_0 &= \beta_0, \\
b_1 + \alpha_1 b_0 &= \beta_1, \\
b_2 + \alpha_1 b_1 + \alpha_2 b_0 &= \beta_2, \\
&\vdots \\
b_{p-1} + \alpha_1 b_{p-2} + \cdots + \alpha_{p-1} b_0 &= \beta_{p-1} \\
b_{p+k} + \alpha_1 b_{p+k-1} + \cdots + \alpha_p b_k &= 0, \qquad (k \geq 0).
\end{aligned} \tag{2.27}
$$

For simplicity, we assumed that $p > q$; adding extra coefficients $\alpha_j = 0$ this can always be achieved. If $p - 1 > q$, then $\beta_{q+1}, \ldots, \beta_{p-1}$ are 0.

Several statements go similarly now as in the case of AR(p). For completeness, we state here the stability condition for ARMA(p,q) processes that can be proved exactly the same way as Proposition 2.2.

Proposition 2.3. *An ARMA(p, q) process has a unique causal MA(∞) solution* (2.14)*, convergent in mean square, if and only if the corresponding AR polynomial $\alpha(z)$ has no zeros in the closed unit disc $\bar{D} = \{z : |z| \leq 1\}$. In this case the ARMA(p,q) process is called* stable.

Stability implies that the covariance function $c(h)$ tends to 0 exponentially fast as $h \to \infty$. With a stable ARMA process we may introduce the *transfer function* $H(z)$:

$$H(z) := \sum_{j=0}^{\infty} b_j z^j = \frac{\beta(z)}{\alpha(z)}, \qquad z \in \bar{D}.$$

The system of linear equations (2.27) for the coefficients $\{b_j\}_{j\geq 0}$ is exactly the same as that obtained from the equation

$$\alpha(z)H(z) = \beta(z), \qquad z \in \bar{D},$$

equating the terms with the same power z^j $(j = 0, 1, 2, \dots)$ on the two sides. Transfer functions will be treated in more detail in Chapters 3 and 4. Using this notation we may write that

$$X_t = H(z)\xi_t = \alpha^{-1}(z)\beta(z)\xi_t, \text{ or } X_t = H(L)\xi_t = \alpha^{-1}(L)\beta(L)\xi_t, \quad t \in \mathbb{Z},$$

compare with (2.26).

The Yule–Walker equations now are more complex than (2.20). For example, it is no longer true that from them one can get a linear system of equations for the unknown parameters α_j $(j = 0, \dots, p)$ and β_ℓ $(\ell = 0, \dots, q)$ if the covariance function $c(h)$ $(h \in \mathbb{Z})$ is known. In fact, multiply the complex conjugate of (2.25) by X_{t-k}, $k \geq 0$, and take expectation:

$$\sum_{j=0}^{p} \bar{\alpha}_j \mathbb{E}(X_{t-k}\bar{X}_{t-j}) = \sum_{\ell=0}^{q} \bar{\beta}_\ell \mathbb{E}(X_{t-k}\bar{\xi}_{t-\ell}).$$

Using the MA(∞) solution (2.14) as well, we obtain

$$\sum_{j=0}^{p} c(j-k)\bar{\alpha}_j = \sum_{\ell=k}^{q} b_{l-k}\bar{\beta}_\ell, \qquad k \geq 0. \tag{2.28}$$

(The right hand side here is 0 if $k > q$.) If we use (2.27) to express the coefficients $b_{k-\ell}$ in terms of the parameters α_j and β_ℓ recursively, then the right hand side of the first q equations in (2.28) will contain polynomial expressions of the parameters.

Using the method of (2.2), the *spectral density of an ARMA process* $\{X_t\}$ is

$$f^X(\omega) = \frac{1}{2\pi}\frac{|\beta(e^{-i\omega})|^2}{|\alpha(e^{-i\omega})|^2}, \quad \beta(e^{-i\omega}) = \sum_{\ell=0}^{q} \beta_\ell e^{-i\ell\omega}, \quad \alpha(e^{-i\omega}) = \sum_{j=0}^{p} \alpha_j e^{-ij\omega}, \tag{2.29}$$

for $\omega \in [-\pi, \pi]$.

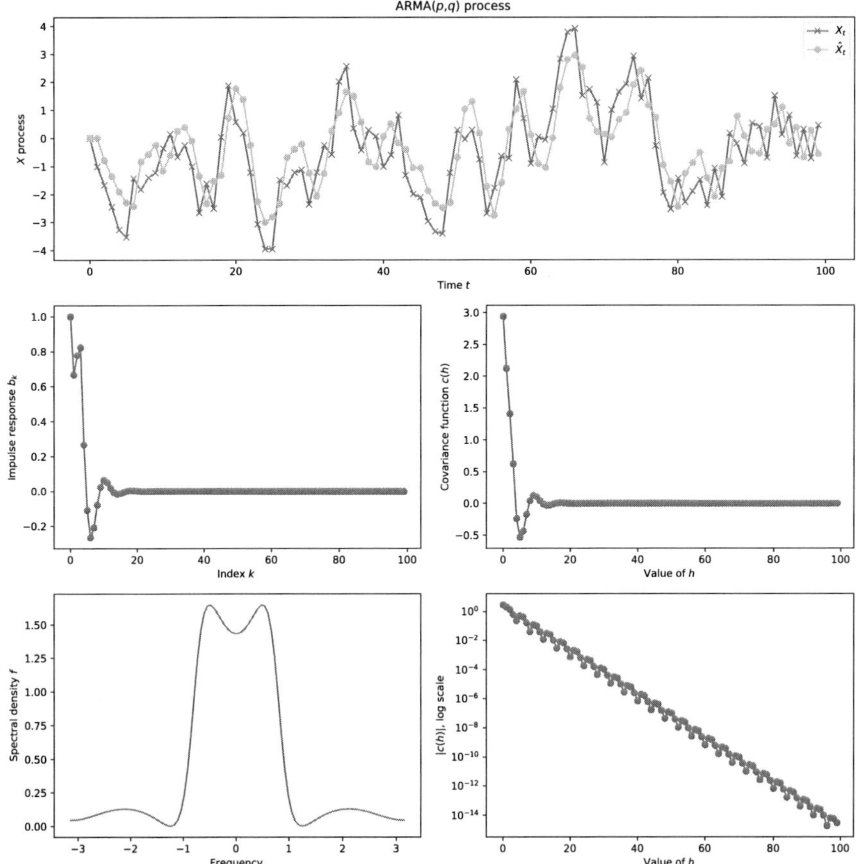

FIGURE 2.9
A typical trajectory and its prediction of a stable ARMA(4,4) process, with its b_k coefficients, covariance function, and spectral density in Example 2.5.

Example 2.5. Figure 2.9 shows a typical trajectory of a stable ARMA$(4, 4)$ process and its prediction based on the finite past $X_0, \ldots, X_{n-1}$, see Subsection 5.2.1. The second, third, and fourth panel show its impulse response b_k coefficients, covariance function, and spectral density. The AR polynomial is $\alpha(z) = 1 - \frac{7}{6}z + \frac{1}{2}z^2 + \frac{1}{12}z^3 - \frac{1}{12}z^4$ and the MA polynomial is $\beta(z) = 1 - \frac{1}{2}z + \frac{1}{2}z^2 + \frac{1}{3}z^3 - \frac{1}{3}z^4$. The last panel shows the exponential decrease of the covariance function.

The first panel on Figure 2.10 shows the mean square prediction error as a function of n as X_n is predicted. The prediction error goes to 1, which is the non-zero impulse response b_0, since it is a regular process, see Section 2.6. The last panel shows $\det(\boldsymbol{C}_n)$.

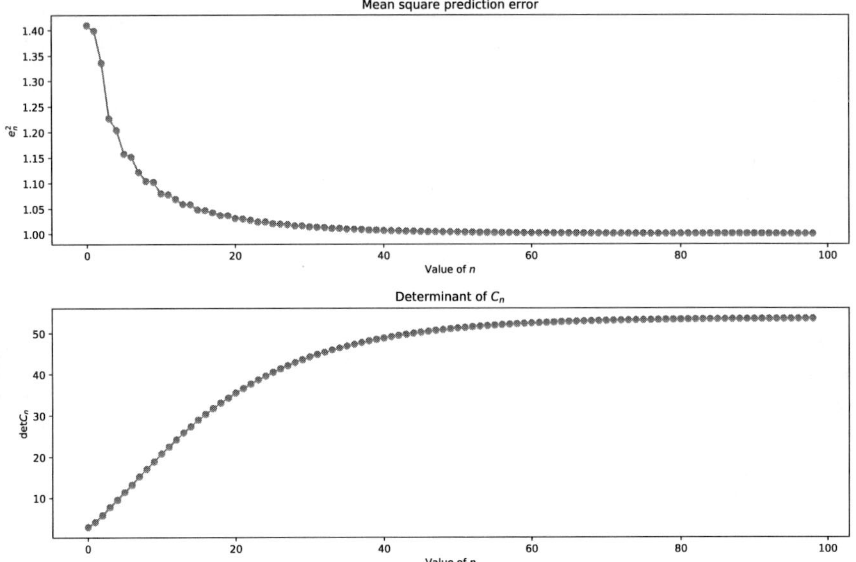

FIGURE 2.10
The mean square prediction error and $\det(\boldsymbol{C}_n)$ of an ARMA(p, q) process in Example 2.5.

2.6 Wold decomposition in 1D

Recall that we use the notation $\overline{\text{span}}\{X_t : t \in A\}$ for the closed linear span of the random variables $\{X_t : t \in A\} \subset L^2(\Omega, \mathcal{F}, \mathbb{P})$. Then for a weakly stationary time series $\{X_t\}_{t\in\mathbb{Z}}$, we define

$$H_n^-(X) = \overline{\text{span}}\{X_t : t \leq n\} \quad (n \in \mathbb{Z}), \quad H_{-\infty}(X) = \bigcap_{n\in\mathbb{Z}} H_n^-(X).$$

If it is clear from the context, we simply write H_n^- instead of $H_n^-(X)$ and $H_{-\infty}$ instead of $H_{-\infty}(X)$. H_n^- is called the *past of* $\{X_t\}$ *until* n and $H_{-\infty}$ is the *remote past of* $\{X_t\}$. Clearly, $H(X) = \overline{\text{span}}\{H_n^- : n \in \mathbb{Z}\}$ and by the weak stationarity of $\{X_t\}_{t\in\mathbb{Z}}$, $H_n^- = S^n H_0^-$ $(n \in \mathbb{Z})$, where S is the unitary operator of right time-shift. The time series is called *singular* if $H_{-\infty} = H(X)$, or equivalently, if $H_k^- = H_{k+1}^-$ for some $k \in \mathbb{Z}$, that is, if $H_k^- = H_{k+m}^-$ for any $k, m \in \mathbb{Z}$; otherwise, it is *non-singular*. The time series is *regular* if $H_{-\infty} = \{0\}$. We are going to see that in general, a weakly stationary time series can be written as an orthogonal sum of a regular and a singular time series.

Assume that $\{X_t\}_{t\in\mathbb{Z}}$ is *non-singular*, so $H_{-1}^- \neq H_0^- = \overline{\text{span}}\{H_{-1}^-, X_0\}$. Let

$$X_0 = X_0^- + X_0^+, \quad X_0^- \in H_{-1}^-, \quad X_0^+ \perp H_{-1}^-.$$

Define the random variable $\xi_0 := X_0^+/\|X_0^+\|$. Then $\xi_0 \in H_0^-$, $\|\xi_0\| = 1$, $\xi_0 \perp H_{-1}^-$. Define $\xi_n := S^n \xi_0$ $(n \in \mathbb{Z})$. Clearly, $\xi_n \in H_n^-$, $\xi_n \perp H_{n-1}^-$, and $H_n^- = \overline{\text{span}}\{H_{n-1}^-, \xi_n\}$. Thus $\{\xi_n\}_{n\in\mathbb{Z}}$ is an orthonormal sequence and $\xi_n \perp H_{-\infty}$ for each n. This procedure resembles a Gram–Schmidt orthogonalization.

Now let us expand X_0 into its orthogonal series w.r.t. $\{\xi_n\}_{t\in\mathbb{Z}}$:

$$X_0 = \sum_{k=0}^{\infty} b_k \xi_{-k} + Y_0. \tag{2.30}$$

Here $b_k = \langle X_0, \xi_{-k}\rangle$, $\sum_k |b_k|^2 < \infty$, and $b_k = 0$ for $k < 0$ because $\xi_k \perp H_0^-$ when $k \geq 1$. In particular,

$$b_0 = \langle X_0, \xi_0\rangle = \langle X_0^+, X_0^+/\|X_0^+\|\rangle = \|X_0^+\| > 0. \tag{2.31}$$

The vector Y_0 is simply the remainder term, which of course is 0 if $\{\xi_n\}_{t\in\mathbb{Z}}$ span $H(X)$, but not in general. It is not hard to see that $Y_0 \in H_{-\infty}$.

Now apply the operator S^t to (2.30):

$$X_t = \sum_{k=0}^{\infty} b_k \xi_{t-k} + Y_t =: R_t + Y_t \quad (t \in \mathbb{Z}), \tag{2.32}$$

where we have defined $Y_t := S^t Y_0$ $(t \in \mathbb{Z})$. This way we have proved the important *Wold decomposition* of $\{X_t\}$.

Theorem 2.1. *Assume that* $\{X_t\}_{t\in\mathbb{Z}}$ *is a* non-singular *weakly stationary time series. Then we can decompose* $\{X_t\}$ *in the form* (2.32), *where* $\{R_t\}_{t\in\mathbb{Z}}$ *is a regular time series (that is, a causal* MA(∞) *process) and* $\{Y_t\}_{t\in\mathbb{Z}}$ *is a singular time series,* $Y_t \in H_{-\infty}$ *for all* t*; the two processes are orthogonal to each other.*

The *best linear prediction* $\hat{X}_h$ of X_h $(h \geq 1)$ is by definition the projection of X_h to the past until 0, that is, to H_0^-. Time 0 is considered the present and $\hat{X}_h$ is called a *h-step ahead prediction.* By the Wold decomposition (2.32) and the projection theorem, the best prediction is

$$\hat{X}_h = \sum_{k=h}^{\infty} b_k \xi_{h-k} + Y_h = \sum_{j=-\infty}^{0} b_{h-j} \xi_j + Y_h \quad (h \geq 1), \tag{2.33}$$

since the right hand side of (2.33) is in H_0^- and the difference $(X_h - \hat{X}_h)$ is orthogonal to H_0^-. Hence, the *prediction error* (the mean-square error) of the h-step ahead prediction is given by

$$\sigma_h^2 := \|X_h - \hat{X}_h\|^2 = \sum_{k=0}^{h-1} |b_k|^2. \tag{2.34}$$

This implies that

$$\lim_{h\to\infty} \sigma_h^2 = \sum_{k=0}^{\infty} |b_k|^2 = \left\| \sum_{k=0}^{\infty} b_k \xi_{\ell-k} \right\|^2 \quad (\ell \in \mathbb{Z}), \qquad \lim_{h\to\infty} \sum_{k=h}^{\infty} b_k \xi_{h-k} = 0.$$

Clearly, the MA part $\{R_t\}_{t\in\mathbb{Z}}$ of $\{X_t\}_{t\in\mathbb{Z}}$ is a regular time series. For, the past $H_n^-(R) := \overline{\text{span}}\{R_t : t \le n\}$ of $\{R_t\}$ is spanned by the vectors $\{\xi_t : t \le n\}$, so

$$H_{-\infty}(R) := \bigcap_{n\in\mathbb{Z}} H_n^-(R) = \lim_{n\to-\infty} H_n^-(R) = \{0\}.$$

If the process $\{X_t\}_{t\in\mathbb{Z}}$ itself is regular, then $Y_t = 0$ for each $t \in \mathbb{Z}$,

$$\lim_{t\to\infty} \sigma_t^2 \to \|X_\ell\|^2 \quad (\ell \in \mathbb{Z}), \text{ and } \lim_{t\to\infty} \hat{X}_t = 0.$$

It is also easy to see that $\{Y_t\}_{t\in\mathbb{Z}}$ is a singular process and spans $H_{-\infty}$. Clearly, if a process $\{X_t\}_{t\in\mathbb{Z}}$ is singular, then $X_t \in H_0^-$ for any $t \in \mathbb{Z}$, so perfect linear prediction is possible: $\hat{X}_t = X_t$ for any $t \in \mathbb{Z}$.

2.7 Spectral form of the Wold decomposition

The regular part of a non-singular time series $\{X_t\}$ is a causal MA(∞) process, so by (2.11) it has an absolutely continuous spectral measure with density

$$f(\omega) = \frac{1}{2\pi}\left| \sum_{j=0}^{\infty} b_j e^{-ij\omega} \right|^2 = \frac{1}{2\pi}|H(e^{-i\omega})|^2, \qquad \sum_{j=0}^{\infty} |b_j|^2 < \infty, \tag{2.35}$$

where $H(z)$ is the transition function of the process. Starting from a different perspective, we may introduce the function $\Phi \in L^2(T)$, called a *spectral factor*, by the Fourier series

$$\Phi(e^{-i\omega}) := \sum_{j=0}^{\infty} b_j e^{-ij\omega}, \qquad \omega \in [-\pi, \pi]. \tag{2.36}$$

It is essential that this Fourier series is one-sided, its coefficients are zero for $j < 0$. Then this definition can be extended to the open unit disc D as an analytic function

$$\Phi(z) := \sum_{j=0}^{\infty} b_j z^j, \qquad z \in D. \tag{2.37}$$

Recall that this power series must be convergent in D, since its radius of convergence R is given by (A.3), and $\limsup_{j\to\infty} |b_j|^{1/j} = 1/R$ cannot be greater than 1, because then there would be infinitely many coefficients b_j with absolute value greater than 1 and that would contradict to the fact that $\sum_{j=0}^{\infty} |b_j|^2$ is convergent.

Therefore Φ is analytic in the open unit disc D and is in L^2 on the unit circle T, with zero Fourier coefficients when $j < 0$. It means that $\Phi \in H^2$, see Section A.3 about Hardy spaces. We mention that the negative sign in the exponents in (2.36) is only a matter of convention, usual in the theory of time series. An H^2 function which is not identically 0, vanishes only on a set of measure 0 on T, so the spectral density f in (2.35) is positive a.e. on T, see the last sentence of Theorem A.13.

Remark 2.2. The next lemma shows that the Wold decomposition (2.32) is a spectral decomposition in the sense too that the support of the spectral measure of the singular process $\{Y_t\}$ must be disjoint from the set $\{\omega \in [-\pi, \pi] : f(\omega) \neq 0\}$. Consequently, the spectral measure of $\{Y_t\}$ is a singular measure w.r.t. Lebesgue measure. This way we see that the Wold decomposition (2.32) of $\{X_t\}$ is equivalent to a decomposition of the spectral measure of a non-singular process into an absolutely continuous and a singular part w.r.t. Lebesgue measure on $[-\pi, \pi]$.

Lemma 2.1. *Assume that $\{X_t\}$ is a stationary time series, $X_t = U_t + V_t$, where $U_t, V_t \in H(X)$ for all $t \in \mathbb{Z}$, $\{U_t\}$ and $\{V_t\}$ are stationary processes, and $\langle U_t, V_s \rangle = 0$ for all t, s. Then there is a decomposition $A \cup B = [-\pi, \pi]$, $A \cap B = \emptyset$, A and B are measurable, and*

$$X_t = \int_{-\pi}^{\pi} e^{it\omega} dZ_\omega = \int_{-\pi}^{\pi} e^{it\omega} \mathbf{1}_A dZ_\omega + \int_{-\pi}^{\pi} e^{it\omega} \mathbf{1}_B dZ_\omega = U_t + V_t \quad (t \in \mathbb{Z}).$$

Proof. We know that there is a unique unitary operator S on $H(X)$ such that $X_t = S^t X_0$ $(t \in \mathbb{Z})$. The assumptions imply that the Hilbert subspaces $H(U)$ and $H(V)$ of $H(X)$ are orthogonal and $H(X) = H(U) \oplus H(V)$, where $\oplus$ denotes orthogonal direct sum. There exist unitary operators S_U and S_V on $H(U)$ and $H(V)$, respectively, such that $U_t = S_U^t U_0$ and $V_t = S_V^t V_0$. Then the direct sum $S_U \oplus S_V$ is a unitary operator on $H(X)$ that maps X_t into X_{t+1}. Thus we conclude that $S = S_U \oplus S_V$ and S_U and S_V are the restrictions of S to U and V, respectively.

Since $U_0, V_0 \in H(X)$, there exist functions $u, v \in L^2([-\pi, \pi], \mathcal{B}, dF)$ such that

$$U_0 = \int_{-\pi}^{\pi} u(\omega) dZ_\omega, \quad V_0 = \int_{-\pi}^{\pi} v(\omega) dZ_\omega,$$

and so

$$U_t = \int_{-\pi}^{\pi} e^{it\omega} u(\omega) dZ_\omega, \quad V_t = \int_{-\pi}^{\pi} e^{it\omega} v(\omega) dZ_\omega \quad (t \in \mathbb{Z}).$$

But the orthogonality of $\{U_t\}$ and $\{V_t\}$ implies that for any t and s,

$$\int_{-\pi}^{\pi} e^{i(t-s)\omega} u(\omega) \overline{v(\omega)} d\omega) = \langle U_t, V_s \rangle = 0.$$

It implies that the function $h := u\bar{v}$ is orthogonal to all trigonometric polynomials in $L^2([-\pi,\pi],\mathcal{B},dF)$. Since in general $h \in L^1([-\pi,\pi],\mathcal{B},dF)$ only, we want to show that this still implies that $h = 0$ a.e. with respect to dF. For any continuous function g such that $g(-\pi) = g(\pi)$, define the bounded linear functional

$$Kg := \int_{-\pi}^{\pi} g(\omega)h(\omega)dF(\omega).$$

Since K vanishes for all trigonometric polynomials, a dense set in the space of continuous functions g with $g(-\pi) = g(\pi)$, it follows that $Kg = 0$ for any such g as well. By the Riesz representation theorem, K can be represented by a complex measure μ so that

$$d\mu(\omega) = h(\omega)dF(\omega).$$

Thus μ must be identically 0 and $h(\omega) = 0$ for a.e. ω with respect to dF.

On the other hand, $u(\omega) + v(\omega) = 1$ a.e. with respect to dF, since

$$\int_{-\pi}^{\pi} \{u(\omega) + v(\omega)\}dZ_\omega = \int_{-\pi}^{\pi} 1\, dZ_\omega = X_0,$$

and the spectral representation is unique.

Thus $uv = 0$ and $u + v = 1$ a.e. (dF), so u and v are indicator functions on disjoint sets A and B, $A \cup B = [-\pi,\pi]$, except on a set of dF measure 0, which does not influence the integrals. This proves the lemma. □

Assume now that $\{X_t\}$ is regular, with spectral representation

$$X_t = \int_{-\pi}^{\pi} e^{it\omega} dZ_\omega. \tag{2.38}$$

We want to show that under not too restrictive assumptions we can find the causal MA(∞) representation

$$X_t = \sum_{k=0}^{\infty} b_k \xi_{t-k}, \quad t \in \mathbb{Z},$$

by factoring the spectral density f of $\{X_t\}$. This method then automatically extends to the regular part of any non-singular time series.

Since $\xi_0 \in H(X)$ by its definition, there is a function $\psi(\omega) \in L^2([-\pi,\pi],\mathcal{B},dF)$ which represents it; we then have

$$\xi_n = S^n \xi_0 = \int_{-\pi}^{\pi} e^{in\omega}\psi(\omega)dZ_\omega. \tag{2.39}$$

We want to show that one can find the function $\psi(\omega)$ and so the orthonormal sequence $\{\xi_n\}$, by suitably factoring the spectral density $f(\omega)$ of the regular part of $\{X_t\}$.

There are three conditions that $\{\xi_n\}$ must satisfy:

1. orthonormality,
2. $\xi_n \perp X_{n-k}$ for $k > 0$,
3. $\xi_n \in H_n^-$ for $n \in \mathbb{Z}$.

By (1),

$$\delta_{nm} = \langle \xi_n, \xi_m \rangle = \int_{-\pi}^{\pi} e^{in\omega} \psi(\omega) e^{-im\omega} \overline{\psi(\omega)} f(\omega) d\omega,$$

which implies

$$|\psi(\omega)|^2 f(\omega) = \frac{1}{2\pi} \quad \text{(a.e.)}$$

because of the uniqueness theorem for Fourier series. Then we can factor the spectral density

$$f(\omega) = \frac{1}{2\pi} \frac{1}{\psi(\omega)} \cdot \frac{1}{\overline{\psi(\omega)}} =: \frac{1}{2\pi} \phi(\omega) \cdot \overline{\phi(\omega)},$$

where the *spectral factor* ϕ is a complex valued square integrable function on $[-\pi, \pi]$. There is a simple one-to-one correspondence of this spectral factor ϕ defined on $[-\pi, \pi]$ and Φ defined by (2.36) on the unit circle T, as explained in Remark 1.2.

Next, condition (2) requires that $\langle \xi_n, X_{n-k} \rangle = 0$, $k > 0$. But

$$\langle \xi_n, X_{n-k} \rangle = \int_{-\pi}^{\pi} e^{in\omega} \psi(\omega) e^{-i(n-k)\omega} f(\omega) d\omega = \frac{1}{2\pi} \int_{-\pi}^{\pi} e^{ik\omega} \overline{\phi(\omega)} d\omega = 0,$$

for all $k > 0$, that is, ϕ has to be orthogonal to $e^{-ik\omega}$ for all $k < 0$. It means that the Fourier series of ϕ has to be of 'power series type':

$$\phi(\omega) = \sum_{k=0}^{\infty} b_k e^{-ik\omega}, \qquad \omega \in [-\pi, \pi],$$

that is, $\phi \in H^2$, where H^2 is the closed linear span of $\{e^{-ik\omega} : k \geq 0\}$ in $L^2(T)$, see Subsection A.3.2 in the Appendix about the Hardy space H^2. Comparing with (2.36) and (2.37), it shows that $\phi(\omega)$ is the boundary value of a function $\Phi(z)$ which is regular in D: $\phi(\omega) = \Phi(e^{-i\omega})$, $\omega \in [-\pi, \pi]$.

Finally, by (3), we must have $\xi_n \in H_n^-$; for this $\xi_0 \in H_0^-$ is enough. But by (2.39),

$$\xi_0 = \int_{-\pi}^{\pi} \psi(\omega) dZ_\omega = \int_{-\pi}^{\pi} \frac{1}{\phi(\omega)} dZ_\omega.$$

The requirement $\xi_0 \in H_0^-$ means that there exist complex numbers $\{\gamma_k\}_{k=0}^{\infty}$, $\sum_{k=0}^{\infty} |\gamma_k|^2 < \infty$, such that

$$\xi_0 = \sum_{k=0}^{\infty} \gamma_k X_{-k}, \text{ and so } \psi(\omega) = \sum_{k=0}^{\infty} \gamma_k e^{-ik\omega}$$

by the spectral representation of the time series $\{X_t\}$. It means that we must have $1/\phi = \psi(\omega) \in H^2$. This way we have proved:

Theorem 2.2. *Assume that the spectral density f of a regular stationary time series $\{X_t\}$ has a factorization $f = \frac{1}{2\pi}\phi \cdot \overline{\phi}$, where $\phi \in H^2$ and $1/\phi \in H^2$.*

(a) Then the orthonormal sequence appearing in the Wold decomposition is given by the random variables

$$\xi_t = \int_{-\pi}^{\pi} \frac{e^{it\omega}}{\phi(\omega)} dZ_\omega \quad (t \in \mathbb{Z}). \tag{2.40}$$

(b) We have the Fourier series

$$\phi(\omega) = \sum_{k=0}^{\infty} b_k e^{-ik\omega}, \quad b_k = \frac{1}{2\pi}\int_{-\pi}^{\pi} e^{ik\omega}\phi(\omega)d\omega, \quad \sum_{k=0}^{\infty} |b_k|^2 < \infty, \tag{2.41}$$

and this defines the coefficients of the Wold representation:

$$X_t = \sum_{k=0}^{\infty} b_k \xi_{t-k}, \quad t \in \mathbb{Z}. \tag{2.42}$$

Proof. We have seen (2.40) above. On the other hand, if $\phi \in H^2$, then by (2.38), (2.40), and (2.41),

$$X_t = \int_{-\pi}^{\pi} e^{it\omega} dZ_\omega = \int_{-\pi}^{\pi} e^{it\omega} \sum_{k=0}^{\infty} b_k e^{-ik\omega} \frac{1}{\phi(\omega)} dZ_\omega = \sum_{k=0}^{\infty} b_k \xi_{t-k},$$

using the properties of spectral representation. □

Comparing (2.35), (2.36) and (2.41), we see that $\phi(\omega) = \Phi(e^{-i\omega}) = H(e^{-i\omega})$, where $\Phi(z)$ is the spectral factor and $H(z)$ is the transition function of the process.

Theorem 2.2 gives an explicit solution to causal MA(∞) representation of regular time series under conditions that are relatively mild. If $\{X_t\}$ is an arbitrary non-singular time series, then Theorem 2.2 is still valid, with the difference that (2.42) gives the regular part of the process, since by Remark 2.2, the spectral density of a non-singular time series is the same as the spectral density of its regular part.

In the next section we are going to discuss two special cases, where the factorization of the spectral density is rather straightforward. Remark 2.6 later will give an explicit formula for the factoring the spectral density of a general regular time series in 1D.

2.8 Factorization of rational and smooth spectral densities

2.8.1 Rational spectral density

We saw in (2.29) that the spectral density f of a stationary ARMA process is a rational function of $e^{-i\omega}$.

Remark 2.3. If a spectral density f is rational function of $z = e^{-i\omega}$, then the rational function cannot have poles on the unit circle T. For, the spectral measure dF is a finite measure, while a rational function of $z = e^{-i\omega}$ that has one or more poles on T cannot have finite integral $\int_{-\pi}^{\pi} f(\omega)d\omega$.

Lemma 2.2. *(Fejér–Riesz lemma [17])*

If the spectral density f is a rational function of $z = e^{-i\omega}$ and $f(\omega) \geq 0$ for $\omega \in [-\pi, \pi]$, then it can be written in the form

$$f(\omega) = \frac{1}{2\pi} \frac{|\beta(e^{-i\omega})|^2}{|\alpha(e^{-i\omega})|^2} \quad (-\pi \leq \omega \leq \pi), \tag{2.43}$$

where the polynomials

$$\alpha(z) = \sum_{j=0}^{p} \alpha_j z^j, \quad \beta(z) = \sum_{j=0}^{q} \beta_j z^j$$

are relative prime, α has no zeros in the closed unit disc $\{z : |z| \leq 1\}$, and β has no zeros in the open unit disc D.

If, moreover, $f(-\omega) = f(\omega)$, then the coefficients of the polynomials α and β can be chosen real.

Proof. By our assumptions we can write that

$$f(\omega) = ce^{-i\ell\omega} \frac{\prod_{k=1}^{n}(e^{-i\omega} - v_k)}{\prod_{j=1}^{m}(e^{-i\omega} - u_j)},$$

where $c \in \mathbb{C}$, $\ell \in \mathbb{Z}$, $m, n \geq 0$, and the sets of complex nonzero numbers $\{u_j\}_{j=1,\ldots,m}$ and $\{v_k\}_{k=1,\ldots,n}$ are disjoint.

Since $f(\omega)$ is real,

$$f(\omega) = \bar{c}e^{i\ell\omega} \frac{\prod_{k=1}^{n}(e^{i\omega} - \bar{v}_k)}{\prod_{j=1}^{m}(e^{i\omega} - \bar{u}_j)} = c' e^{i\omega(\ell+n-m)} \frac{\prod_{k=1}^{n}(\bar{v}_k^{-1} - e^{-i\omega})}{\prod_{j=1}^{m}(\bar{u}_j^{-1} - e^{-i\omega})}, \tag{2.44}$$

where $c' \in \mathbb{C}$.

Since $f(\omega) \geq 0$, it coincides with the absolute values of all the above expressions written for it. Considering that

$$\begin{aligned} |e^{-i\omega} - \bar{w}^{-1}| &= |w|^{-1} \left|\bar{w}e^{-i\omega} - 1\right| = r^{-1}|re^{-i(\theta+\omega)} - 1| \\ &= r^{-1}|1 - re^{i(\theta+\omega)}| = |w|^{-1} \left|e^{-i\omega} - w\right|, \qquad w = re^{i\theta}, \end{aligned}$$

it follows that for each root u_j and v_k, where $|u_j| > 1$ and $|v_k| > 1$, there exists another root $u_{j'}$ and $v_{k'}$, respectively, such that $u_j = \bar{u}_{j'}^{-1}$ and $v_k = \bar{v}_{k'}^{-1}$.

Also because of $f(\omega) \geq 0$, in case of any factor $(e^{-i\omega} - v_k)$, where $v_k = e^{-i\omega_0}$ is on the unit circle T, the root must be double:

$$\overline{(e^{-i\omega} - e^{-i\omega_0})} = (e^{i\omega} - e^{i\omega_0}) = -e^{i(\omega+\omega_0)}(e^{-i\omega} - e^{-i\omega_0}).$$

These prove (2.43).

If $f(-\omega) = f(\omega)$, then by the first equality of (2.44), for every root u_j and v_k there corresponds a root $u_{j'} = \bar{u}_j$ and $v_{k'} = \bar{v}_k$, respectively. Thus α and β can be chosen real. □

Corollary 2.2. *If the spectral density f is a rational function of $z = e^{-i\omega}$ and $f(\omega) > 0$ for $\omega \in [-\pi, \pi]$, the factorization $f = \frac{1}{2\pi}\phi \cdot \overline{\phi}$ is straightforward by* (2.43)*:*

$$f(\omega) = K \frac{\prod_{k=1}^{q}(e^{-i\omega} - v_k)(e^{i\omega} - \bar{v}_k)}{\prod_{j=1}^{p}(e^{-i\omega} - u_j)(e^{i\omega} - \bar{u}_j)},$$

$$\phi(\omega) = \sqrt{2\pi K} \frac{\prod_{k=1}^{q}(e^{-i\omega} - v_k)}{\prod_{j=1}^{p}(e^{-i\omega} - u_j)} = \frac{\beta(e^{-i\omega})}{\alpha(e^{-i\omega})}, \tag{2.45}$$

where $K > 0$ is a constant and each $|v_k| > 1$, $|u_j| > 1$. In this case Theorem 2.2 can be applied to find an explicit causal MA(∞) *representation of the time series since both ϕ and $1/\phi$ are continuous, so bounded, on $[-\pi, \pi]$ and this way they belong to the Hardy space H^2. When $p > 1$, it may be useful to apply a partial fraction expansion of ϕ in order to obtain the Fourier series explicitly.*

Equations (2.10) and (2.45) show that a process with spectral density satisfying the assumptions of Corollary 2.2 is an MA(q) process when $p = 0$, that is, the denominator is 1. Also, (2.24) and (2.45) give that the process is an AR(p) process when $q = 0$, that is, the numerator is 1. Finally, (2.29) and (2.45) show that the process is a proper ARMA(p, q) process otherwise. It simplifies the discussion if from now on, we consider AR and MA processes as special ARMA processes.

Remark 2.4. By Lemma 2.2 it follows that any 1D stationary time series with spectral density f which is a rational function of $z = e^{-i\omega}$ can be represented with a *stable* ARMA process. Also, any ARMA process satisfies the assumptions of Lemma 2.2, so it can be represented as a stable ARMA process.

2.8.2 Smooth spectral density

Now assume that $f > 0$ and f is continuously differentiable on $[-\pi, \pi]$. Then $\log f$ is also continuously differentiable and it can be expanded into a uniformly convergent Fourier series:

$$\log f(\omega) = \sum_{n=-\infty}^{\infty} \beta_n e^{in\omega}, \quad \beta_{-n} = \bar{\beta}_n.$$

Now we write

$$\log f(\omega) = Q(\omega) + \overline{Q(\omega)} := \left(\frac{1}{2}\beta_0 + \sum_{n=-\infty}^{-1} \beta_n e^{in\omega}\right) + \left(\frac{1}{2}\beta_0 + \sum_{n=1}^{\infty} \beta_n e^{in\omega}\right).$$

Then by Theorem 2.2

$$f(\omega) = e^{Q(\omega)} \cdot e^{\overline{Q(\omega)}} =: \frac{1}{2\pi}\phi(\omega) \cdot \overline{\phi(\omega)}$$

is the correct factorization. For, then both $\bar{\phi}$ and $1/\bar{\phi}$ are continuous on $[-\pi, \pi]$ and the Fourier series of $\bar{\phi}$ contains only non-negative powers of $e^{i\omega}$.

This leads to new formulas for the prediction error. By Theorem 2.2, we need the Fourier series of

$$\phi(\omega) = \sqrt{2\pi} e^{Q(\omega)} = \sqrt{2\pi} \sum_{k=0}^{\infty} \frac{1}{k!} \left(\frac{1}{2}\beta_0 + \sum_{j=1}^{\infty} \bar{\beta}_j e^{-ij\omega}\right)^k.$$

In particular, the constant term is

$$b_0 = \sqrt{2\pi} \sum_{k=0}^{\infty} \frac{1}{k!} \left(\frac{\beta_0}{2}\right)^k = \sqrt{2\pi} e^{\beta_0/2}.$$

But β_0 is just the 0th Fourier coefficient of $\log f$, so by formula (2.34) the one-step ahead prediction error is given by

$$\sigma_1^2 = b_0^2 = 2\pi e^{\beta_0} = 2\pi \exp \int_{-\pi}^{\pi} \log f(\omega) \frac{d\omega}{2\pi}. \tag{2.46}$$

The expressions for the several step-ahead prediction errors $\sigma_2^2, \sigma_3^2, \ldots$ are similar, but more complicated.

2.9 Classification of stationary time series in 1D

The theorems and their proofs in this section are based on the seminal paper "Stationary sequences in Hilbert space" [34] by Kolmogorov from 1941.

Wold decomposition in Section 2.6 showed that a stationary time series

$\{X_t\}_{t\in\mathbb{Z}}$ is regular if and only if it can be represented as a causal MA(∞) process

$$X_t = \sum_{j=0}^{\infty} b_j \xi_{t-j}, \tag{2.47}$$

with an orthonormal sequence $\{\xi_t\}_{t\in\mathbb{Z}}$. Moreover, in (2.35) and (2.36) we saw that the spectral density function f^X of a regular time series $\{X_t\}$ can be written as $f^X(\omega) = \frac{1}{2\pi}|\Phi(e^{-i\omega})|^2$, where $\Phi(z)$ is analytic in the open unit disc D and is in L^2, more accurately, is in H^2, on the unit circle T. The next lemma gives an even more precise and interesting description of Φ.

Lemma 2.3. *[34, Theorem 21] For any regular stationary time series $\{X_t\}_{t\in\mathbb{Z}}$, the analytic function $\Phi(z)$ has no zeros in the open unit disc D.*

Proof. Indirectly, assume that there exists a $z_0 \in D$, where $\Phi(z_0) = 0$. Then consider the linear fractional transformation

$$\Phi(z) := \frac{z - z_0}{1 - \bar{z}_0 z},$$

which maps the unit circle T and the open unit disc D onto itself in a one-to-one way, respectively. We introduce a new stationary time series $\{\eta_t\}_{t\in\mathbb{Z}}$ by linear filtration from the orthonormal series $\{\xi_t\}_{t\in\mathbb{Z}}$ of (2.47). As we saw in Section 2.2, such a filter can be determined by the Fourier series of its weights $\{\phi_j\}_{j\in\mathbb{Z}}$, the conditional spectral density of η w.r.t. ξ:

$$f^{\eta|\xi}(\omega) := \hat{\phi}(\omega) := \sum_{j=-\infty}^{\infty} \phi_j e^{-ij\omega}. \tag{2.48}$$

In the present case we choose

$$f^{\eta|\xi}(\omega) = \hat{\phi}(\omega) = \Phi(e^{-i\omega}) = \frac{e^{-i\omega} - z_0}{1 - \bar{z}_0 e^{-i\omega}}. \tag{2.49}$$

Since $f^{\eta|\xi}$ is a bounded function,

$$|f^{\eta|\xi}(\omega)|^2 = 1, \tag{2.50}$$

it belongs to $L^2([-\pi, \pi], \mathcal{B}, dF^\xi)$, where

$$dF^\xi(\omega) = \frac{1}{2\pi} d\omega,$$

and it shows that the time series $\{\eta_k\}_{t\in\mathbb{Z}}$ is well-defined by the above filtering. Moreover, by (2.2),

$$dF^\eta(\omega) = |f^{\eta|\xi}(\omega)|^2 dF^\xi(\omega) = \frac{1}{2\pi} d\omega,$$

so $\{\eta_k\}_{t\in\mathbb{Z}}$ is also an orthonormal sequence.

By (2.7),

$$dF^{X,\eta}(\omega) = f^{X|\xi}(\omega)\overline{f^{\eta|\xi}(\omega)}dF^{\xi}(\omega) = f^{X|\xi}(\omega)\overline{f^{\eta|\xi}(\omega)}\frac{1}{2\pi}d\omega.$$

Since by (2.50), $\overline{f^{\eta|\xi}(\omega)} = 1/f^{\eta|\xi}(\omega)$, and by (2.6), $f^{X|\eta} = dF^{X,\eta}/dF^{\eta}$, we obtain that

$$f^{X|\eta}(\omega) = \frac{f^{X|\xi}(\omega)}{f^{\eta|\xi}(\omega)}.$$

Comparing (2.2) and (1.28) it follows that $f^{X|\xi}(\omega) = \Phi(e^{-i\omega})$, thus by (2.49) we get the result

$$f^{X|\eta}(\omega) = \Phi(e^{-i\omega})\frac{1-\bar{z}_0 e^{-i\omega}}{e^{-i\omega}-z_0} =: \tilde{\Phi}(e^{-i\omega}).$$

Also,

$$\tilde{\Phi}(z) = \Phi(z)\frac{1-\bar{z}_0 z}{z-z_0} \quad (|z| \leq 1),$$

which is an analytic function in D, since $z_0 \in D$ is a zero of $\Phi(z)$. Moreover, $\tilde{\Phi}(z)$ is in L^2 (in fact, in H^2) on T. Thus the properties of

$$\tilde{\Phi}(z) = \sum_{j=0}^{\infty} \tilde{b}_j z^j \quad (z \in D)$$

guarantees that we may use the series $\{\eta_t\}_{t\in\mathbb{Z}}$ as an orthonormal sequence in $H(X)$ instead of $\{\xi_t\}_{t\in\mathbb{Z}}$:

$$X_t = \sum_{j=0}^{\infty} \tilde{b}_j \eta_{t-j} \quad (t \in \mathbb{Z}).$$

Formula (2.48) implies that

$$\eta_t = \sum_{j=0}^{\infty} \phi_j \xi_{t-j} \quad (t \in \mathbb{Z}).$$

This shows that $\eta_t \in H_t^-(\xi) := \overline{\text{span}}\{\xi_j : j \leq t\}$ for any t, and so $\eta_t \in H_t^-(X) := \overline{\text{span}}\{X_j : j \leq t\}$ for any t. Thus the procedure of the Wold decomposition described in Section 2.6 implies that

$$\tilde{b}_0 \eta_t = b_0 \xi_t \quad \Rightarrow \quad \eta_t = \frac{b_0}{\tilde{b}_0}\xi_t \quad (t \in \mathbb{Z}).$$

Then it would follow that

$$f^{\eta|\xi}(\omega) = \frac{b_0}{\tilde{b}_0},$$

which contradicts (2.49), and this proves the lemma. □

Theorem 2.3. *[34, Theorem 22] A stationary time series* $\{X_t\}_{t \in \mathbb{Z}}$ *is regular if and only if the following three conditions hold on* $[-\pi, \pi]$*:*

(1) its spectral measure dF *is absolutely continuous w.r.t. Lebesgue measure;*

(2) its spectral density f *is positive a.e.;*

(3) (Kolmogorov's condition) $\log f$ *is integrable.*

Proof. Every regular time series can be represented as a causal MA(∞) process (2.47), so the necessity of conditions (1) and (2) follows from this. Let us show the necessity of condition (3). By (2.35) and (2.36) it follows that its spectral density function f can be written as

$$f(\omega) = \frac{1}{2\pi}|\Phi(e^{-i\omega})|^2 = \frac{1}{2\pi}\left|\sum_{j=0}^{\infty} b_j e^{-ij\omega}\right|^2,$$

where $\Phi(z)$ is analytic in the open unit disc D and is in L^2 on the unit circle T. By Lemma 2.3 we know that $\Phi(z)$ has no zeros in D. Denote by $Q(z)$ that branch of $\log(\Phi(z)/\sqrt{2\pi})$ which at $z = 0$ takes the real value

$$Q(0) = \log\frac{\Phi(0)}{\sqrt{2\pi}} = \log\frac{b_0}{\sqrt{2\pi}} \in \mathbb{R}, \tag{2.51}$$

since $b_0 > 0$ by (2.31). The fact that $\Phi(z)$ has no zeros in D implies that the above choice uniquely defines the analytic function $Q(z)$ in D and

$$\Phi(z) = \sqrt{2\pi}\, e^{Q(z)} \quad (z \in D). \tag{2.52}$$

Clearly,

$$\operatorname{Re} Q(z) = \log|\Phi(z)| - \frac{1}{2}\log(2\pi). \tag{2.53}$$

Denote by $\operatorname{Re}^+ z := \max(\operatorname{Re}(z), 0)$ and $\log^+ x := \max(\log x, 0)$ $(x \geq 0)$. Then

$$\operatorname{Re}^+ Q(z) < \log^+|\Phi(z)| \leq |\Phi(z)|,$$

and so by the maximum modulus theorem, for any $0 \leq \rho < 1$ we have

$$\int_{-\pi}^{\pi} \operatorname{Re}^+ Q(\rho e^{-i\omega}) d\omega < \int_{-\pi}^{\pi} |\Phi(\rho e^{-i\omega})| d\omega \leq \int_{-\pi}^{\pi} |\Phi(e^{-i\omega})| d\omega =: K < \infty.$$

Equation (2.53) shows that $\operatorname{Re} Q(z)$ is a continuous *subharmonic function* in D, see Theorem A.8. It means that

$$\operatorname{Re} Q(0) \leq \frac{1}{2\pi}\int_{-\pi}^{\pi} \operatorname{Re} Q(\rho e^{-i\omega}) d\omega \quad (0 \leq \rho < 1).$$

These imply that

$$\int_{-\pi}^{\pi} |\mathrm{Re}\, Q(\rho e^{-i\omega})| d\omega = \int_{-\pi}^{\pi} \{2\mathrm{Re}^{+} Q(\rho e^{-i\omega}) - \mathrm{Re}\, Q(\rho e^{-i\omega})\} d\omega$$
$$\leq 2K - 2\pi \mathrm{Re}\, Q(0) \qquad (0 \leq \rho < 1). \tag{2.54}$$

By Theorem A.10, the radial limit

$$\lim_{\rho \to 1} \Phi(\rho e^{-i\omega}) = \Phi(e^{-i\omega})$$

exist a.e. on T, and (2.53) and (2.54) imply that $\Phi(e^{-i\omega}) \neq 0$ a.e. on T. Hence

$$\lim_{\rho \to 1} \log \frac{\Phi(\rho e^{-i\omega})}{\sqrt{2\pi}} = \lim_{\rho \to 1} Q(\rho e^{-i\omega}) = Q(e^{-i\omega})$$

also exist a.e. on T. Then (2.54) implies that the boundary value of $\mathrm{Re}\, Q(z)$ on T:

$$\mathrm{Re}\, Q(e^{-i\omega}) = \log \frac{|\Phi(e^{-i\omega})|}{\sqrt{2\pi}} = \frac{1}{2} \log \frac{|\Phi(e^{-i\omega})|^2}{2\pi} = \frac{1}{2} \log f(\omega) \tag{2.55}$$

is integrable w.r.t. Lebesgue measure on $[-\pi, \pi]$. This proves the necessity of condition (3) of the theorem.

Now we turn to the proof of sufficiency of the conditions of the theorem. Thus we may define the harmonic real function $\mathrm{Re}\, Q(z)$ in D by the definition

$$\mathrm{Re}\, Q(\rho e^{-i\omega}) := \frac{1}{4\pi} \int_{-\pi}^{\pi} \log f(t) P_{\rho}(\omega - t) dt. \tag{2.56}$$

(See the definition and properties of Poisson integrals in Section A.2.) There exists a unique analytic function $Q(z)$ in D whose real part is $\mathrm{Re}\, Q(z)$, with its harmonic conjugate $\mathrm{Im}\, Q(z)$, if we assume that $Q(0) = \mathrm{Re}\, Q(0) \in \mathbb{R}$, and consequently, $\mathrm{Im}\, Q(0) = 0$. Then we may define the analytic function $\Phi(z)$ in D by formula (2.52):

$$\Phi(z) = \sqrt{2\pi} \exp \left\{ \frac{1}{4\pi} \int_{-\pi}^{\pi} \log f(t) \frac{e^{it} + z}{e^{it} - z} dt \right\}. \tag{2.57}$$

By Jensen's inequality, see e.g. [50, p. 63], if g is an a.e. positive function and μ is a probability measure on a set A, then

$$\exp \int_A \log g \, d\mu \leq \int_A g d\mu.$$

Apply this to (2.56):

$$\frac{|\Phi(\rho e^{-i\omega})|^2}{2\pi} = \exp\left(2\mathrm{Re}\, Q(\rho e^{-i\omega})\right) \leq \frac{1}{2\pi} \int_{-\pi}^{\pi} f(t) P_{\rho}(\omega - t) dt.$$

Hence by Fubini's theorem,

$$\frac{1}{2\pi}\int_{-\pi}^{\pi}|\Phi(\rho e^{-i\omega})|^2 d\omega \le \int_{-\pi}^{\pi} f(t)\frac{1}{2\pi}\int_{-\pi}^{\pi} P_\rho(\omega-t)d\omega = \int_{-\pi}^{\pi} f(t)dt < \infty$$

for any $0 \le \rho < 1$. This shows that the boundary value $\Phi(e^{-i\omega})$ exists, it is in $L^2(T)$, and $\Phi(z)$ belongs to the Hardy space H^2, see Section A.3. It follows that

$$\Phi(e^{-i\omega}) = \sum_{j=0}^{\infty} b_j e^{-ij\omega} \quad (b_j \in \mathbb{C}). \tag{2.58}$$

By (A.9), for the boundary values we have

$$|\Phi(e^{-i\omega})|^2 = 2\pi \exp\left(2\mathrm{Re}\, Q(e^{-i\omega})\right) = 2\pi f(\omega) \tag{2.59}$$

for a.e. $\omega \in [-\pi, \pi]$. Let us use linear filtering of the stationary time series $\{X_t\}_{t\in\mathbb{Z}}$ by the conditional spectral density

$$f^{\xi|X}(\omega) = \frac{1}{\Phi(e^{-i\omega})}.$$

By (2.59),

$$\int_{-\pi}^{\pi} |f^{\xi|X}(\omega)|^2 f(\omega)d\omega = 1,$$

so $f^{\xi|X} \in L^2(-\pi,\pi], \mathcal{B}, dF^X)$, the filter is well-defined. By (2.2),

$$dF^{\xi}(\omega) = |f^{\xi|X}(\omega)|^2 f(\omega)d\omega = \frac{1}{2\pi},$$

so the resulting process: $\{\xi_k\}_{t\in\mathbb{Z}}$ is an orthonormal sequence. Since

$$f^{X|\xi}(\omega) = 1/f^{\xi|X}(\omega) = \Phi(e^{-i\omega}),$$

by (2.58) it follows that

$$X_t = \sum_{j=0}^{\infty} b_j \xi_{k-j} \quad (t \in \mathbb{Z}),$$

which shows that $\{X_t\}$ is a regular process and so it completes the proof of the theorem. □

Remark 2.5. If $\{X_t\}_{t\in\mathbb{Z}}$ is regular, then $Q(z)$ defined in the previous theorem is analytic in D, so we can write that

$$Q(z) = \sum_{k=0}^{\infty} \gamma_k z^k \quad (z \in D).$$

By condition (3) of the theorem and by formula (2.55), $\operatorname{Re} Q(e^{-i\omega}) \in L^1(T)$, so it can be expanded in a Fourier series:

$$\operatorname{Re} Q(e^{-i\omega}) = \frac{1}{2}\log f(\omega) \sim \sum_{k=0}^{\infty} \operatorname{Re}(\gamma_k e^{ik\omega}) = \sum_{k=0}^{\infty} (\alpha_k \cos k\omega - \beta_k \sin k\omega) \tag{2.60}$$

for $\omega \in [-\pi, \pi]$, where $\gamma_k = \alpha_k + i\beta_k$. Then we also have

$$\operatorname{Im} Q(e^{-i\omega}) \sim \sum_{k=1}^{\infty} \operatorname{Im}(\gamma_k e^{ik\omega}) = \sum_{k=1}^{\infty} (\alpha_k \sin k\omega + \beta_k \cos k\omega),$$

since $\gamma_0 \in \mathbb{R}$ by (2.51).

Remark 2.6. Equation (2.57) gives an explicit formula for the factorization of the spectral density of an arbitrary one-dimensional regular time series:

$$f(\omega) = \frac{1}{2\pi}|\Phi(e^{-i\omega}|^2, \quad \Phi(z) = \sqrt{2\pi}\exp\left\{\frac{1}{4\pi}\int_{-\pi}^{\pi} \log f(t)\frac{e^{it}+z}{e^{it}-z}dt\right\},$$

where $z \in D$, $\Phi(e^{-i\omega})$ is the boundary value of the analytic function $\Phi(z)$, by the proof of the previous theorem.

We mention that according to the definition in Subsection A.3.1 in the Appendix, $\Phi^2(z)/(2\pi)$ is an outer function. Its absolute value at $z = e^{-i\omega}$ is the spectral density $f(\omega)$.

Remark 2.7. (Kolmogorov–Szegő formula)

If $\{X_t\}_{t\in\mathbb{Z}}$ is regular, by (2.51), (2.52), and (2.60) we obtain that

$$b_0 = \Phi(0) = \sqrt{2\pi}e^{Q(0)} = \sqrt{2\pi}e^{\alpha_0} = \sqrt{2\pi}\exp\left(\frac{1}{4\pi}\int_{-\pi}^{\pi}\log f(\omega)d\omega\right),$$

so

$$\sigma_1^2 = b_0^2 = 2\pi\exp\int_{-\pi}^{\pi}\log f(\omega)\frac{d\omega}{2\pi},$$

see Remark 2.6 as well. Thus for any regular time series we have got the same prediction error formula as obtained earlier in the special case of smooth spectral densities, see (2.46). As follows from the next theorem, see Corollary 2.3, this formula is valid for any non-singular time series as well.

The next theorem gives the different classes of singular time series.

Theorem 2.4. *[34, Theorem 23] Assume that $\{X_t\}_{t\in\mathbb{Z}}$ is a stationary time series with spectral measure dF on $[-\pi, \pi]$. Then there exists a unique Lebesgue decomposition*

$$dF = dF_a + dF_s, \quad dF_a \ll d\omega, \quad dF_s \perp d\omega, \tag{2.61}$$

where $d\omega$ denotes the Lebesgue measure, $dF_a(\omega) = f_a(\omega)d\omega$ is the absolutely continuous part of dF with density f_a and dF_s is the singular part of dF, which is concentrated on a zero Lebesgue measure subset of $[-\pi, \pi]$.

The following three cases are distinguished:

(1) $f_a(\omega) = 0$ *on a set of positive Lebesgue measure on* $[-\pi, \pi]$*;*

(2) $f_a(\omega) > 0$ *a.e., but* $\int_{-\pi}^{\pi} \log f_a(\omega) d\omega = -\infty$*;*

(3) $f_a(\omega) > 0$ *a.e. and* $\int_{-\pi}^{\pi} \log f_a(\omega) d\omega > -\infty$*.*

Then in cases (1) and (2), the time series X is singular. In case (3), the time series $\{X_t\}$ is non-singular, so we may apply the Wold decomposition to write

$$X_t = R_t + Y_t = \sum_{j=0}^{\infty} b_j \xi_{t-j} + Y_t (t \in \mathbb{Z}), \tag{2.62}$$

where $\{R_t\}$ is a regular time series with absolutely continuous spectral measure $dF_a(\omega) = f_a(\omega) d\omega$ and $\{Y_t\}$ is a singular time series with singular spectral measure dF_s, as described by (2.61).

Proof. Assume first that $\{X_t\}$ is non-singular. Then, as we saw earlier, in the Wold decomposition (2.62) $\{R_t\}$ and $\{Y_t\}$ are orthogonal to each other, $\{R_t\}$ is a regular process with absolutely continuous spectrum $dF_a(\omega) = f_a(\omega) d\omega$ and $\{Y_t\}$ is a singular process with singular spectrum dF_s w.r.t. Lebesgue measure. Theorem 2.3 shows that for the density f_a of the regular part case (3) of the present theorem holds.

Conversely, we want to show that in case (3) $\{X_t\}$ is non-singular. Our starting point will be the Lebesgue decomposition (2.61). We want to find two mutually orthogonal stationary time series $\{R_t\}$ and $\{Y_t\}$ both subordinated to $\{X_t\}$, $R_t + Y_t = X_t$, with spectral measure dF_a and dF_s, respectively. Since $\{R_t\}$ and $\{Y_t\}$ are orthogonal to each other, their covariance function $C^{R,Y}(k) := \mathbb{E}(R_{t+k} Y_t) = 0$ for any $k \in \mathbb{Z}$. Thus

$$C^{R,X}(k) = \mathbb{E}(R_{t+k} X_t) = \mathbb{E}(R_{t+k} R_t) = C^R(k), \quad C^{Y,X}(k) = C^Y(k),$$

for $k \in \mathbb{Z}$ and so we have the same relationship for spectral measures: $dF^{R,X} = dF^R = dF_a$ and $dF^{Y,X} = dF^Y = dF_s$. By (2.2) and (2.6) it implies for the conditional densities that

$$f^{R|X} = |f^{R|X}|^2, \quad f^{Y|X} = |f^{Y|X}|^2, \quad |f^{R|X}|^2 + |f^{Y|X}|^2 = \frac{dF^R}{dF} + \frac{dF^Y}{dF} = 1.$$

It means that $f^{R|X}$ and $f^{Y|X}$ may take only 0 and 1 values, complementarily,

$$f^{R|X} = \frac{dF_a}{dF}, \quad f^{Y|X} = \frac{dF_s}{dF},$$

and they belong to $L^2([-\pi, \pi], \mathcal{B}, dF)$. Moreover, since $\{R_t\}$ and $\{Y_t\}$ should belong to $H(X)$, by (2.1) we may define them as

$$R_t = \int_{-\pi}^{\pi} e^{it\omega} f^{R|X}(\omega) dZ_\omega, \quad Y_t = \int_{-\pi}^{\pi} e^{it\omega} f^{Y|X}(\omega) dZ_\omega, \quad X_t = R_t + Y_t$$

for $t \in \mathbb{Z}$. Thus $\{R_t\}$ has spectral measure $dF_a(\omega) = f_a(\omega)d\omega$ and $\{Y_t\}$ has a singular spectral measure dF_s. Then by Theorem 2.3, case (3) of the present theorem implies that $\{R_t\}$ is a regular time series. Therefore,

$$X_t = R_t + Y_t = \sum_{j=0}^{\infty} b_j \xi_{t-j} + Y_t \quad (t \in \mathbb{Z}), \tag{2.63}$$

where $\{\xi_t\}_{t\in\mathbb{Z}}$ is an orthonormal sequence.

Let $H_k^-(R) := \overline{\text{span}}\{R_j : j \le k\}$, the past of $\{R_t\}$ until k, and $H(Y) := \overline{\text{span}}\{Y_j : j \in \mathbb{Z}\}$, the closed space spanned by $\{Y_t\}$. Formula (2.63) shows that

$$H_k^-(X) := \overline{\text{span}}\{X_j : j \le k\} \subset (H_k^-(R) \oplus H(Y)).$$

(The right hand side is the closed direct sum, that is, the smallest closed space spanned by $H_k^-(R)$ and H_Y.) Now we write

$$X_{k+1} = X_{k+1}^- + X_{k+1}^+ := \left\{ \sum_{j=1}^{\infty} b_j \xi_{k+1-j} + Y_{k+1} \right\} + b_0 \xi_{k+1}.$$

Clearly, $X_{k+1}^- \in (H_k^-(R) \oplus H_Y)$, while $X_{k+1}^+ = b_0 \xi_{k+1} \perp (H_k^-(R) \oplus H_Y)$. Moreover, by the condition in case (3) and the Kolmogorov–Szegő formula (Remark 2.7) we see that $b_0 > 0$, hence $X_{k+1} \notin H_k^-(X)$. This proves that X is non-singular.

Thus we have shown that in case (3) the time series is non-singular and for any non-singular time series the case (3) holds. It implies that in the cases (1) and (2) the time series is singular and this completes the proof of the theorem. □

Corollary 2.3. *In the proof of the previous theorem we saw that for any non-singular time series $\{X_t\}$ the Wold decomposition $X_t = R_t + Y_t$ gives a regular process $\{R_t\}$ whose spectral density function f^R is a.e. the same as the spectral density function f_a of $\{X_t\}$. Thus the Kolmogorov–Szegő formula (Remark 2.7) is valid for any non-singular process $\{X_t\}$ and its spectral density function f_a.*

2.10 Examples for singular time series

2.10.1 Type (0) singular time series

In the Lebesgue decomposition (2.61) one can further decompose the singular spectral measure:

$$dF_s = dF_d + dF_c,$$

where dF_d is the discrete spectrum corresponding to at most countable many jumps of the spectral distribution function F, while dF_c is the continuous singular spectrum.

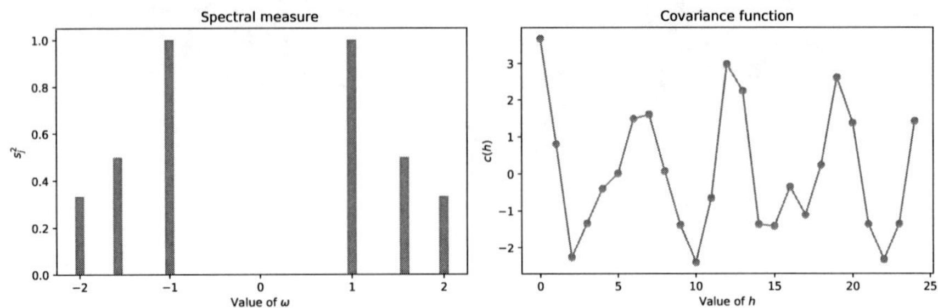

FIGURE 2.11
Spectral measure and covariance function of a Type(0) singular process in Example 2.6.

(a) A typical example for a process with discrete spectrum:

$$X_t = \sum_{j=1}^{n} A_j e^{it\omega_j}, \qquad t \in \mathbb{Z}, \tag{2.64}$$

where $-\pi < \omega_1 < \cdots < \omega_n \leq \pi$; $A_1, \dots, A_n$ are uncorrelated random variables with mean 0 and variance s_j^2 $(j = 1, \dots, n)$. (The A_j's can be e.g. independent Gaussian random variables.)

This process is weakly stationary with

$$c(k) = \mathbb{E}(X_{t+k}\overline{X_t}) = \sum_{j=1}^{n} \mathbb{E}(|A_j|^2)e^{ik\omega_j} = \sum_{j=1}^{n} s_j^2 e^{ik\omega_j} = \int_{\pi}^{\pi} e^{ik\omega}\, dF(\omega)$$

for $k \in \mathbb{Z}$, where

$$F(\omega) = \sum_{\omega_j \leq \omega} s_j^2, \qquad \omega \in [-\pi, \pi].$$

Observe that the covariance function does not tend to 0 as $k \to \infty$.

Example 2.6. Figure 2.11 shows the spectral measure and the covariance function of a Type (0) process (2.64) with $n = 6$. The frequencies ω_j are $\{-2, -\pi/2, -1, 1, \pi/2, 2\}$ and the variances s_j^2 are $\{1/3, 1/2, 1, 1, 1/2, 1/3\}$. Since in this example the covariance function is real valued, by the construction in Subsection 1.4.1 it is possible to construct a real valued process $\{X_t\}$ with this covariance function.

The first panel on Figure 2.12 shows the mean square prediction error e_n^2 when predicting X_n based on the finite past $X_0, \dots, X_{n-1}$, see Subsection 5.2.1. If $n \geq 5$, square error $e_n^2 = 0$. On the second panel $\det(\boldsymbol{C}_n)$ is shown, which is also 0 if $n \geq 6$, see Remark 5.5.

(b) The standard example for a continuous singular function on $[0, 1]$ is the Cantor function γ, "the devil's ladder." Suppose $\mathcal{C}$ is the Cantor set in $[0, 1]$,

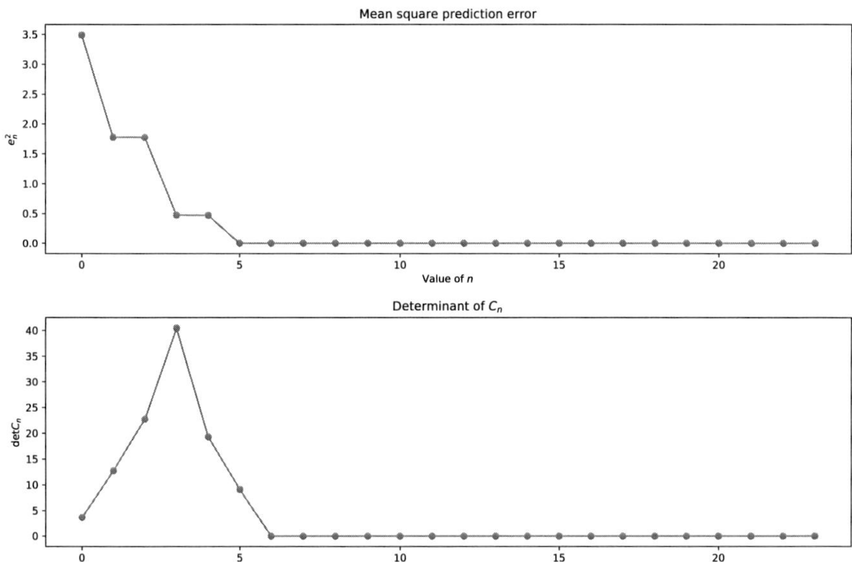

FIGURE 2.12
Prediction error and $\det(\boldsymbol{C}_n)$ of a Type(0) singular process in Example 2.6.

that is, $x \in \mathcal{C}$ if and only if in base 3 expansion. While in (a) above we discussed a singular process with pure discrete spectrum, now we consider a singular process with continuous singular spectrum.

$$x = \sum_{n=1}^{\infty} a_n 3^{-n}, \quad a_n = 0 \text{ or } 2.$$

Then the Cantor function $\gamma : [0,1] \to [0,1]$ can be defined as

$$\gamma(x) = \begin{cases} \sum_{n=1}^{\infty} \frac{1}{2} a_n 2^{-n}, & x = \sum_{n=1}^{\infty} a_n 3^{-n} \in \mathcal{C}; \\ \sup\{\gamma(y) : y \leq x, \ y \in \mathcal{C}\}, & x \in [0,1] \setminus \mathcal{C}. \end{cases}$$

Then $\gamma(0) = 0$, $\gamma(1) = 1$, γ is non-decreasing on $[0,1]$, and $\gamma'(x) = 0$ for a.e. $x \in [0,1]$.

By the case $d = 1$ of Corollary 1.6, the definition

$$F(\omega) = \gamma\left(\frac{\omega + \pi}{2\pi}\right), \quad \omega \in [-\pi, \pi],$$

gives the spectral distribution function of a singular stationary time series X. Heuristically, the spectrum of X consists of the points of an uncountable but zero Lebesgue measure Cantor set, with infinitesimally small amplitudes.

2.10.2 Type (1) singular time series

A simple example for a singular time series corresponding to case (1) of Theorem 2.4 is the one with spectral density function

$$f(\omega)=\begin{cases}\frac{1}{2}, & |\omega|\le 1;\\ 0, & 1<|\omega|\le\pi.\end{cases}$$

By the case $d=1$ of Corollary 1.6, one can construct a singular stationary time series with this spectral density.

Its covariance function is

$$c(k)=\int_{-1}^{1}e^{ik\omega}\frac{1}{2}d\omega=\begin{cases}1, & k=0;\\ \frac{\sin k}{k}, & k\neq 0.\end{cases}$$

It is an example for an absolutely non-summable covariance function which still corresponds to an absolutely continuous spectral measure. On the other hand, $\sum_k c(k)$ converges conditionally; also, $\sum_k |c(k)|^2<\infty$.

Generalizing the previous example, observe the following interesting phenomenon. For any $\delta>0$ fixed, a time series with spectral density

$$f(\omega)=\begin{cases}\frac{1}{2(\pi-\delta)}, & |\omega|\le\pi-\delta;\\ 0, & \pi-\delta<|\omega|\le\pi.\end{cases}$$

is still singular, like the one above. On the other hand, if we take $\delta=0$, that is, $f(\omega)=\frac{1}{2\pi}$ for any $\omega\in[-\pi,\pi]$, then the time series becomes a regular, orthonormal series.

Example 2.7. Figure 2.13 shows the spectral density and the covariance function of the above described Type (1) singular process.

The first panel on Figure 2.14 shows the mean square prediction error e_n^2 when predicting X_n based on the finite past $X_0,\dots,X_{n-1}$, see Subsection 5.2.1. As n is growing, the prediction error is going to 0, since this is a singular process. On the second panel $\det(\boldsymbol{C}_n)$ is shown, which goes to 0 as well, see Remark 5.5.

2.10.3 Type (2) singular time series

An example for case (2) of Theorem 2.4 is the following spectral density:

$$f(\omega)=e^{-\frac{1}{|\omega|}},\quad \omega\in[-\pi,\pi]\setminus\{0\},\quad f(0)=0.$$

Then $f(\omega)>0$ a.e. and f is continuous everywhere on $[-\pi,\pi]$,

$$\int_{-\pi}^{\pi}f(\omega)d\omega<\infty,\quad \int_{\pi}^{\pi}\log f(\omega)d\omega=\int_{-\pi}^{\pi}-\frac{1}{|\omega|}=-\infty.$$

By the case $d=1$ of Corollary 1.6, one can construct a singular stationary

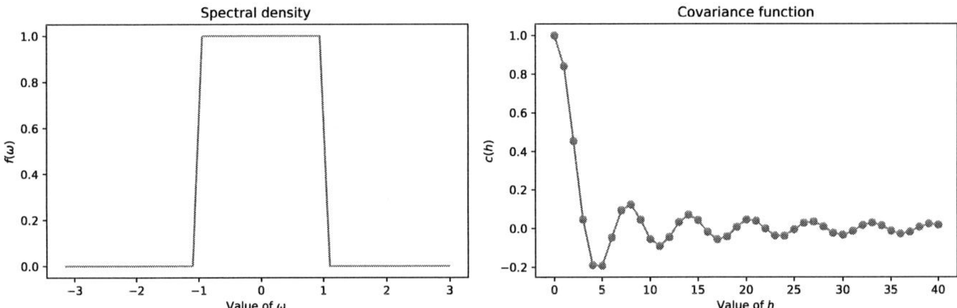

FIGURE 2.13
Spectral density and covariance function of a Type (1) singular process in Example 2.7.

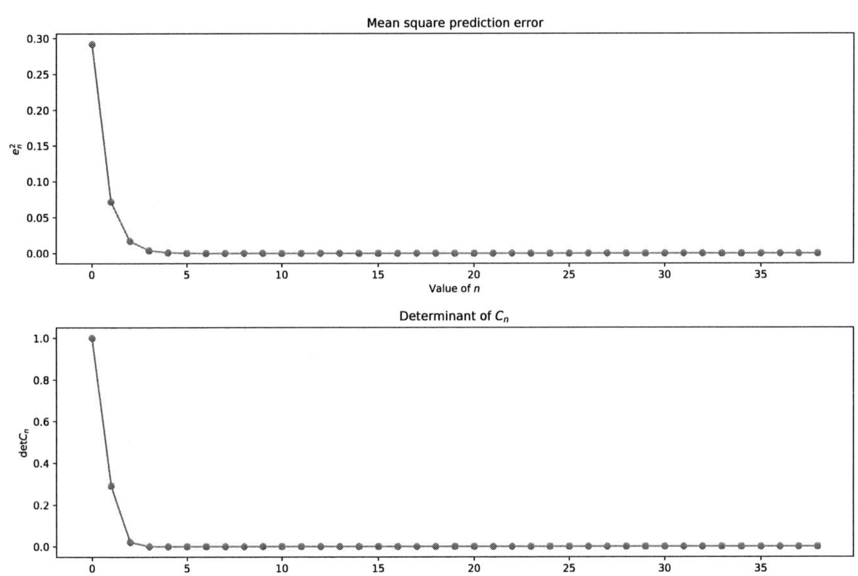

FIGURE 2.14
Prediction error and $\det(\boldsymbol{C}_n)$ of a Type (1) singular process in Example 2.7.

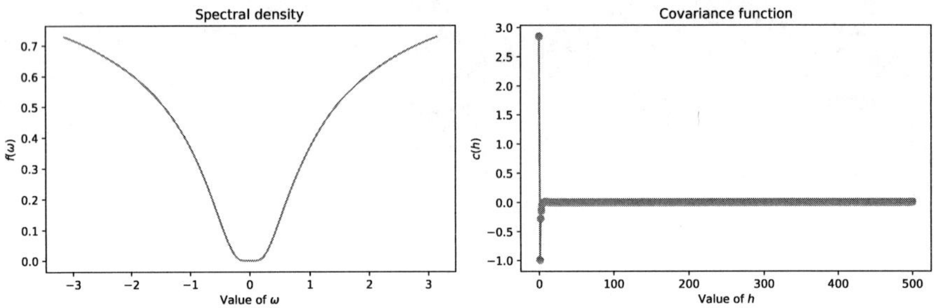

FIGURE 2.15
Spectral density and covariance function of a Type (2) singular process in Example 2.8.

time series $\{X_t\}$ with this spectral density. Theorem 4.1 later shows that a time series can be represented as a two-sided infinite MA (a sliding summation) if and only if it has constant rank, that is, in 1D its spectral density is positive a.e., like in the case of this $\{X_t\}$. However, since this $\{X_t\}$ is singular, it cannot be represented as a causal (one-sided) infinite MA. In general, in 1D the same is true for any singular time series of Type (2) and only for these ones.

Example 2.8. Figure 2.15 shows the spectral density and the covariance function of the above described Type (2) singular process. The last few values of $c(h)$ are not used because of the numerical distortion of the IDFT (FFT) procedure used to compute the covariance function from the spectral density.

The first panel on Figure 2.16 shows the mean square prediction error e_n^2 when predicting X_n based on the finite past $X_0, \dots, X_{n-1}$, see Subsection 5.2.1. As n is growing, the mean square prediction error is eventually going to 0, since this is a singular process. The second panel shows $\det(\boldsymbol{C}_n)$, which goes to 0 as well, see Remark 5.5.

Corollary 2.4. *Theorem 2.4 has the following interesting corollary. Any non-singular process may contain a Type (0) singular component, but cannot contain a Type (1) or (2) singular part. For, if case (1) or (2) of Theorem 2.4 holds for a time series then that process must be singular. Adding an orthogonal regular process to a Type (1) or (2) singular process results in a regular process.*

For, let $\{X_t\}$ be a regular and $\{Y_t\}$ be a Type (1) or (2) singular stationary time series, $\text{Cov}(X_s, Y_t) = 0$ *for all $s, t \in \mathbb{Z}$. Then*

$$\text{Cov}(X_{t+h} + Y_{t+h}, X_t + Y_t) = c^X(h) + c^Y(h) = \int_{-\pi}^{\pi} e^{ih\omega}(f^X(\omega) + f^Y(\omega))d\omega,$$

where c^X and c^Y are the covariance functions and f^X and f^Y are the spectral

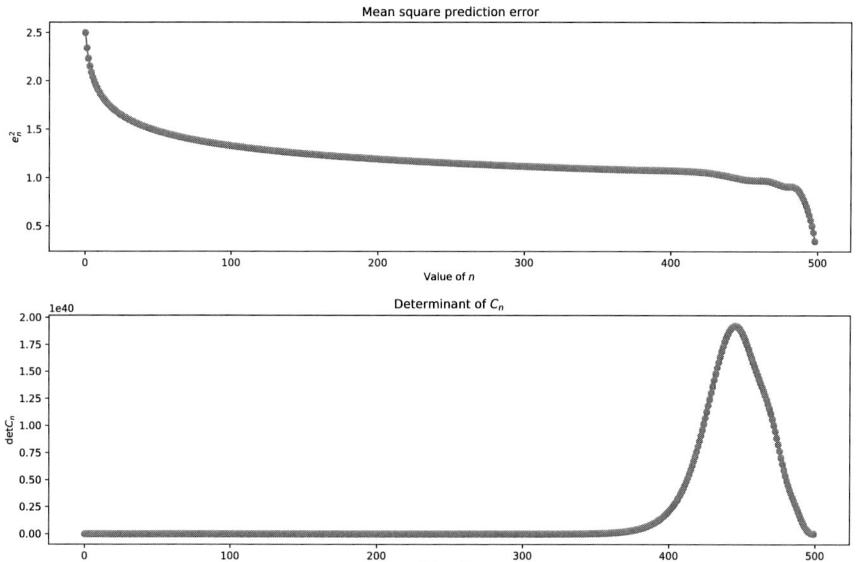

FIGURE 2.16
Prediction error and $\det(\boldsymbol{C}_n)$ of a Type (2) singular process in Example 2.8.

density functions of X and Y, respectively. Consequently,

$$\int_{-\pi}^{\pi} \log(f^X(\omega) + f^Y(\omega))d\omega \geq \int_{-\pi}^{\pi} \log f^X(\omega)d\omega > -\infty.$$

This shows that $\{X_t + Y_t\}$ is regular.

2.11 Summary

Linear filtering of a stationary time series $\{X_t\}_{t\in\mathbb{Z}}$ means applying a TLF (time-invariant linear filter) to it:

$$Y_t := \sum_{j=-\infty}^{\infty} c_{tj}X_j = \sum_{k=-\infty}^{\infty} b_k X_{t-k} \quad (t \in \mathbb{Z}).$$

Time-invariance means that the coefficient c_{tj} depends only on $t - j$, i.e. $c_{tj} = b_{t-j}$, giving the final form of a TLF. The so obtained $\{Y_t\}$ is also a weakly stationary sequence, and we use the wording that it is *subordinated* to the process $\{X_t\}$. A sufficient condition for the almost sure and mean square convergence of the above filtration is that $\sum_{j=-\infty}^{\infty} |b_j| < \infty$.

The second order, weakly stationary 1D time series $\{Z_t\}_{t\in\mathbb{Z}}$ is called *white noise sequence* if its autocovariances are $c(0) = \sigma^2$ and $c(h) = 0$ for $h = \pm 1, \pm 2, \dots$. It is denoted by $\mathrm{WN}(\sigma^2)$. In other words, Z_ts are uncorrelated and have variance σ^2. In particular, if Z_ts are i.i.d., with finite variance, they constitute a white noise sequence, and in the Gaussian case, the two notions are the same. We call the WN(1) sequence *orthonormal sequence.* Its spectral density is constant ($\frac{1}{2\pi}$), like the spectrum of the white light. With the $\{\xi_t\} \sim$ WN(1) sequence, the following special processes are introduced.

The *two-sided MA* (*sliding summation*) is defined by

$$X_t = \sum_{k=-\infty}^{\infty} b_k \xi_{t-k} = \sum_{j=-\infty}^{\infty} b_{t-j}\xi_j, \qquad t \in \mathbb{Z},$$

where for the sequence of non-random complex coefficients $\{b_k\}$ we assume that $\sum_k |b_k|^2 < \infty$. (Then the Riesz–Fischer theorem implies the mean square convergence of the above infinite series.) If $\xi_j = 0$ for any $j \neq j_0$, then $X_t = b_{t-j_0}\xi_{j_0}$, $t \in \mathbb{Z}$. This is why the sequence $\{b_k\}_{k\in\mathbb{Z}}$ is called the *impulse response function.* Since the single impulse ξ_{j_0} (random shock) at time j_0 can create nonzero responses not only for times $t \geq j_0$, but also for times $t < j_0$, a sliding summation process $\{X_t\}$ is *non-causal* in general. In contrast, when $b_k = 0$ whenever $k < 0$, we obtain a *causal (one-sided, future-independent) MA(∞) process*:

$$X_t = \sum_{k=0}^{\infty} b_k \xi_{t-k} = \sum_{j=0}^{\infty} b_{t-j}\xi_j, \qquad t \in \mathbb{Z},$$

with the covariance function $c(h) = \sum_{k=0}^{\infty} b_{k+h}\bar{b}_k \quad (h \geq 0), \quad c(-h) = \bar{c}(h)$.

It is customary to introduce the *lag operator or left (backward) shift* L, and the formal power series $H(L) := \sum_k b_k L^k$, so we can write the sliding summation as $X_t = H(L)\xi_t$, $t \in \mathbb{Z}$. Further, it is also customary to denote the operator L by the indeterminate z as well and to call the power series

$$H(z) = \sum_{k=-\infty}^{\infty} b_k z^k$$

the *transfer function* of the sliding summation $\{X_t\}$ or the *z-transform* of $H(L)$. Since by our assumption $\sum_k |b_k|^2 < \infty$, we have $H(z) \in L^2(T)$, where T denotes the unit circle of the complex plane $\mathbb{C}$.

The finite order MA processes are generated with finitely many random shocks. The *qth order moving average process*, denoted by MA(q), is defined by

$$X_t = \beta(L)\xi_t = \sum_{k=0}^{q} \beta_k \; \xi_{t-k},$$

where $\{\xi_t\} \sim$ WN(1), L is the lag operator, and

$$\beta(z) = \sum_{k=0}^{q} \beta_k z^k$$

is the *MA polynomial.* The covariance function of the MA(q) process is

$$c(h) = \sum_{k=h}^{q} \beta_k \bar{\beta}_{k-h} \quad (0 \le h \le q), \quad c(-h) = \bar{c}(h), \quad c(h) = 0 \text{ if } |h| > q.$$

Conversely, if the autocovariance function of a zero mean stationary process is such that $c(h) = 0$ for $|h| > q$ and $c(q) \neq 0$, then it is a MA(q) process. The spectral density of an MA(q) process is $f(\omega) = \frac{1}{2\pi}|\beta(e^{-ik\omega})|^2$ which is a $2q$ degree polynomial of $e^{-i\omega}$.

A *pth order autoregressive process* AR(p) is defined by

$$\alpha(L)X_t = \beta\xi_t,$$

where $\{\xi_t\} \sim \text{WN}(1)$, L is the lag operator, and

$$\alpha(z) = \sum_{k=0}^{q} \alpha_k z^k$$

is the *AR polynomial,* $\alpha_0 = 1$, $\beta \neq 0$.

A stationary causal MA(∞) solution of it exists if and only if the AR(p) polynomial is *stable*, i.e. it has no roots in the closed unit disc ($|z| \le 1$) of $\mathbb{C}$. If $\boldsymbol{C}_p$ of Chapter 1 is positive definite, then this condition holds. Between the autocovariances and the constants of the AR(p) process we have the following *Yule–Walker equations*:

$$c(-k) + \sum_{j=1}^{p} c(j-k)\bar{\alpha}_j = \delta_{0k}|\beta|^2, \qquad k \ge 0.$$

Taking them for $0 \le k \le p$, one obtains a system of linear equations for the unknowns $\alpha_1, \dots, \alpha_p$, $|\beta|^2$, if the covariances $c(0), c(1), \dots, c(p)$ are known. It has a unique solution if and only if the aforementioned matrix $\boldsymbol{C}_p$ is positive definite, see also Chapter 5. In this case the AR(p) process is also stable.

The ARMA(p, q) processes ($p \ge 0$, $q \ge 0$) are generalizations of both AR(p) and MA(q) processes: in the $p = 0$ case we get a MA(q), while in the $q = 0$ case we get an AR(p) process. With the aforementioned polynomials, an ARMA(p, q) process is defined by

$$\alpha(L)X_t = \beta(L)\xi_t.$$

Again, we want to find a causal stationary MA(∞) solution of this equation. Assume that the polynomials $\alpha(z)$ and $\beta(z)$ have no common zeros. Then $\{X_t\}$ is causal if and only if $\alpha(z) \neq 0$ for all $|z| \le 1$ (the same stability condition as in case of the AR process). The coefficients (impulse responses) of the MA(∞) process $X_t = \sum_{j=0}^{\infty} b_j \xi_{t-j}$ are obtainable by the power series expansion of the transfer function

$$H(z) = \sum_{j=0}^{\infty} b_j z^j = \alpha^{-1}(z)\beta(z), \quad |z| \le 1.$$

The spectral density of the ARMA(p, q) process is

$$f(\omega) = \frac{1}{2\pi} \frac{|\beta(e^{-i\omega})|^2}{|\alpha(e^{-i\omega})|^2}.$$

This is a rational function (ratio of polynomials) of $e^{-i\omega}$. Rational spectral densities play an important role in state space models of Chapter 3. It can also be shown that any continuous spectral density can be approximated with the spectral density of a MA(q) and/or AR(p) process with any small error, albeit with possibly large q and/or p.

So processes of rational spectral density are special MA(∞) processes. One-sided moving average processes are also called *regular* (also called *purely non-deterministic*) as they have no remote past at all (they are entirely governed by random shocks). In contrast, *singular* (also called *deterministic*) processes are completely determined by their remote past. As a mixture of them, *non-singular processes* cannot be completely predicted based on their past values, but there are added values (innovations) of the newcoming observations.

The Wold decomposition in 1D guarantees that any non-singular weakly stationary time series (not completely determined by the remote past) can be decomposed into a regular (MA(∞)) and a singular process that are orthogonal to each other. The original proof of Wold is based on the one-step ahead predictions with longer and longer past, where the error terms (innovations) give the shocks, and prediction errors converge to an optimum value (that is not zero in the presence of a regular part), see Chapter 5. The spectral form of the Wold decomposition is related to the transfer function and discussed in details in the multivariate situation in Chapter 4.

By a theorem of Kolmogorov, a 1D stationary time series is regular if and only if the following three conditions hold on $[-\pi, \pi]$:

1. its spectral measure dF is absolutely continuous w.r.t. Lebesgue measure;
2. its spectral density f is positive almost everywhere;
3. *(Kolmogorov's condition)* $\log f$ is integrable.

Note that we have different classes of 1D singular time series as follows. The spectral measure of a stationary time series has a unique Lebesgue decomposition

$$dF = dF_a + dF_s, \quad dF_a \ll d\omega, \quad dF_s \perp d\omega,$$

where $d\omega$ denotes Lebesgue measure, $dF_a(\omega) = f_a(\omega)d\omega$ is the absolutely continuous part of dF with density f_a and dF_s is the singular part of dF, which is concentrated on a zero Lebesgue measure subset of $[-\pi, \pi]$. We call it Type (0) singularity. Apart from this, we distinguish between the following three cases:

(1) $f_a(\omega) = 0$ on a set of positive Lebesgue measure on $[-\pi, \pi]$;

(2) $f_a(\omega) > 0$ a.e. , but $\int_{-\pi}^{\pi} \log f_a(\omega) d\omega = -\infty$;

(3) $f_a(\omega) > 0$ a.e. and $\int_{-\pi}^{\pi} \log f_a(\omega) d\omega > -\infty$.

In cases (1) and (2), the time series is singular, whereas in case (3), the time series is non-singular and has the Wold decomposition. So a regular process can coexist only with a Type (0) singular one, while adding an orthogonal regular process to Type (1) or (2) singularity results in a regular process.

3

Linear system theory, state space models

3.1 Introduction

We apply state space models to multidimensional stationary processes, the same approach as that of the seminal Kálmán's filtering. We consider the reachability and observability of a system, defined by an inner description. Power series and extended input/output maps are also considered together with an external description of the system.

Algebraic tools of modules are intensively used. By means of these, we investigate when and how one can get an internal description from an external one, more exactly, from a reduced input/output map. We also introduce the notion of minimal polynomial of a state space if it is finite dimensional.

Next, we turn to realizations that use power series techniques. The transfer function $H(z)$ of a linear system is a power series containing only non-negative powers of z and its coefficients are $p \times q$ complex matrices. Equivalently, $H(z)$ is a $p \times q$ matrix, whose entries are complex power series with only non-negative powers of z. The transfer function $H(z)$ of a linear system with finite dimensional state space is a rational matrix. Equivalences are proved about the rationality of the transfer function and the bounded rank of the infinite block Hankel matrix formed by its coefficients.

The McMillan degree of the transfer function is the dimension of the state space in a minimal realization. It is uniquely defined. More general linear time-invariant dynamical systems are also treated. Relation of Hankel operators and realization theory is investigated. Finally, stochastic time-invariant linear systems are introduced that are driven by two multidimensional random stationary time series. Multidimensional ARMA (VARMA) processes are special cases of them.

3.2 Restricted input/output map

Suppose that we are given a quadruple of bounded linear operators (A, B, C, D), $A : X \to X$, $B : U \to X$, $C : X \to Y$, and $D : U \to Y$, where U is a q-dimensional, Y is a p-dimensional and X (the *state space*) is an n-dimensional complex vector space. The corresponding *linear time invariant dynamical system* is described by

$$\begin{aligned} x_{k+1} &= Ax_k + Bu_k \\ y_k &= Cx_k + Du_k \qquad (k \in \mathbb{Z}). \end{aligned} \tag{3.1}$$

We may write that $U = \mathbb{C}^q$, $X = \mathbb{C}^n$, $Y = \mathbb{C}^p$, $A \in \mathbb{C}^{n \times n}$, $B \in \mathbb{C}^{n \times q}$, $C \in \mathbb{C}^{p \times n}$, and $D \in \mathbb{C}^{p \times q}$.

Eventually, the input sequence $\{u_k\}$ is going to be a random stationary time series, that generates the dynamics. The resulting random stationary time series $\{x_k\}$ describes the consecutive, directly unobservable, internal states of the system, while the random stationary time series $\{y_k\}$ is the observable output sequence. However, except for the last section of this chapter, everything will be deterministic. Interestingly, for discussing many important properties of linear dynamical systems it is immaterial whether the input process is deterministic or stochastic.

In this section we consider a very simple case. We assume that

1. $u_j = 0$ when $j \geq 1$;
2. the input sequence $\{u_j : j \leq 0\}$ contains only finitely many nonzero terms;
3. the system was at rest when the first nonzero input arrived, say at time $-j_0 \leq 0$ (which is arbitrary, but finite): $x_{-j_0} = 0$.

If $k \leq 0$, then by the first equation of (3.1), progressing from time $-j_0$ forward step-by-step, we get that

$$x_k = Bu_{k-1} + ABu_{k-2} + A^2Bu_{k-3} + \cdots + A^{j_0+k-1}Bu_{-j_0}.$$

Then remembering the above assumptions (1)–(3), we may write that

$$x_k = \sum_{j=0}^{\infty} A^{j+k-1}Bu_{-j} \qquad (k \geq 1). \tag{3.2}$$

This sum is convergent since it contains only finitely many non-zero terms. Further, we consider the output sequence for times $k \geq 1$ only, where $u_k = 0$:

$$y_k = \sum_{j=0}^{\infty} CA^{j+k-1}Bu_{-j} \quad (k \geq 1). \tag{3.3}$$

This sum is convergent as well, since contains only finitely many non-zero terms.

Then we define the *restricted input/output map* ϕ_0 of the system by

$$\phi_0(\{u_j\}_{j=-\infty}^{0}) = \{y_k\}_{k=1}^{\infty}, \tag{3.4}$$

as described by (3.3). (It is necessary writing the input sequence as a sequence

$\{u_j\}_{j=-\infty}^{0}$, because the starting time $j_0 \leq 0$ of the input sequence can be arbitrary.)

Observe that ϕ_0 does not depend on the operator D, therefore in this chapter, except when it is explicitly stated otherwise, we may assume that $D = 0$. It is useful to write ϕ_0 in matrix form:

$$\begin{bmatrix} y_1 \\ y_2 \\ y_3 \\ \vdots \end{bmatrix} = \mathcal{H} \begin{bmatrix} u_0 \\ u_{-1} \\ u_{-2} \\ \vdots \end{bmatrix} = \begin{bmatrix} CB & CAB & CA^2B & \cdots \\ CAB & CA^2B & \cdot & \cdots \\ CA^2B & \cdot & \cdot & \cdots \\ \vdots & \vdots & \vdots & \end{bmatrix} \begin{bmatrix} u_0 \\ u_{-1} \\ u_{-2} \\ \vdots \end{bmatrix} \tag{3.5}$$

The infinite coefficient matrix $\mathcal{H}$ here is a *block Hankel matrix*; each of its blocks is a $p \times q$ complex matrix (see Appendix B).

3.3 Reachability and observability

Let $\Sigma = (A, B, C)$ be a finite dimensional linear system. We say that the *state $x \in X$ is reachable* if there is a $k \geq 0$ and a sequence of inputs $\{u_0, u_1, \ldots, u_{k-1}\}$ of length k that drives the system from the state $x_0 = 0$ to the state $x_k = x$:

$$x = \sum_{j=0}^{k-1} A^{k-1-j} B u_j = [B, AB, \ldots, A^{k-1}B] \begin{bmatrix} u_{k-1} \\ u_{k-2} \\ \vdots \\ u_0 \end{bmatrix}.$$

By the Cayley–Hamilton theorem, if $k \geq n$, A^k can be expressed in terms of $I, A, \ldots, A^{n-1}$. Thus

$$\text{Range}[B, AB, \ldots, A^{k-1}B] = \text{Range}[B, AB, \ldots, A^{n-1}B] \quad (k \geq n).$$

The matrix

$$\mathcal{R} := [B, AB, \ldots, A^{n-1}B] \tag{3.6}$$

is the *reachability matrix* of the system and its range is the *subspace of reachable states.* We say that the *system Σ is reachable* if any $x \in X$ is reachable, that is, if $\text{rank}(\mathcal{R}) = \dim(\text{Range}(\mathcal{R})) = n$, equivalently, if $\mathcal{R}$ has n linearly independent columns, equivalently, if $\text{Ker}(\mathcal{R}^*) = \{0\}$. Clearly, $\text{Range}(\mathcal{R})$ is an A-invariant subspace in X: $A\,\text{Range}(\mathcal{R}) \subset \text{Range}(\mathcal{R})$.

We say that the *state $x \in X$ is observable* if starting from the state $x_0 = x$ at time 0 and having a constant zero input sequence $u_j = 0$ $(j \geq 0)$, there is a time instant $k \geq 0$ such that

$$y_k = CA^k x \neq 0.$$

The reason for the name 'observable' is that the state x is observable if and only if for any $x_{(1)}, x_{(2)} \in X$, $x_{(2)} - x_{(1)} = x$, starting at time 0 with initial state $x_{(1)}$ and $x_{(2)}$, respectively, there exists a time instant $k \geq 0$ such that the output y_k is different in the two cases. This fact follows from the linearity of the system.

Again, by the Cayley–Hamilton theorem, if

$$\mathcal{O} := \begin{bmatrix} C \\ CA \\ \vdots \\ CA^{n-1} \end{bmatrix}, \text{ then Ker} \begin{bmatrix} C \\ CA \\ \vdots \\ CA^{k} \end{bmatrix} = \text{Ker}(\mathcal{O}) \text{ for each } k \geq n-1, \quad (3.7)$$

where $\mathcal{O}$ is called the *observability matrix* of the system. The orthogonal complement of its kernel is the *subspace of observable states.* The *system* Σ *is called observable*, if any $x \in X$ is observable, that is, if $\text{Ker}(\mathcal{O}) = \{0\}$, equivalently, if $\mathcal{O}$ has n linearly independent rows, equivalently, if $\text{rank}(\mathcal{O}) = n$. Clearly, $\text{Ker}(\mathcal{O})$ is an A-invariant subspace in X.

3.4 Power series and extended input/output maps

An important tool in system theory is the application of power series. The set of integers $\mathbb{Z}$, that is, the set of time instants can be mapped in a one-to-one way onto powers of the indeterminate z which is the operator of the *left (backward) time shift in* $\mathbb{Z}$. So an input u at time $-j$ can be written as uz^j and an output y at time k can be written as yz^{-k}. Correspondingly, the *right (forward) time shift* is denoted by z^{-1}.

Now we consider more general input/output sequences than in Section 3.2. Thus we suppose that

1. the input sequence contains only finitely many terms with negative indices;
2. the system was at rest when the first nonzero input arrived, say at time $-j_0 \leq 0$ (which is arbitrary): $x_{-j_0} = 0$.

So the input sequence can be written as a formal power series

$$u(z) := \sum_{j=-j_0}^{\infty} u_j z^{-j} = \sum_{j=-\infty}^{\infty} u_j z^{-j} \quad (3.8)$$

Recall that here the coefficients $u_j \in U$ are q-dimensional complex vectors and in the last sum there are only finitely many terms with negative index j.

The formal power series $u(z)$ is also called the *z-transform* of the sequence $\{\dots, u_{-2}, u_{-1}, u_0, u_1, u_2, \dots\}$.

Let $\mathcal{U}$ denote the set of all input sequences $\{u_j : j \in \mathbb{Z}\}$ that contain only finitely many nonzero terms with negative indices. Equivalently, $\mathcal{U}$ is the vector space of all formal power series $u(z)$ as given by (3.8) with any $-j_0 \in \mathbb{Z}$.

Likewise, a corresponding output sequence $(y_{-j_0+1}, y_{-j_0+2}, \ldots)$ is the following formal power series (z-transform) with p-dimensional complex vector coefficients $y_k \in Y$:

$$y(z) := \sum_{k=-j_0+1}^{\infty} y_k z^{-k} = \sum_{k=-\infty}^{\infty} y_k z^{-k}. \tag{3.9}$$

Remember that in the last sum there are only finitely many terms with negative indices k.

Let $\mathcal{Y}$ denote all output sequences $\{y_k : k \in \mathbb{Z}\}$ that contain only finitely many nonzero terms with negative indices. Equivalently, $\mathcal{Y}$ is the vector space of all formal power series $y(z)$ as given by (3.9) with any $-j_0 \in \mathbb{Z}$.

With a finite dimensional linear system $\Sigma = (A, B, C)$, if $x_{-j_0} = 0$ and $k \geq -j_0 + 1$, then

$$y_k = \sum_{j=-j_0}^{k-1} CA^{k-j-1}Bu_j = \sum_{j=-\infty}^{k-1} CA^{k-j-1}Bu_j. \tag{3.10}$$

(The last series converges, because it contains only finitely many non-zero terms.)

Now we can consider the *extended input/output map* $\phi : \mathcal{U} \to \mathcal{Y}$ of the finite dimensional linear system $\Sigma = (A, B, C)$, which is defined by

$$\phi(u(z)) = y(z), \quad \phi\left(\sum_{j=-\infty}^{\infty} u_j z^{-j}\right) = \sum_{k=-\infty}^{\infty} y_k z^{-k}, \tag{3.11}$$

where y_k is given by (3.10).

Also, we define the *left shift* L that works in the ordinary way on the input/output sequences defined on $\mathbb{Z}$:

$$Lu(z) := zu(z), \quad Ly(z) := zy(z).$$

Then simply we have

$$\phi(u(z)) = y(z) \quad \Rightarrow \quad \phi(Lu(z)) = Ly(z). \tag{3.12}$$

It is very important that ϕ is not only a complex linear function from $\mathcal{U}$ into $\mathcal{Y}$, but a homomorphism (see Appendix D) for left shifts as well, as is shown by (3.12). Let $\mathbb{C}[z]$ denote the ring of all polynomials in the indeterminate z, with complex coefficients (see Appendix D). Let

$$p(z) = a_m z^m + a_{m-1} z^{m-1} + \cdots + a_1 z + a_0 \in \mathbb{C}[z]$$

be an arbitrary complex polynomial. The effect of $p(z)$ on an input sequence $u(z)$ or on an output sequence $y(z)$ is simply the corresponding combination of left shifts, additions and complex multiplications:

$$L_{p(z)}u(z) := p(z)u(z) = \sum_{r=0}^{m} a_r z^r u(z) = \sum_{j=-j_0-m}^{\infty} z^{-j} \sum_{r=0}^{m} a_r u_{j+r},$$

$$L_{p(z)}y(z) := p(z)y(z) = \sum_{r=0}^{m} a_r z^r y(z) = \sum_{k=-j_0-m+1}^{\infty} z^{-k} \sum_{r=0}^{m} a_r y_{k+r}.$$

Algebraically, it means that ϕ is a $\mathbb{C}[z]$-*module homomorphism* from $\mathcal{U}$ to $\mathcal{Y}$ (see Appendix D):

$$\phi(u(z)) = y(z) \quad \Rightarrow \quad \phi(L_{p(z)}u(z)) = L_{p(z)}y(z),$$

or briefly, $\phi L_p = L_p \phi$.

Let us introduce the *transfer function* $H(z)$ by the formal power series

$$H(z) := \sum_{\ell=0}^{\infty} H_\ell z^\ell := D + \sum_{\ell=1}^{\infty} CA^{\ell-1}Bz^\ell, \tag{3.13}$$

the coefficients $H_\ell := CA^{\ell-1}B$ ($\ell \geq 1$) being $p \times q$ complex matrices. The coefficient $H_0 = D$ has been assumed to be a $p \times q$ zero matrix so far, but eventually it can be nonzero.

Proposition 3.1. *We have the following properties of a transfer function:*

(1) If z is considered a complex variable, then (3.13) *converges for $|z| < 1/\rho(A)$, where $\rho(A)$ denotes the* spectral radius of the matrix *A:*

$$\rho(A) := \max\{|\lambda| : \lambda \in \mathbb{C},\, \det(\lambda I_n - A) = 0\},$$

see Lemma B.2. Then $H(z)$ is an analytic matrix function in the disc $\{z : |z| < 1/\rho(A)\}$, meaning that each entry of the matrix $H(z)$ is an analytic complex function there. In particular, if $\rho(A) = 1$, then $H(z)$ is analytic in the open unit disc and if $\rho(A) < 1$, then $H(z)$ is analytic in the closed unit disc.

(2)

$$H(z) = D + C(I_n - Az)^{-1}zB, \qquad |z| < 1/\rho(A), \tag{3.14}$$

$$H(z^{-1}) = D + C(I_n z - A)^{-1}B, \qquad |z| > \rho(A). \tag{3.15}$$

(3) The z-transform of the input/output map ϕ is $H(z^{-1})$:

$$y(z) = \phi(u(z)) = H(z^{-1})u(z), \qquad |z| > \rho(A). \tag{3.16}$$

(4) The transfer function $H(z)$ gives the output in terms of the input:

$$y_k = H(z)u_k, \qquad k \in \mathbb{Z}.$$

(5)

$$\lim_{z\to\infty} H(z^{-1}) = D.$$

Proof. By Lemma B.2, for any $\epsilon > 0$,

$$\|A^k\| \le (\rho(A)+\epsilon)^k,$$

so

$$\left\| \sum_{\ell=1}^{\infty} CA^{\ell-1}Bz^{\ell} \right\| \le \sum_{\ell=1}^{\infty} \|C\| \cdot \|A^{\ell-1}\| \cdot \|B\| \cdot |z|^{\ell}$$

$$\le \|C\| \cdot \|B\| \cdot |z| \sum_{\ell=1}^{\infty} [|z|(\rho(A)+\epsilon)]^{\ell-1} < \infty,$$

if $|z| < 1/(\rho(A)+\epsilon)$. This proves (1).

Simple algebra shows that we have the following relationship for formal power series:

$$(I_n - Az)\sum_{\ell=1}^{\infty} A^{\ell-1}z^{\ell} = I_n z.$$

If $|z| < 1/\rho(A)$, then this implies that $I_n - Az$ is invertible and

$$(I_n - Az)^{-1}z = \sum_{\ell=1}^{\infty} A^{\ell-1}z^{\ell}. \tag{3.17}$$

By definition (3.13) this proves (3.14). Equation 3.15 follows from this.

By (3.9) and (3.10),

$$y_k = \sum_{j=-\infty}^{k-1} H_{k-j}u_j \quad (k \in \mathbb{Z}), \quad y(z) = \sum_{k=-\infty}^{\infty} y_k z^{-k}.$$

On the other hand,

$$H(z^{-1})u(z) = \sum_{\ell=1}^{\infty} H_{\ell}z^{-\ell} \sum_{j=-\infty}^{\infty} u_j z^{-j} = \sum_{k=-\infty}^{\infty} z^{-k} \sum_{j=-\infty}^{k-1} H_{k-j}u_j.$$

These prove (3).

Statement (4) follows from

$$H(z)u_k = \sum_{j=1}^{\infty} H_j u_{k-j} = y_k.$$

Statement (5) is a simple consequence of (3.15). □

If all $u_j = 0$ when $j \neq t$, then by (3.10),

$$y_k = H_{k-t}u_t, \quad (k > t).$$

This shows that the sequence $\{H_{\ell} : \ell = 1, 2, \dots\}$ represents the impact of a single impulse u_t at time t on the future values of the process. That is why it is called the *impulse response function.*

Equations (3.10) and (3.11) imply the important *strict causality* of the input/output map:

$$u(z) = \sum_{j=0}^{\infty} u_j z^{-j} \quad \Rightarrow \quad y(z) = \phi(u(z)) = \sum_{k=1}^{\infty} y_k z^{-k}. \tag{3.18}$$

By words: An input sequence beginning at time 0 may only influence the output sequence from time 1.

While (3.1) is called the *internal description* of the linear system, (3.16) is the *external description*: if we can determine (e.g. measure or estimate) the coefficients H_ℓ, then we can compute the output sequence from the input sequence, without using (or knowing) the matrices (A, B, C).

In Section 3.2 we considered special input and output sequences. Consistently with that, let $\mathcal{U}_-$ denote the set of all input sequences $\{u_j : j \leq 0\}$ that contain only finitely many nonzero terms. Also, let $\mathcal{Y}_+$ denote all output sequences $\{y_k : k \geq 1\}$. Then the *restricted input/output map* $\phi_0 : \mathcal{U}_- \to \mathcal{Y}_+$ was defined by (3.4) and (3.5):

$$\phi_0(u(z)) = y(z) \in \mathcal{Y}_+ \quad \text{if} \quad u(z) \in \mathcal{U}_-. \tag{3.19}$$

In this case the *left shift* L should be adapted to the restricted situation:

$$L(\dots, u_{-2}, u_{-1}, u_0) = (\dots, u_{-2}, u_{-1}, u_0, 0),$$
$$L(y_1, y_2, y_3, \dots) = (y_2, y_3, y_4, \dots).$$

Then it follows that

$$\phi_0(L(\dots, u_{-2}, u_{-1}, u_0)) = L\phi_0(\dots, u_{-2}, u_{-1}, u_0) \tag{3.20}$$

for any $(\dots, u_{-2}, u_{-1}, u_0) \in \mathcal{U}_-$.

Using the power series notation introduced above, we may write that

$$u(z) \in \mathcal{U}_- \quad \Leftrightarrow \quad u(z) = \sum_{j=0}^{j_0} u_{-j} z^j \quad (j_0 \geq 0, \text{ arbitrary}), \tag{3.21}$$

$$y(z) \in \mathcal{Y}_+ \quad \Leftrightarrow \quad y(z) = \sum_{k=1}^{\infty} y_k z^{-k}.$$

We can identify the left shift L in the space $\mathcal{U}_-$ with the multiplication by the identity function $\chi(z) = z$, and in the space $\mathcal{Y}_+$ with the operation on formal power series

$$Ly(z) = [zy(z)]_-, \text{ where } \left[\sum_{j=-\infty}^{r} a_j z^j\right]_- := \sum_{j=-\infty}^{-1} a_j z^j, \tag{3.22}$$

whenever $r \geq 0$. Another notation for the operator $[\cdot]_-$ is the projection π_-. Then (3.20) can be expressed in the form

$$\phi_0(u(z)) = y(z) \quad \Rightarrow \quad \phi_0(Lu(z)) = Ly(z).$$

It implies that ϕ_0 is also a $\mathbb{C}[z]$-*module homomorphism*:

$$\phi_0(u(z)) = y(z) \quad \Rightarrow \quad \phi_0(L_{p(z)}u(z)) = L_{p(z)}y(z), \tag{3.23}$$

or briefly, $\phi_0 L_p = L_p \phi_0$.

Let $\iota_+ : \mathcal{U}_- \to \mathcal{U}$ denote the inclusion map and $\pi_- : \mathcal{Y} \to \mathcal{Y}_+$ denote the projection map. Then by (3.16), (3.19), (3.22), and (3.23) the following diagram is commutative:

$$\begin{array}{ccc} \mathcal{U}_- & \xrightarrow{\phi_0} & \mathcal{Y}_+ \\ \iota_+ \downarrow & & \uparrow \pi_- \\ \mathcal{U} & \xrightarrow{\phi} & \mathcal{Y} \end{array} \tag{3.24}$$

That is,

$$\phi_0 = \pi_- \, \phi \, \iota_+,$$

or with power series,

$$y(z) = \phi_0(u(z)) = [H(z^{-1})u(z)]_- \in \mathcal{Y}_+ \text{ if } u(z) \in \mathcal{U}_-. \tag{3.25}$$

Proposition 3.2. *Given the restricted input/output map ϕ_0, there exists a unique strictly causal input/output map ϕ which makes diagram (3.24) commutative. Namely, we can define*

$$\phi\left(\sum_{j=-\infty}^{\infty} u_j z^{-j}\right) := \sum_{j=-\infty}^{\infty} \phi_0(u_j) z^{-j}.$$

Proof. By (3.25),

$$\phi_0(u_j) = \sum_{r=1}^{\infty} u_j C A^{r-1} B z^{-r}.$$

Thus by the above definition,

$$\phi\left(\sum_{j=-\infty}^{\infty} u_j z^{-j}\right) = \sum_{j=-\infty}^{\infty} \phi_0(u_j) z^{-j} = \sum_{k=-\infty}^{\infty} y_k z^{-k}, \tag{3.26}$$

where

$$y_k = \sum_{j=-\infty}^{k-1} C A^{k-j-1} B u_j,$$

exactly as (3.10) and (3.11) demand. Remember that in each sum there are only finitely many terms with negative indices. The uniqueness of ϕ follows from the fact that ϕ has to be linear, so (3.26) has to hold.

Then the strict causality of ϕ and the commutativity of the diagram (3.24) follow the same way as before, see (3.18) and (3.22). □

3.5 Realizations

In this section we investigate when and how one can get an internal description $\Sigma = (A, B, C)$ from an external description, more exactly, from a reduced input/output map $\phi_0 : \mathcal{U}_- \to \mathcal{Y}_+$. Realizations are not unique, but Theorem 3.1(b) below shows that in the finite dimensional case *minimal realizations* are isomorphic. First we are looking for a finite dimensional complex vector space X describing the space of internal states x, and linear maps $F : \mathcal{U}_- \to X$ and $G : X \to \mathcal{Y}_+$ such that

$$\phi_0 = G\,F.$$

Such a decomposition of ϕ_0 will be called a *factorization*.

To explain our search for such a factorization, let us review some earlier results. The Hankel matrix of the system $\Sigma = (A, B, C)$ defined by (3.5) can be written as an infinite block dyadic product:

$$\mathcal{H} = \begin{bmatrix} C \\ CA \\ CA^2 \\ \vdots \end{bmatrix} [B, AB, A^2B, \dots]. \tag{3.27}$$

Clearly, the system is observable if and only if the first factor in (3.27) that maps X to $\mathcal{Y}_+$ is one-to-one. Also, the system is reachable if and only if the second factor that maps $\mathcal{U}_-$ to X is onto. By (3.2), the image of the second factor is determined by x_1, the internal state at time 1, because that uniquely determines the whole output sequence $(y_1, y_2, y_3, \dots) \in \mathcal{Y}_+$, since now the input u_j is supposed to be zero if $j \geq 1$:

$$x_1 = \sum_{j=0}^{j_0} A^j B u_{-j} = \sum_{j=0}^{\infty} A^j B u_{-j}, \quad y_k = C x_k = C A^{k-1} x_1 \quad (k \geq 1).$$

Theorem 3.1. *We have the following important properties of factorization.*

(a) *Suppose that we have two factorizations:* $\phi_0 = G_1\,F_1 = G_2\,F_2$, *where* $F_j : \mathcal{U}_- \to X_j$ *and* $G_j : X_j \to \mathcal{Y}_+$ $(j = 1, 2)$. *If* F_1 *is onto and* G_2 *is one-to-one, then there exists a linear map* $L : X_1 \to X_2$ *such that the following diagram is commutative:*

$$\begin{array}{ccccc} & \overset{F_1}{\nearrow} & X_1 & \overset{G_1}{\searrow} & \\ \mathcal{U}_- & & L\downarrow & & \mathcal{Y}_+ \\ & \underset{F_2}{\searrow} & X_2 & \underset{G_2}{\nearrow} & \end{array}$$

(b) *Suppose that in statement (a) both* F_1 *and* F_2 *are onto and both* G_1 *and* G_2 *are one-to-one. Then the linear map* L *is invertible, so the two factorizations are equivalent.*

(c) *Suppose that we have a factorization $\phi_0 = G_1\, F_1$ with a finite dimensional state space X_1. Then dim(X_1) is minimal if and only if F_1 is onto and G_1 is one-to-one.*

Proof. (a) Let $x^1 \in X_1$ be arbitrary. Since F_1 is onto, there exists a $u(z) \in \mathcal{U}_-$ such that $F_1(u(z)) = x^1$. Then $\phi_0(u(z)) = G_1(x^1) \in \mathcal{Y}_+$, and since G_2 is one-to-one, there exists a unique $x^2 \in X_2$ such that $G_2(x^2) = G_1(x^1)$. Then we can set $L(x^1) := x^2$.

(b) It follows from (a) that in this case $L = G_2^{-1}G_1$ and $L^{-1} = G_1^{-1}G_2$.

(c) We may assume that the state space X_1 is such that our description is reachable and observable. For, otherwise we may omit those subspaces from X_1 that are either not reachable or not observable. In turn, it means that F_1 is onto and G_1 is one-to-one.

Suppose that $\phi_0 = G_2\, F_2$ is an arbitrary factorization with a state space X_2. Then

$$\begin{aligned} \dim(X_2) &\geq \dim(\text{Range}(G_2)) \geq \dim(\text{Range}(G_2\, F_2)) \\ &= \dim(\text{Range}(\phi_0)) = \dim(X_1), \end{aligned} \tag{3.28}$$

because G_1 is one-to-one.

By statement (b), if $\phi_0 = G_2\, F_2$ is an arbitrary factorization with a state space X_2 such that F_2 is onto and G_2 is one-to-one, then there exists an invertible linear map $L : X_1 \to X_2$, thus $\dim(X_2) = \dim(X_1)$.

Conversely, if $\dim(X_2)$ is minimal, then in (3.28) there is equality everywhere, so $\dim(X_2) = \dim(\text{Range}(G_2))$. This means that G_2 is one-to-one. Further, if F_2 were not onto, then $\dim(F_2) < \dim(X_2)$ would hold, which would contradict the equality $\dim(X_2) = \dim(\text{Range}(G_2\, F_2))$. □

Theorem 3.2. *For any $k, j \geq 1$, let*

$$\mathcal{H}_{k,j} = \begin{bmatrix} H_1 & H_2 & H_3 & \cdots & H_j \\ H_2 & H_3 & \cdot & \cdots & \cdot \\ H_3 & \cdot & \cdot & \cdots & \cdot \\ \vdots & \vdots & \vdots & & \vdots \\ H_k & \cdot & \cdot & \cdots & H_{k+j-1} \end{bmatrix}$$

be the $k \times j$ upper left sub-matrix of the infinite block Hankel matrix $\mathcal{H}$ defined in (3.5). *(Each block H_r is a $p \times q$ complex matrix.) The system of equations*

$$H_r = CA^{r-1}B \quad (r \geq 1)$$

has a linear system solution $\Sigma = (A, B, C)$ (as described in Section 3.2) if and only if

$$\sup_{k,j} \text{rank}(\mathcal{H}_{k,j}) < \infty.$$

Moreover, then the construction in the proof gives a linear system that is both reachable and observable, so its state space X has minimal dimension.

Proof. If $\Sigma = (A, B, C)$ is a linear system as described in Section 3.2 and $H_r = CA^{r-1}B$ $(r \geq 1)$, then, similarly as in (3.27), we can write that

$$\mathcal{H}_{k,j} = \begin{bmatrix} C \\ CA \\ \vdots \\ CA^{k-1} \end{bmatrix} [B, AB, \ldots, A^{j-1}B]. \tag{3.29}$$

Thus $\mathrm{rank}(\mathcal{H}_{k,j}) \leq n$, since the second factor has n rows.

Conversely, suppose that $\mathrm{rank}(\mathcal{H}_{k,j})$ is a bounded function of k and j. Since the rank is a non-decreasing, integer valued function, it must be constant if $k \geq k_0$ and $j \geq j_0$. Consider the matrices $\mathcal{H}_{k_0+1,j_0+1} \in \mathbb{C}^{(k_0+1)p \times (j_0+1)q}$. Choose a basis in the range of $\mathcal{H}_{k_0+1,j_0+1}$; let us denote the number of the basis vectors by n. From these n column vectors create the matrix $M_{k_0+1} \in \mathbb{C}^{(k_0+1)p \times n}$. Each column vector of $\mathcal{H}_{k_0+1,j_0+1}$ is a linear combination of the columns of M_{k_0+1}, so there exists a matrix $V_{j_0+1} \in \mathbb{C}^{n \times (j_0+1)q}$ such that

$$\mathcal{H}_{k_0+1,j_0+1} = M_{k_0+1} V_{j_0+1}. \tag{3.30}$$

Define the matrix $C \in \mathbb{C}^{p \times n}$ as the first p rows of the matrix M_{k_0+1} and define the matrix $B \in \mathbb{C}^{n \times q}$ as the first q columns of the matrix V_{j_0+1}. It is clear from the construction that C and B may remain the same for any $\mathcal{H}_{k,j}$ if $k \geq k_0$ and $j \geq j_0$.

Let us denote the columns of matrix V_{j_0+1} by $v_1, v_2, \ldots, v_{(j_0+1)q}$. Define the matrix $A \in \mathbb{C}^{n \times n}$ as a solution — if there exists one — of the system of linear equations

$$Av_r = v_{q+r} \qquad (r = 1, 2, \ldots, j_0 q). \tag{3.31}$$

Denote

$$V_{j_0} := [v_1, v_2, \ldots, v_{j_0 q}], \qquad V_{j_0}^s := [v_{q+1}, v_{q+2}, \ldots, v_{(j_0+1)q}].$$

Obviously, to have a solution A of (3.31) it is necessary that for $c \in \mathbb{C}^{j_0 q}$,

$$V_{j_0} c = 0 \quad \Rightarrow \quad V_{j_0}^s c = 0. \tag{3.32}$$

Moreover, (3.32) is also sufficient for the solvability of (3.31). For, writing down the augmented matrix of the system (3.31):

$$\begin{bmatrix} v_1 & v_2 & \cdots & v_{j_0 q} \\ --- & --- & --- & --- \\ v_{q+1} & v_{q+2} & \cdots & v_{(j_0+1)q} \end{bmatrix},$$

we see that whenever elementary column operations result a zero column in the upper matrix V_{j_0}, the same operations give a zero column in the lower matrix $V_{j_0}^s$ as well, so the rows of $V_{j_0}^s$ depend linearly on the rows of V_{j_0}, which is equivalent to the solvability of (3.31).

Thus we have to check that (3.32) holds. Suppose that $V_{j_0}c = 0$. Then $M_{k_0+1} V_{j_0} c = 0$ as well. Because of the block Hankel matrix structure of $\mathcal{H}_{k,j}$, we see that

$$M_{k_0+1} V_{j_0} = \begin{bmatrix} H_1 & H_2 & \cdots & H_{j_0} \\ H_2 & H_3 & \cdots & H_{j_0+1} \\ \vdots & \vdots & & \vdots \\ H_{k_0+1} & H_{k_0+2} & \cdots & H_{k_0+j_0} \end{bmatrix}$$

and

$$M_{k_0} V_{j_0}^s = \begin{bmatrix} H_2 & H_3 & \cdots & H_{j_0+1} \\ \vdots & \vdots & & \vdots \\ H_{k_0+1} & H_{k_0+2} & \cdots & H_{k_0+j_0} \end{bmatrix},$$

which is the same as $M_{k_0+1} V_{j_0}$, deleting its first row. Thus $M_{k_0} V_{j_0}^s c = 0$ holds too. Since the columns of M_{k_0} are independent, it follows that $V_{j_0}^s c = 0$, which was to show and which implies that the system (3.31) is solvable for A.

By our assumptions, $\mathrm{rank}(\mathcal{H}_{k,j}) = \mathrm{rank}(M_k) = n$ whenever $k \geq k_0$ and $j \geq j_0$. Hence $\mathrm{rank}(V_j) = n$ too when $j \geq j_0$. It implies that the columns of V_{j_0} span the n-dimensional complex vector space on which A acts, hence the solution A of the system (3.31) is unique and is the same for any $k \geq k_0$ and $j \geq j_0$.

Writing down the first block row of (3.30) using (3.31), it follows that

$$H_r = CA^{r-1}B \quad (r = 1, 2, \ldots j_0 + 1).$$

Since it is true for any $j \geq j_0 + 1$ as well, we have this representation for any $r \geq 1$.

Also, we saw that in the factorization (3.29), equivalently in (3.30), the two factors have rank n, it follows that the linear system $\Sigma = (A, B, C)$ is both reachable and observable, so its state space X has minimal dimension. □

Next we introduce the notion of minimal polynomial of a state space X. Suppose that ϕ_0 is a restricted input/output map of a linear system and $\phi_0 = GF$ is a factorization with a finite dimensional state space X; suppose as well that F is onto and G is one-to-one. Then we can define the effect of left shift L on each state $x \in X$ as a multiplication by the indeterminate z. Since F is onto, there exists a $u(z) \in \mathcal{U}_-$ such that $F(u(z)) = x$. Then

$$z \cdot x := F(zu(z)) \in X \quad (x \in X).$$

The vector $z \cdot x$ is uniquely defined, because if $x = F(u_1(z)) = F(u_2(z))$, then

$$\phi_0(u_1(z)) = G\,F(u_1(z)) = G\,F(u_2(z)) = \phi_0(u_2(z))$$

and applying left shift,

$$\phi_0(zu_1(z)) = \phi_0(zu_2(z));$$

since G is one-to-one, it follows that

$$z \cdot x = F(zu_1(z)) = F(zu_2(z)).$$

If $p(z) = a_m z^m + \cdots + a_1 z + a_0 \in \mathbb{C}[z]$ is an arbitrary polynomial in the indeterminate z, and $x = F(u(z)))$, then by definition

$$p(z) \cdot x := \sum_{j=0}^{m} a_j z^j \cdot x = F(p(z)u(z)) \in X.$$

If $y(z) = \phi_0(u(z))$, then by (3.25) and (3.23),

$$G(p(z) \cdot x) = G\,F(p(z)u(z)) = \phi_0(p(z)u(z)) = [H(z^{-1})p(z)u(z)]_- = L_{p(z)}y(z).$$

The *annihilator* Ann_X of the state space X is an ideal in $\mathbb{C}[z]$ defined as

$$\mathrm{Ann}_X := \{p(z) \in \mathbb{C}[z] : p(z) \cdot x = 0 \quad \forall x \in X\}.$$

Since $\mathbb{C}[z]$ is a principal ideal domain, if $\mathrm{Ann}_X \neq \{0\}$, there exists a non-zero polynomial $\psi_X \in \mathrm{Ann}_X$ such that

$$\mathrm{Ann}_X = \psi_X \mathbb{C}[z].$$

For uniqueness, we may assume that ψ_X is *monic*, that is, its leading coefficient is 1. Then ψ_X is called the *minimal polynomial* of X.

Proposition 3.3. *If the state space X is finite dimensional, there exists a minimal polynomial ψ_X of X.*

Proof. Since $\mathbb{C}[z]$ is a principal ideal domain, it is enough to show that $\mathrm{Ann}_X \neq \{0\}$. Take a basis $(e_1, \ldots, e_n)$ in X. Then the sequence of vectors $\{z^j \cdot e_1 : j = 0, 1, \ldots, n\}$ must be linearly dependent. Thus there exists a nonzero polynomial $p_1(z)$ (of degree at most n) such that $p_1(z) \cdot e_1 = 0$. Similarly, for any e_k $(k = 1, \ldots, n)$ there exists a nonzero polynomial $p_k(z)$ such that $p_k(z) \cdot e_k = 0$. Then we can take the *least common multiple* (*lcm*):

$$\mathrm{lcm}(p_1, \ldots, p_n) \in \mathrm{Ann}_X.$$

□

Next, we turn to realizations that use power series techniques, introduced in Section 3.4. The transfer function $H(z) \in \mathbb{C}^{p \times q}(z)$ of a linear system was defined in (3.13). Its definition shows that $H(z)$ is a power series and its coefficients are $p \times q$ complex matrices. Equivalently, $H(z)$ is a $p \times q$ matrix, whose entries are complex power series. $H(z)$ is called *rational* if it is a rational matrix, that is, there exists a nonzero polynomial $\phi(z) \in \mathbb{C}[z]$ such that $\phi(z)H(z)$ is a polynomial matrix: $\phi(z)H(z) \in \mathbb{C}^{p \times q}[z]$.

Theorem 3.3. *The transfer function $H(z^{-1})$ of a linear system $\Sigma = (A, B, C)$, defined by* (3.13), (3.14), *and (3.15), with finite dimensional state space X, is a rational matrix*

$$H(z^{-1}) = \frac{P(z)}{\psi_X(z)}, \quad P(z) \in \mathbb{C}[z]^{p\times q},$$

where $\psi_X(z)$ is the minimal polynomial of the state space X.

Proof. Let us denote the basis of coordinate unit vectors in the input space $U = \mathbb{C}^{q\times 1}$ by $(e_1, \ldots, e_q)$. Consider q different input sequences from $\mathcal{U}_-$:

$$u_j^r := \begin{cases} e_r & j = 0 \\ 0 & j \neq 0 \end{cases} \qquad (r = 1, \ldots, q).$$

Then the corresponding polynomials are

$$u^r(z) = e_r \quad (r = 1, \ldots, q). \tag{3.33}$$

By (3.25), the corresponding output polynomials from $\mathcal{Y}_+$ are

$$y^r(z) = [H(z^{-1})u^r(z)]_- = H(z^{-1})u^r(z) = H(z^{-1})e_r,$$

because now $H(z^{-1})u^r(z)$ does not have terms with non-negative powers of z.

Let us apply the polynomial left shift $\psi_X(z)$ to the input sequences. Then by (3.23),

$$\phi_0(L_{\psi_X} u^r(z)) = G\,F(L_{\psi_X} u^r(z)) = G(\psi_X(z) \cdot F(u^r(z))) = 0,$$

since $\psi_X(z) \cdot x = 0$ for any $x \in X$ as $\psi_X(z) \in \mathrm{Ann}_X$. It means that

$$[H(z^{-1})\psi_X(z)u^r(z)]_- = 0,$$

so $H(z^{-1})\psi_X(z)u^r(z)$ contains only non-negative powers of z, that is,

$$H(z^{-1})\psi_X(z)u^r(z) = P_r(z) \in \mathbb{C}[z]^{p\times 1} \quad (r = 1, \ldots, q),$$

with certain $p\times 1$ polynomial matrices $P_r(z)$. Thus by (3.33), $\psi_X(z)H(z^{-1})e_r = P_r(z)$ $(r = 1, \ldots, q)$,

$$\psi_X(z)H(z^{-1}) = P(z) = [P_1(z), \ldots, P_q(z)],$$

which proves the statement of the theorem. □

Observe that $H(z)$ is a rational matrix if and only if $H(z^{-1})$ is a rational matrix.

Theorem 3.4. *The following properties of a linear system are equivalent:*

(1) The reduced input/output map $\phi_0 : \mathcal{U}_- \to \mathcal{Y}_+$ defined by (3.19) *has an internal description $\Sigma = (A, B, C)$ with finite dimensional state space X.*

(2) If $\mathcal{H}_{k,j}$ denotes the $k \times j$ upper left minor of the infinite Hankel block matrix $\mathcal{H}$ defined in (3.5), *then $\sup_{k,j} \mathrm{rank}(\mathcal{H}_{k,j}) < \infty$.*

(3) $\mathrm{Range}(\phi_0) = \mathrm{Range}(\mathcal{H})$ *is finite dimensional.*

(4) The transfer function $H(z)$ *defined by* (3.13) *or* (3.14) *is a rational matrix.*

(5) The transfer function $H(z)$ *has a* matrix fraction description (MFD) *defined by Theorem D.1 in the Appendix.*

Proof. Theorem 3.2 proved the equivalence of (1) and (2) and Theorem 3.3 proved that (1) implies (4).

Supposing condition (4), $H(z^{-1}) = P(z)/\psi(z)$, where $P(z) \in \mathbb{C}[z]^{p\times q}$ and $\psi(z) \in \mathbb{C}[z]$. By (3.21), $u(z) \in \mathcal{U}_-$ if and only if $u(z) \in \mathbb{C}[z]^{q\times 1}$. Thus for any $u(z) \in \mathcal{U}_-$, $H(z^{-1})\psi(z)u(z) = P(z)u(z)$ contains only non-negative powers of z. Thus we get that

$$\phi_0(L_\psi u(z)) = [H(z^{-1})\psi(z)u(z)]_- = 0. \tag{3.34}$$

Then $\Psi := \psi(z)\mathbb{C}[z]^{q\times 1}$ is a submodule in the $\mathbb{C}[z]$-module $\mathbb{C}[z]^{q\times 1}$, whose elements ϕ_0 maps to 0. For any $u(z) \in \mathcal{U}_-$ we can perform standard polynomial division by the ordinary polynomial $\psi(z)$ to obtain

$$u(z) = Q(z)\psi(z) + R(z), \quad Q(z), R(z) \in \mathbb{C}[z]^{q\times 1}, \quad \deg(R(z)) < \deg(\psi(z)),$$

where $\deg(R(z))$ denotes the degree of a polynomial $R(z)$. Thus by (3.34), it follows that

$$\phi_0(u(z)) = \phi_0(\psi(z)Q(z)) + \phi_0(R(z)) = \phi_0(R(z)),$$

which implies that

$$\dim(\mathrm{Range}(\phi_0)) \le \dim(\mathbb{C}[z]^{q\times 1}/\Psi) = q\deg(\psi(z)).$$

This proves (3).

If (3) holds, then, since each $\mathcal{H}_{k,j}$ is a minor of $\mathcal{H}$,

$$\mathrm{rank}(\mathcal{H}_{k,j}) \le \mathrm{rank}(\mathcal{H}) < \infty,$$

so (2) follows.

Finally, Theorem D.1 and Remark D.3 show the equivalence of (4) and (5). □

The *McMillan degree* of the transfer function $H(z)$ is the dimension of the state space X in a minimal realization. Theorem 3.1(b) implies that the McMillan degree is uniquely defined.

3.6 Stochastic linear systems

3.6.1 Stability

Now we consider a *stochastic time invariant linear system*:

$$\begin{aligned} \mathbf{X}_{t+1} &= A\mathbf{X}_t + B\mathbf{U}_t \\ \mathbf{Y}_t &= C\mathbf{X}_t + D\mathbf{V}_t \qquad (t \in \mathbb{Z}), \end{aligned} \tag{3.35}$$

which is driven by the white noise processes $\{\mathbf{U}_t\}_{t\in\mathbb{Z}} \sim \mathrm{WN}(\boldsymbol{\Sigma})$ and $\{\mathbf{V}_t\}_{t\in\mathbb{Z}} \sim \mathrm{WN}(\boldsymbol{S})$. The matrices $A \in \mathbb{C}^{n\times n}$, $B \in \mathbb{C}^{n\times q}$, $C \in \mathbb{C}^{p\times n}$, and $D \in \mathbb{C}^{p\times s}$ give the internal description of the system. The covariance matrices $\boldsymbol{\Sigma} \in \mathbb{C}^{q\times q}$ and $\boldsymbol{S} \in \mathbb{C}^{s\times s}$ are given self-adjoint non-negative definite matrices. It is assumed that $\mathbb{E}(\mathbf{U}_t\mathbf{V}_\tau^*) = \delta_{t\tau}\boldsymbol{R}$, where $\boldsymbol{R} \in \mathbb{C}^{q\times s}$ represents the cross-covariance between the two driving processes. An important special case is when $\mathbf{V}_t = \mathbf{U}_t$ for each $t \in \mathbb{Z}$.

As before, $\{\mathbf{X}_t\}_{t\in\mathbb{Z}}$ is the sequence of the directly not observable $\mathbb{C}^n$-valued internal states and $\{\mathbf{Y}_t\}_{t\in\mathbb{Z}}$ is the sequence of the observable $\mathbb{C}^p$-valued output terms. These are stationary time series now.

We assume that the system is at rest at the remote past, that is,

$$\mathbf{X}_1 = \sum_{j=0}^{\infty} A^j B\mathbf{U}_{-j},$$

where the sum is convergent in mean square. (The convergence of the sum follows from stability, see next.) It implies that $\mathbb{E}\mathbf{X}_t = 0$ for each t, and each $\mathbf{U}_t$ and $\mathbf{V}_t$ are orthogonal to the past:

$$\mathbb{E}(\mathbf{U}_t\mathbf{X}_\tau^*) = 0, \qquad \mathbb{E}(\mathbf{V}_t\mathbf{X}_\tau^*) = 0 \qquad (\forall\, \tau \le t).$$

The system (3.35) is called *stable* if all the eigenvalues of the matrix A are in the open unit disc $\{z \in \mathbb{C} : |z| < 1\}$, that is, $\det(zI - A) \ne 0$ if $|z| \ge 1$. The *spectrum* $\sigma(A)$ of the matrix A is the set of all its eigenvalues and the *spectral radius* of A is

$$\rho(A) = \max\{|\lambda| : \lambda \in \sigma(A)\}.$$

See the properties of spectral radius in Lemma B.2 in the Appendix.

Theorem 3.5. *If a system* (3.35) *is stable, that is,* $\rho(A) < 1$*, then it has a unique causal stationary* MA(∞) *solution both for* $\{\mathbf{X}_t\}$ *and* $\{\mathbf{Y}_t\}$*:*

$$\mathbf{X}_t = \sum_{j=1}^{\infty} A^{j-1}B\mathbf{U}_{t-j}, \quad \mathbf{Y}_t = D\mathbf{V}_t + \sum_{j=1}^{\infty} CA^{j-1}B\mathbf{U}_{t-j} \quad (t \in \mathbb{Z}).$$

These series converge with probability 1 and in mean square.

Proof. This proof is similar to the one given for AR(1) processes in Section 2.4 above. Suppose that $\rho(A) < 1$. First consider the convergence of the claimed MA solution for $\{\mathbf{X}_t\}$. Since $\{\mathbf{U}_t\}$ is an orthogonal series with constant covariance matrix $\mathbf{\Sigma}$, the Cauchy–Schwarz inequality implies that

$$\mathbb{E}|\mathbf{U}_{t-j}| \leq \left(\mathbb{E}(|\mathbf{U}_{t-j}|^2)\right)^{1/2} = (\operatorname{tr}(\mathbf{\Sigma}))^{1/2} = \text{constant}.$$

Here $\operatorname{tr}(\mathbf{\Sigma})$ denotes the *trace* of matrix $\mathbf{\Sigma}$, $\operatorname{tr}(\mathbf{\Sigma}) = \sum_{j=1}^{q} \sigma_{jj}$. Thus by Lemma B.2(2) in the Appendix we have

$$\begin{aligned}\mathbb{E}\left|\sum_{j=1}^{\infty} A^{j-1} B\mathbf{U}_{t-j}\right| &\leq \sum_{j=1}^{\infty} \|A^{j-1}\| \, \|B\| \, \mathbb{E}|\mathbf{U}_{t-j}| \\ &\leq (\operatorname{tr}(\mathbf{\Sigma}))^{1/2} \|B\| \sum_{j=1}^{\infty} K c^{j-1} < \infty \quad (\rho(A) < c < 1),\end{aligned}$$

which shows that the proposed MA solution for $\{\mathbf{X}_t\}$ converges with probability 1. It is clear then that $\{\mathbf{X}_t\}$ satisfies the first equation of (3.35). The statement of the theorem for $\{\mathbf{Y}_t\}$ follows from this and from the second equation of (3.35).

The weak stationarity of $\{\mathbf{X}_t\}$ follows directly:

$$\begin{aligned}\boldsymbol{C}_X(k) := \mathbb{E}(\mathbf{X}_{t+k}\mathbf{X}_t^*) &= \sum_{j,\ell=1}^{\infty} A^{j-1} B \, \mathbb{E}(\mathbf{U}_{t+k-j}\mathbf{U}_{t-\ell}^*) \, B^* A^{*(\ell-1)} \\ &= \sum_{\ell=1}^{\infty} A^{k+\ell-1} B \, \mathbf{\Sigma} B^* A^{*(\ell-1)} = A^k \boldsymbol{C}_X(0) \qquad (k \geq 0) \end{aligned} \tag{3.36}$$

and $\boldsymbol{C}_X(-k) = \boldsymbol{C}_X^*(k)$, independently of t.

Now we show that the proposed MA solution is convergent in mean square and it is the only stationary solution of (3.35). Iterate the first equation of (3.35), shifted by one unit backward, $(k-1)$ times, :

$$\mathbf{X}_t = B\mathbf{U}_{t-1} + AB\mathbf{U}_{t-2} + A^2 B\mathbf{U}_{t-3} + \cdots + A^{k-1} B\mathbf{U}_{t-k} + A^k \mathbf{X}_{t-k}.$$

Then using the properties of a trace, we obtain

$$\begin{aligned}\mathbb{E}\left|\mathbf{X}_t - \sum_{j=1}^{k} A^{j-1} B\mathbf{U}_{t-j}\right|^2 &= \mathbb{E}\left|A^k \mathbf{X}_{t-k}\right|^2 \\ &= \mathbb{E}\left(\mathbf{X}_{t-k}^* A^{*k} A^k \mathbf{X}_{t-k}\right) = \mathbb{E} \operatorname{tr}\left(\mathbf{X}_{t-k}^* (A^* A)^k \mathbf{X}_{t-k}\right) \\ &= \mathbb{E} \operatorname{tr}\left(\mathbf{X}_{t-k}\mathbf{X}_{t-k}^* (A^* A)^k\right) = \operatorname{tr}\left(\mathbb{E}(\mathbf{X}_{t-k}\mathbf{X}_{t-k}^*)(A^* A)^k\right) \\ &= \operatorname{tr}\left(\boldsymbol{C}_X(0)(A^* A)^k\right) = \operatorname{tr}\left(A^k \boldsymbol{C}_X(0) A^{*k}\right),\end{aligned} \tag{3.37}$$

since we are looking for a stationary solution $\{\mathbf{X}_t\}$ whose covariance matrix $\boldsymbol{C}_X(0)$ is constant. Under the condition $\rho(A) < 1$, Lemma B.2 and (3.37) imply that $\sum_{j=1}^{\infty} A^{j-1} B\mathbf{U}_{t-j}$ converges in mean square to $\{\mathbf{X}_t\}$. This completes the proof of the theorem. □

Remark 3.1. Applying the indeterminate z as the operator of left (backward) shift, one can write that $z^j\mathbf{U}_t = \mathbf{U}_{t-j}$ $(j \geq 1)$. Using this and (3.17), the solutions given in Theorem 3.5 can be written as

$$\mathbf{X}_t = (I_n - Az)^{-1}zB\mathbf{U}_t, \quad \mathbf{Y}_t = D\mathbf{V}_t + C(I_n - Az)^{-1}zB\mathbf{U}_t \quad (t \in \mathbb{Z}).$$

Looking at the transfer function $\boldsymbol{H}(z)$ defined in (3.13) and (3.14), in the case $\mathbf{V}_t \equiv \mathbf{U}_t$ we can write that

$$\mathbf{Y}_t = \boldsymbol{H}(z)\mathbf{U}_t = \sum_{j=0}^{\infty} \boldsymbol{H}_j\mathbf{U}_{t-j}, \quad t \in \mathbb{Z}.$$

3.6.2 Prediction, miniphase condition, and covariance

Now we study the prediction in the case of a stochastic linear system

$$\begin{aligned} \mathbf{X}_{t+1} &= A\mathbf{X}_t + B\mathbf{U}_t \\ \mathbf{Y}_t &= C\mathbf{X}_t + D\mathbf{U}_t \qquad (t \in \mathbb{Z}). \end{aligned} \tag{3.38}$$

Assuming stability: $\rho(A) < 1$, it is easy to find the best linear prediction. By Theorem 3.5, the solution series

$$\mathbf{X}_t = \sum_{j=1}^{\infty} A^{j-1}B\mathbf{U}_{t-j}, \quad \mathbf{Y}_t = D\mathbf{U}_t + \sum_{j=1}^{\infty} CA^{j-1}B\mathbf{U}_{t-j}, \quad t \in \mathbb{Z}$$

converge in mean square. In mean square the *best linear h-step ahead prediction* ($h = 1, 2, \dots$) of $\mathbf{Y}_{t+h}$ is

$$\hat{\mathbf{Y}}_{t+h} := CA^h\mathbf{X}_t = \sum_{j=1}^{\infty} CA^{h+j-1}B\mathbf{U}_{t-j}, \quad t \in \mathbb{Z}. \tag{3.39}$$

This follows from the projection theorem, since

$$\mathbf{Y}_{t+h} - \hat{\mathbf{Y}}_{t+h} = D\mathbf{U}_{t+h} + \sum_{j=1}^{h} CA^{j-1}B\mathbf{U}_{t+h-j}, \quad t \in \mathbb{Z},$$

is orthogonal to $\hat{\mathbf{Y}}_{t+h}$. The *mean square prediction error*, c.f. (2.34), is

$$\|\mathbf{Y}_{t+h} - \hat{\mathbf{Y}}_{t+h}\|^2 = D\boldsymbol{\Sigma}D^* + \sum_{j=1}^{h} CA^{j-1}B\boldsymbol{\Sigma}B^*A^{*(j-1)}C^*.$$

In many applications one cannot observe the driving white noise process $\{\mathbf{U}_t\}$, only the output process $\{\mathbf{Y}_t\}$, so it is important if one can express the former in terms of the latter. Then this can be used in the best linear prediction (3.39).

One can describe an 'inverse' linear system of (3.38) that uses the observable process $\{\mathbf{Y}_t\}$ as input and the not observable process $\{\mathbf{U}_t\}$ as output. We assume that $m = n$ and $D = \boldsymbol{I}_n$. Starting with

$$\begin{aligned}\mathbf{X}_{t+1} &= A\mathbf{X}_t + B\mathbf{U}_t \\ \mathbf{Y}_t &= C\mathbf{X}_t + \mathbf{U}_t, \qquad t \in \mathbb{Z},\end{aligned}$$

we obtain that

$$\mathbf{U}_t = -C\mathbf{X}_t + \mathbf{Y}_t,$$

and so

$$\mathbf{X}_{t+1} = (A - BC)\mathbf{X}_t + B\mathbf{Y}_t.$$

The last two equations describe the 'inverse' system where the matrix $(A-BC)$ plays the role of the matrix A in the original system: the transition matrix in the state space. The stability condition $\rho(A) < 1$ of Theorem 3.5 is then replaced by the so-called *strict miniphase condition*:

$$\rho(A - BC) < 1 \tag{3.40}$$

of the 'inverse' system. Under this condition the 'inverse' system has a solution

$$\mathbf{U}_t = \sum_{j=0}^{\infty} \tilde{\boldsymbol{H}}_j \mathbf{Y}_{t-j} = \mathbf{Y}_t - \sum_{j=1}^{\infty} C(A - BC)^{j-1} B\mathbf{Y}_{t-j} \quad (t \in \mathbb{Z}),$$

which converges almost surely and in mean square. Here $\tilde{\boldsymbol{H}}_j$, $j \geq 0$, is the impulse response function of the 'inverse' system, whose transfer function is

$$\boldsymbol{H}^{-1}(z) = \sum_{j=0}^{\infty} \tilde{\boldsymbol{H}}_j z^j, \qquad |z| \leq 1.$$

Remark 3.2. By the first equation of (3.35) we have

$$\boldsymbol{C}_X(0) = \mathbb{E}(\mathbf{X}_{t+1}\mathbf{X}_{t+1}^*) = \mathbb{E}\left\{(A\mathbf{X}_t + B\mathbf{U}_t)(\mathbf{U}_t^* B^* + \mathbf{X}_t^* A^*)\right\},$$

and it gives a linear equation, a so-called *Lyapunov equation* for the covariance matrix $\boldsymbol{C}_X(0)$:

$$\boldsymbol{C}_X(0) = A\boldsymbol{C}_X(0)A^* + B\boldsymbol{\Sigma}B^*. \tag{3.41}$$

If $\rho(A) < 1$, by (3.36) this Lyapunov equation has a unique solution:

$$\boldsymbol{C}_X(0) = \sum_{j=0}^{\infty} A^j B\boldsymbol{\Sigma}B^* A^{*j}. \tag{3.42}$$

From the second equation of (3.35) we get the following result for the autocovariance function of $\{\mathbf{Y}_t\}$:

$$\begin{aligned}\boldsymbol{C}_Y(k) &:= \mathbb{E}(\mathbf{Y}_{t+k}\mathbf{Y}_t^*) = \mathbb{E}\left\{(C\mathbf{X}_{t+k} + D\mathbf{V}_{t+k})\,(\mathbf{X}_t^* C^* + \mathbf{V}_t^* D^*)\right\} \\ &= C\boldsymbol{C}_X(k)C^* + \mathbf{1}_{\{k=0\}} D\boldsymbol{S}D^* + \mathbf{1}_{\{k\geq 1\}} CA^{k-1}B\boldsymbol{R}D^* \quad (k \geq 0),\end{aligned} \tag{3.43}$$

since

$$\mathbb{E}(\mathbf{X}_{t+k}\mathbf{V}_t^*) = \sum_{j=1}^{\infty} A^{j-1}B\,\mathbb{E}(\mathbf{U}_{t+k-j}\mathbf{V}_t^*) = \mathbf{1}_{\{k\geq 1\}}A^{k-1}B\boldsymbol{R}.$$

Also, $\boldsymbol{C}_Y(-k) = \boldsymbol{C}_Y^*(k)$. In particular, we have

$$\boldsymbol{C}_Y(0) = C\boldsymbol{C}_X(0)C^* + D\boldsymbol{S}D^*.$$

Remark 3.3. Lemma B.2(3) and (3.36) show that if $\rho(A) < 1$, the covariance function $\boldsymbol{C}_X(k)$ decays exponentially as $|k| \to \infty$:

$$\|\boldsymbol{C}_X(k)\| \leq \|\boldsymbol{C}_X(0)\|\; Kc^k,$$

where $\rho(A) < c < 1$.

By (3.43), $\boldsymbol{C}_Y(k)$ also decays exponentially as $|k| \to \infty$.

In the important special case $\mathbf{V}_t = \mathbf{U}_t$, $\forall t \in \mathbb{Z}$, and using (3.36) too, (3.43) becomes

$$\boldsymbol{C}_Y(k) = \mathbf{1}_{\{k=0\}}D\boldsymbol{\Sigma}D^* + \mathbf{1}_{\{k\geq 1\}}CA^{k-1}BD^* + \sum_{\ell=1}^{\infty} CA^{k+\ell-1}B\,\boldsymbol{\Sigma}B^*A^{*(\ell-1)}C^*,$$

when $k \in \mathbb{Z}$. At this point we may apply the transfer function

$$H(z) = \sum_{\ell=0}^{\infty} H_\ell z^\ell = D + \sum_{\ell=1}^{\infty} CA^{\ell-1}Bz^\ell;$$

see (3.13), with $H_0 := D$. Thus we get

$$\boldsymbol{C}_Y(k) = \sum_{\ell=0}^{\infty} H_{k+\ell}\,\boldsymbol{\Sigma}H_\ell^* \in \mathbb{C}^{p\times p} \qquad (k \in \mathbb{Z}).$$

In particular,

$$\boldsymbol{C}_Y(0) = \sum_{\ell=0}^{\infty} H_\ell\,\boldsymbol{\Sigma}H_\ell^*.$$

Based on this, now we can introduce the *block Hankel matrix of the covariance function* $\boldsymbol{C}_Y(k)$:

$$\mathcal{H}(\boldsymbol{C}_Y) := \begin{bmatrix} \boldsymbol{C}_Y(1) & \boldsymbol{C}_Y(2) & \boldsymbol{C}_Y(3) & \cdots \\ \boldsymbol{C}_Y(2) & \boldsymbol{C}_Y(3) & \boldsymbol{C}_Y(4) & \cdots \\ \boldsymbol{C}_Y(3) & \boldsymbol{C}_Y(4) & \boldsymbol{C}_Y(5) & \cdots \\ \vdots & \vdots & \vdots & \end{bmatrix}$$

$$= \begin{bmatrix} H_1 & H_2 & H_3 & \cdots \\ H_2 & H_3 & H_4 & \cdots \\ H_3 & H_4 & H_5 & \cdots \\ \vdots & \vdots & \vdots & \end{bmatrix} \begin{bmatrix} \boldsymbol{\Sigma}H_0^* & 0 & 0 & \cdots \\ \boldsymbol{\Sigma}H_1^* & \boldsymbol{\Sigma}H_0^* & 0 & \cdots \\ \boldsymbol{\Sigma}H_2^* & \boldsymbol{\Sigma}H_1^* & \boldsymbol{\Sigma}H_0^* & \cdots \\ \vdots & \vdots & \vdots & \ddots \end{bmatrix}.$$

Here the first factor is $\mathcal{H}$, the block Hankel matrix of the linear system defined in (3.5), while the second factor is a lower triangular block Toeplitz matrix based on $\mathcal{H}$ and on the covariance matrix $\mathbf{\Sigma}$ of the driving white noise $\{\mathbf{U}_t\}$.

In Section 3.3 we saw that the *reachability* of a linear system depends on the rank of the block matrix

$$\mathcal{R}_\infty := [B, AB, A^2B, \ldots].$$

In fact, we saw that $\text{rank}(\mathcal{R}_\infty) = \text{rank}(\mathcal{R})$, where $\mathcal{R}$ is the reachability matrix defined by (3.6). Moreover, the linear system (3.35) is reachable if and only if $\text{rank}(\mathcal{R}) = n$, where n denotes the dimension of the state space.

Under the condition of stability $\rho(A) < 1$, let us evaluate the *reachability Gramian* $\mathcal{P}$:

$$\mathcal{P} := \mathcal{R}_\infty \mathcal{R}_\infty^* = \sum_{j=0}^{\infty} A^j BB^* A^{*j} \in \mathbb{C}^{n \times n}.$$

Then $\mathcal{P}$ is the unique solution of the Lyapunov equation

$$\mathcal{P} = A\mathcal{P}A^* + BB^*,$$

compare with (3.41) and (3.42). In fact, if $\mathbf{\Sigma} = I_n$, then $\mathcal{P} = \boldsymbol{C}_X(0)$, the covariance matrix of $\{\mathbf{X}_t\}$. It also follows from the above facts that the linear system (3.35) is reachable if and only if $\text{rank}(\mathcal{P}) = n$, that is, $\mathcal{P}$ is non-singular.

The case of *observability* is similar. Also in Section 3.3 we saw that the observability of a linear system depends on the rank of the block matrix

$$\mathcal{O}_\infty := \begin{bmatrix} C \\ CA \\ CA^2 \\ \vdots \end{bmatrix}.$$

In fact, we saw that $\text{rank}(\mathcal{O}_\infty) = \text{rank}(\mathcal{O})$, where $\mathcal{O}$ is the observability matrix defined by (3.7). Moreover, the linear system (3.35) is observable if and only if $\text{rank}(\mathcal{O}) = n$.

Under the condition of stability $\rho(A) < 1$, let us evaluate the *observability Gramian* $\mathcal{Q}$:

$$\mathcal{Q} := \mathcal{O}_\infty^* \mathcal{O}_\infty = \sum_{j=0}^{\infty} A^{*j} C^* C A^j \in \mathbb{C}^{n \times n}.$$

Then $\mathcal{Q}$ is the unique solution of the Lyapunov equation

$$\mathcal{Q} = A^* \mathcal{Q} A + C^* C.$$

It also follows from the above facts that the linear system (3.35) is observable if and only if $\text{rank}(\mathcal{Q}) = n$, that is, $\mathcal{Q}$ is non-singular.

3.7 Summary

State space model approach of R.E. Kálmán is used. A quadruple of bounded linear operators (A, B, C, D), $A : X \to X$, $B : U \to X$, $C : X \to Y$, and $D : U \to Y$ is given, where $U = \mathbb{C}^q$, $X = \mathbb{C}^n$, $Y = \mathbb{C}^p$, $A \in \mathbb{C}^{n\times n}$, $B \in \mathbb{C}^{n\times q}$, $C \in \mathbb{C}^{p\times n}$, and $D \in \mathbb{C}^{p\times q}$. The corresponding time invariant, linear dynamical system is described by

$$\begin{aligned} x_{k+1} &= Ax_k + Bu_k \\ y_k &= Cx_k + Du_k \quad (k \in \mathbb{Z}). \end{aligned}$$

The *restricted input/output map* ϕ_0 of the system is defined by

$$\phi_0(\{u_j\}_{j=-\infty}^{0}) = \{y_k\}_{k=1}^{\infty},$$

where starting time $j_0 \le 0$ of the input sequence can be arbitrary. Observe that ϕ_0 does not depend on the operator D, therefore we assume that $D = 0$. It is useful to write ϕ_0 in matrix form:

$$\begin{bmatrix} y_1 \\ y_2 \\ y_3 \\ \vdots \end{bmatrix} = \mathcal{H} \begin{bmatrix} u_0 \\ u_{-1} \\ u_{-2} \\ \vdots \end{bmatrix} = \begin{bmatrix} CB & CAB & CA^2B & \cdots \\ CAB & CA^2B & \cdot & \cdots \\ CA^2B & \cdot & \cdot & \cdots \\ \vdots & \vdots & \vdots & \end{bmatrix} \begin{bmatrix} u_0 \\ u_{-1} \\ u_{-2} \\ \vdots \end{bmatrix}.$$

The infinite coefficient matrix $\mathcal{H}$ here is a *block Hankel matrix*; each of its blocks is a $p \times q$ complex matrix.

Let (A, B, C) be a finite dimensional linear system. We say that the *state* $x \in X$ *is reachable* if there is a $k \ge 0$ and a sequence of inputs $u_0, u_1, \ldots, u_{k-1}$ that drives the system from the state $x_0 = 0$ to the state $x_k = x$:

$$x = \sum_{j=0}^{k-1} A^{k-1-j} B u_j = [B, AB, \ldots, A^{k-1}B] \begin{bmatrix} u_{k-1} \\ u_{k-2} \\ \vdots \\ u_0 \end{bmatrix}.$$

By the Cayley–Hamilton theorem,

$$\text{Range}[B, AB, \ldots, A^{k-1}B] = \text{Range}[B, AB, \ldots, A^{n-1}B] \quad (k \ge n).$$

The matrix $\mathcal{R} := [B, AB, \ldots, A^{n-1}B]$ is the *reachability matrix* of the system and its range is the *subspace of reachable states*. We say that the *system is reachable* if any $x \in X$ is reachable, that is, if $\text{rank}(\mathcal{R}) = \dim(\text{Range}(\mathcal{R})) = n$, equivalently, if $\mathcal{R}$ has n linearly independent columns, equivalently, if $\text{Ker}(\mathcal{R}^*) = \{0\}$. Clearly, $\text{Range}(\mathcal{R})$ is an A-invariant subspace in X.

We say that the *state* $x \in X$ *is observable* if starting from the state $x_0 = x$ at time 0 and having a constant zero input sequence $u_j = 0$ $(j \geq 0)$, there is a time instant $k \geq 0$ such that

$$y_k = CA^k x \neq 0.$$

The reason for the name "observable" is that the state x is observable if and only if for any $x_{(1)}, x_{(2)} \in X$, $x_{(2)} - x_{(1)} = x$, starting at time 0 with initial state $x_{(1)}$ and $x_{(2)}$, respectively, there exists a time instant $k \geq 0$ such that the output y_k is different in the two cases. This fact follows from the linearity of the system. Again, by the Cayley–Hamilton theorem,

$$\operatorname{Ker} \begin{bmatrix} C \\ CA \\ \vdots \\ CA^k \end{bmatrix} = \operatorname{Ker} \begin{bmatrix} C \\ CA \\ \vdots \\ CA^{n-1} \end{bmatrix} =: \operatorname{Ker}(\mathcal{O}) \quad (k \geq n-1),$$

where $\mathcal{O}$ is called the *observability matrix* of the system. The orthogonal complement of its kernel is the *subspace of observable states.* The *system is observable*, if any $x \in X$ is observable, that is, if $\operatorname{Ker}(\mathcal{O}) = \{0\}$, equivalently, if $\mathcal{O}$ has n linearly independent rows, equivalently, if $\operatorname{rank}(\mathcal{O}) = n$. Clearly, $\operatorname{Ker}(\mathcal{O})$ is an A-invariant subspace in X.

After we consider more general input/output sequences: assume that the observed input sequence contains only finitely many terms with negative indices; the system was at rest when the first nonzero input arrived, say at time $-j_0 \leq 0$ (which is arbitrary): $x_{-j_0} = 0$. So the input sequence can be written as a formal power series $u(z) := \sum_{j=-j_0}^{\infty} u_j z^{-j}$. A corresponding output sequence $(y_{-j_0+1}, y_{-j_0+2}, \dots)$ is the following formal power series with p-dimensional complex vector coefficients $y_k \in Y$: $y(z) := \sum_{k=-j_0+1}^{\infty} y_k z^{-k}$. Let $\mathcal{Y}$ is the vector space of all formal power series $y(z)$ with this property.

Also, $y(z) = \phi(u(z)) = H(z^{-1})u(z)$ with the *transfer function* $H(z)$. If $|z| > \rho(A)$ (the spectral radius of A), then $Iz - A$ is invertible and $H(z^{-1}) = C(Iz - A)^{-1}B$. Clearly, $\lim_{z\to\infty} H(z^{-1}) = 0$. These facts imply the important *strict causality* of the input/output map, i.e., an input sequence beginning at time 0 may influence only the output sequence from time 1. This gives a link between the *internal* (with state space equations) and the *external description* (with transfer function) of the linear system.

Then we investigate when and how one can get an internal description (A, B, C) from an external description, more exactly, from a reduced input/output map $\phi_0 : \mathcal{U}_- \to \mathcal{Y}_+$. Realizations are not unique, but Theorem 3.1 guarantees that in the finite dimensional case minimal realizations are isomorphic.

The following properties of a linear system are equivalent: the reduced input/output map $\phi_0 : \mathcal{U}_- \to \mathcal{Y}_+$ has an internal description (A, B, C) with finite dimensional state space X; the infinite block Hankel matrix has a bounded rank; the transfer function $H(z)$ is rational and has a matrix fraction description (MFD).

The *McMillan degree* of the transfer function $H(z)$ is the dimension of the state space X in a minimal realization. Theorem 3.1 shows that the McMillan degree is uniquely defined.

Eventually, a *stochastic time-invariant linear system* is considered:

$$\begin{aligned} \mathbf{X}_{t+1} &= A\mathbf{X}_t + B\mathbf{U}_t \\ \mathbf{Y}_t &= C\mathbf{X}_t + D\mathbf{V}_t \qquad (t \in \mathbb{Z}). \end{aligned} \tag{3.44}$$

which is driven by two white noise processes $\{\mathbf{U}_t\}$ and $\{\mathbf{V}_t\}$. These systems will play an important role in the subsequent chapters.

4

Multidimensional time series

4.1 Introduction

In this chapter we investigate properties of multidimensional, weakly stationary time series, similarly to the 1D case. From the literature we use the following important ones: [11, 14, 22, 28, 29, 35, 36, 39, 41, 47, 51, 54, 59, 60]. In contrast to Chapter 2, here we proceed deductively: from the most general constant rank processes, via regular (causal) ones, to the VARMA (vector autoregressive) processes and state space models.

Not surprisingly, here the classification is more complicated; for example, the spectral density matrix can be of reduced rank, but not zero. Linear filtering and constant rank processes are considered, and we prove that the spectral density matrix of a d-dimensional process has a constant rank $r \leq d$ if and only if the process is a two-sided moving average, obtained as a TLF with an r-dimensional white noise process; in this way, the spectral density matrix is also factorized. Special cases are the regular processes that have a causal (future independent) MA representation. Relations between the impulse responses, transfer function, and spectral factor are also discussed.

A non-singular process has a Wold decomposition in multidimension too. Here only the so-called innovation subspaces are unique, whereas their dimension is equal to the constant rank of the process. Further subclass of regular processes are the ones with a rational spectral density matrix. These can be finitely parametrized, and have either a state space, a stable VARMA representation, or an MFD (Matrix Fractional Description). Here, in the factorization of the spectral density matrix, the so-called spectral factors are also rational matrices. This fact has important consequences, for example, in the dynamic factor analysis and relations in the time domain (see Chapter 5).

4.2 Linear transformations, subordinated processes

While we met the notions of *linear filter* and *subordinated process* in Section 2.2 in the case of a one-dimensional time series, their extension to the d-

dimensional case requires some technical tools; these are going to be discussed in this section.

Let $\{\mathbf{X}_t\}_{t\in\mathbb{Z}}$ be an d-dimensional stationary time series with spectral representation

$$\mathbf{X}_t = \int_{-\pi}^{\pi} e^{it\omega} d\mathbf{Z}_\omega$$

and spectral measure matrix $d\mathbf{F} = d\mathbf{F}^X$ as discussed in Chapter 1. Assume that we are given a matrix function

$$\boldsymbol{T}(\omega) = [t_{jk}(\omega)]_{m\times d}, \quad t_{jk} \in L^2([-\pi,\pi], \mathcal{B}, \mathrm{tr}(d\mathbf{F})). \tag{4.1}$$

Here

$$\mathrm{tr}(d\mathbf{F}) = dF^{11} + \cdots + dF^{dd} \tag{4.2}$$

is a non-negative measure which dominates each dF^{jk} by the non-negative definiteness of $d\mathbf{F}$. Indeed, substituting $z_j = e^{it_1}$, $z_k = e^{it_2}$, $z_\ell = 0$ for $\ell \neq j, k$ and fixing α and β in (1.40), we obtain that

$$\Delta F^{jj} + \Delta F^{kk} + \Delta F^{jk} e^{i(t_1-t_2)} + \Delta F^{kj} e^{i(t_2-t_1)} \geq 0 \quad \text{for any} \quad t_1, t_2,$$

which implies that $\Delta F^{jj} + \Delta F^{kk} \geq 2|\Delta F^{jk}|$.

By definition, the m-dimensional process $\{\mathbf{Y}_t\}_{t\in\mathbb{Z}}$ is a *linear transform* of or obtained by a *time invariant linear filter (TLF)* from $\{\mathbf{X}_t\}$ if

$$\mathbf{Y}_t = \int_{-\pi}^{\pi} e^{it\omega} \boldsymbol{T}(\omega)\, d\mathbf{Z}_\omega \quad (t \in \mathbb{Z}). \tag{4.3}$$

It means that with the random measure

$$d\mathbf{Z}_\omega^Y := \boldsymbol{T}(\omega)\, d\mathbf{Z}_\omega$$

we have a representation of the process $\{\mathbf{Y}_t\}$:

$$\mathbf{Y}_t = \int_{-\pi}^{\pi} e^{it\omega} d\mathbf{Z}_\omega^Y \quad (t \in \mathbb{Z}).$$

Similarly to formula (1.28),

$$\begin{aligned}\mathrm{Cov}(\mathbf{Y}_{t+h}, \mathbf{Y}_t) &= \left[\int_{-\pi}^{\pi} e^{ih\omega} \sum_{r,s=1}^{n} t_{pr}(\omega)\overline{t_{qs}(\omega)} dF^{rs}(\omega)\right]_{p,q=1}^{m} \\ &= \int_{-\pi}^{\pi} e^{ih\omega} \boldsymbol{T}(\omega) d\mathbf{F}(\omega) \boldsymbol{T}^*(\omega) \quad (h \in \mathbb{Z}).\end{aligned} \tag{4.4}$$

Thus $\{\mathbf{Y}_t\}$ is also a stationary time series with spectral measure matrix

$$d\mathbf{F}^Y = \boldsymbol{T}\, d\mathbf{F}^X\, \boldsymbol{T}^*. \tag{4.5}$$

Considering $\text{Cov}(\mathbf{Y}_{t+h}\mathbf{X}_t) = \mathbb{E}(\mathbf{Y}_{t+h}\mathbf{X}_t^*)$, one can similarly obtain that $\{\mathbf{X}_t\}$ and $\{\mathbf{Y}_t\}$ are jointly stationary and their joint spectral density matrix is

$$d\mathbf{F}^{Y,X} = \boldsymbol{T}\, d\mathbf{F}^X. \tag{4.6}$$

It is not difficult to show that the last formula is not only necessary, but sufficient for obtaining $\{\mathbf{Y}_t\}$ from $\{\mathbf{X}_t\}$ by linear transformation. Because of (4.6) we may call $\boldsymbol{T}$ the *conditional spectral density* of $\{\mathbf{Y}_t\}$ w.r.t. $\{\mathbf{X}_t\}$ and denote $\boldsymbol{T} = \boldsymbol{f}^{Y|X}$. Compare with the one-dimensional case (2.6).

By Fourier transformation, we rewrite the definition of linear transformation in the time domain, assuming that condition (4.1) holds:

$$\boldsymbol{T}(\omega) = \sum_{j=-\infty}^{\infty} \boldsymbol{\tau}(j)e^{-ij\omega}, \quad \boldsymbol{\tau}(j) = \frac{1}{2\pi}\int_{-\pi}^{\pi} e^{ij\omega}\boldsymbol{T}(\omega)d\omega, \quad \sum_{j=-\infty}^{\infty} \|\boldsymbol{\tau}(j)\|_F^2 < \infty,$$

where the Fourier coefficients $\boldsymbol{\tau}(j)$ are $m \times d$ matrices. Because of the isometry between $H(\mathbf{X})$ and $L^2([-\pi,\pi],\mathcal{B},\text{tr}(d\mathbf{F}))$, definition (4.3) of linear transform is equivalent to

$$\begin{aligned}\mathbf{Y}_t &= \int_{-\pi}^{\pi} e^{it\omega} \sum_{j=-\infty}^{\infty} \boldsymbol{\tau}(j)e^{-ij\omega}\, d\mathbf{Z}_\omega = \sum_{j=-\infty}^{\infty} \boldsymbol{\tau}(j) \int_{-\pi}^{\pi} e^{i(t-j)\omega}\, d\mathbf{Z}_\omega \\ &= \sum_{j=-\infty}^{\infty} \boldsymbol{\tau}(j)\mathbf{X}_{t-j}, \qquad t \in \mathbb{Z}.\end{aligned} \tag{4.7}$$

This shows that the linear transform $\{\mathbf{Y}_t\}$ is really obtained by linear filtering from $\{\mathbf{X}_t\}$, that is, by a sliding summation with given matrix weights $\boldsymbol{\tau}(j)$.

We call an m-dimensional stationary time series $\{\mathbf{Y}_t\}_{t\in\mathbb{Z}}$ *causally subordinated* to an d-dimensional time series $\{\mathbf{X}_t\}_{t\in\mathbb{Z}}$ if $\{\mathbf{Y}_t\}$ is obtained from $\{\mathbf{X}_t\}$ by a linear transform (4.3) or (4.7), and, also,

$$H_k^-(\mathbf{Y}) \subset H_k^-(\mathbf{X}) \text{ for all } k \in \mathbb{Z}.$$

By stationarity, it suffices to require the above condition to hold for a specific value of k, for example, for $k = 0$.

In the causally subordinated case the filtering equation (4.7) modifies as a one-sided infinite summation:

$$\mathbf{Y}_t = \sum_{j=0}^{\infty} \boldsymbol{\tau}(j)\mathbf{X}_{t-j}, \quad \boldsymbol{\tau}(j) = 0 \quad \text{if} \quad j < 0. \tag{4.8}$$

Therefore, the Fourier series of $\boldsymbol{T}$ becomes one-sided:

$$\boldsymbol{T}(\omega) = \sum_{j=0}^{\infty} \boldsymbol{\tau}(j)e^{-ij\omega}, \quad \sum_{j=0}^{\infty} \|\boldsymbol{\tau}(j)\|_F^2 < \infty, \tag{4.9}$$

where $\|\cdot\|_F$ denotes the Frobenius norm, see Appendix B.

4.3 Stationary time series of constant rank

Assume that $\mathbf{X}_t = (X_t^1, \dots, X_t^d)$ $(t \in \mathbb{Z})$ is a d-dimensional complex valued weakly stationary time series with absolutely continuous spectral measure with density matrix $\boldsymbol{f}$ on $[-\pi, \pi]$, see Chapter 1. We say that $\{\mathbf{X}_t\}$ has *constant rank* r, if the matrix $\boldsymbol{f}(\omega)$ has rank r for *almost every* $\omega \in [-\pi, \pi]$. In 1D it means the $f(\omega) > 0$ for a.e. ω.

Theorem 4.1. *([47, Section I.9] We have the following back-and-forth statements.*

(a) Assume that the stationary time series $\mathbf{X}_t = (X_t^1, \dots, X_t^d)$ $(t \in \mathbb{Z})$ has an absolutely continuous spectral measure with density matrix $\boldsymbol{f}$ of constant rank r. Then $\boldsymbol{f}$ can be factored as

$$\boldsymbol{f}(\omega) = \frac{1}{2\pi}\boldsymbol{\phi}(\omega)\boldsymbol{\phi}^*(\omega) \quad \text{for a.e. } \omega \in [-\pi, \pi],$$

where $\boldsymbol{\phi}(\omega) \in \mathbb{C}^{d \times r}$, $r \le d$. Also, $\{\mathbf{X}_t\}$ can be represented as a two-sided infinite MA process (a sliding summation)

$$\mathbf{X}_t = \sum_{j=-\infty}^{\infty} \boldsymbol{b}(j)\boldsymbol{\xi}_{t-j}, \tag{4.10}$$

where $\{\boldsymbol{\xi}_t\}_{t \in \mathbb{Z}}$ is a $\mathrm{WN}(\boldsymbol{I}_r)$ (orthonormal) sequence, $\boldsymbol{b}(j) = [b_{k\ell}(j)] \in \mathbb{C}^{d \times r}$ $(j \in \mathbb{Z})$ is a non-random matrix-valued sequence, the Fourier coefficients of the factor $\boldsymbol{\phi}(\omega)$:

$$\boldsymbol{\phi}(\omega) = \sum_{j=-\infty}^{\infty} \boldsymbol{b}(j)e^{-ij\omega}, \quad \boldsymbol{b}(j) = \frac{1}{2\pi}\int_{-\pi}^{\pi} \boldsymbol{\phi}(\omega)d\omega, \quad \sum_{j=-\infty}^{\infty} \|\boldsymbol{b}(j)\|_F^2 < \infty. \tag{4.11}$$

This Fourier series converges to $\boldsymbol{\phi}$ in L^2 sense.

(b) Conversely, any stationary time series $\{\mathbf{X}_t\}_{t \in \mathbb{Z}}$ represented as a two-sided infinite MA process (4.10) has an absolutely continuous spectral measure with density matrix $\boldsymbol{f}$ of constant rank r, where r is the dimension of the white noise sequence $\{\boldsymbol{\xi}_t\}$.

Proof. First suppose that $\boldsymbol{f}$ has constant rank r. Since $2\pi\boldsymbol{f}$ is self-adjoint and non-negative definite by Corollary 1.1, it has a Gram-decomposition for a.e. ω by Proposition B.1:

$$2\pi\boldsymbol{f}(\omega) = \boldsymbol{\phi}(\omega)\boldsymbol{\phi}^*(\omega), \quad \boldsymbol{\phi}(\omega) \in \mathbb{C}^{d \times r}. \tag{4.12}$$

Now for a.e. ω there exists a matrix $\boldsymbol{\psi}(\omega) \in \mathbb{C}^{r \times d}$ (not unique if $d > r$) such that

$$\boldsymbol{\psi}(\omega)\boldsymbol{\phi}(\omega) = \boldsymbol{I}_r. \tag{4.13}$$

For, fixing $\ell = 1, \dots, r$, the system of equations

$$\sum_{j=1}^{d} \psi_{\ell j} \phi_{jk} = \delta_{\ell k} \quad (k = 1, \dots, r)$$

has a solution $(\psi_{\ell 1}, \dots, \psi_{\ell d})$ for each ω where the coefficient matrix $\boldsymbol{\phi}(\omega)$ has rank r, i.e. for almost every ω.

Then (4.12) and (4.13) imply that for a.e. ω,

$$\boldsymbol{\psi} \boldsymbol{f} \boldsymbol{\psi}^* = \frac{1}{2\pi} \boldsymbol{\psi} \boldsymbol{\phi} \boldsymbol{\phi}^* \boldsymbol{\psi}^* = \frac{1}{2\pi} \boldsymbol{I}_r. \tag{4.14}$$

By the spectral representation (1.19),

$$\mathbf{X}_t = \int_{-\pi}^{\pi} e^{it\omega} d\mathbf{Z}_\omega, \qquad \mathbf{Z}_\omega = (Z_\omega^1, \dots, Z_\omega^d).$$

We are immediately going to show that the following random variables are well-defined and are in $H(\mathbf{X})$ for any Borel subset $B \subset [-\pi, \pi]$:

$$V_B^\ell := \int_B \sum_{j=1}^{d} \psi_{\ell j}(\omega) \, dZ_\omega^j \quad (\ell = 1, \dots, r). \tag{4.15}$$

Set $V_\omega^\ell := V_{(-\pi,\omega]}^\ell$ and $\mathbf{V}_\omega := (V_\omega^1, \dots, V_\omega^r)$ if $\omega \in (-\pi, \pi]$. Then the components of the process $\mathbf{V}_\omega$, $-\pi < \omega \leq \pi$, are orthogonal and have orthogonal increments:

$$\begin{aligned}
\mathbb{E}\left(V_B^\ell \, \overline{V_{B'}^{\ell'}}\right) &= \int_{-\pi}^{\pi} \mathbf{1}_B(\omega) \mathbf{1}_{B'}(\omega) \sum_{j,j'=1}^{d} \psi_{\ell j}(\omega) f_{jj'}(\omega) \overline{\psi_{j'\ell'}(\omega)} \, d\omega \\
&= \frac{1}{2\pi} \delta_{\ell \ell'} \, d\omega(B \cap B') \quad (\ell, \ell' = 1, \dots, r),
\end{aligned} \tag{4.16}$$

where we used (4.14) and the isometric isomorphism between the Hilbert spaces $H(\mathbf{X}) \subset L^2(\Omega, \mathcal{F}, \mathbb{P})$ and $L^2([-\pi, \pi], \mathcal{B}, \operatorname{tr}(d\boldsymbol{F}))$. Here $d\omega$ is Lebesgue measure, $d\boldsymbol{F}$ is the spectral measure matrix of $\{\mathbf{X}_t\}$, and $\operatorname{tr}(d\boldsymbol{F})$ defined by (4.2) is a non-negative measure which dominates each dF^{jk}. At the same time, (4.16) shows that each $\psi_{\ell j} \in L^2([-\pi, \pi], \mathcal{B}, \operatorname{tr}(d\boldsymbol{F}))$, so by (4.1) the integral in (4.15) exists and $V_B^\ell \in H(\mathbf{X})$.

Let us introduce now the stationary time series $\boldsymbol{\xi}_t = (\xi_t^1, \dots, \xi_t^r)$ $(t \in \mathbb{Z})$ by stochastic integrals:

$$\xi_t^\ell := \int_{-\pi}^{\pi} e^{it\omega} dV_\omega^\ell = \int_{-\pi}^{\pi} e^{it\omega} \sum_{j=1}^{d} \psi_{\ell j}(\omega) \, dZ_\omega^j,$$

or briefly,

$$\boldsymbol{\xi}_t = \int_{-\pi}^{\pi} e^{it\omega} d\mathbf{V}_\omega = \int_{-\pi}^{\pi} e^{it\omega} \boldsymbol{\psi}(\omega) \, d\mathbf{Z}_\omega. \tag{4.17}$$

(4.16) implies that $\{\boldsymbol{\xi}_t\}_{t\in\mathbb{Z}}$ is an *orthonormal sequence* in $H(\mathbf{X})$:

$$\mathbb{E}\left(\boldsymbol{\xi}_t\,\overline{\boldsymbol{\xi}_s}\right) = \delta_{ts}\boldsymbol{I}_r \qquad t, s \in \mathbb{Z}.$$

(4.12), (4.13), and (4.14) imply that

$$(\boldsymbol{\phi}\boldsymbol{\psi} - \boldsymbol{I}_d)\boldsymbol{f}(\boldsymbol{\psi}^*\boldsymbol{\phi}^* - \boldsymbol{I}_d) = (\boldsymbol{\phi}\boldsymbol{\psi} - \boldsymbol{I}_d)\frac{1}{2\pi}\boldsymbol{\phi}\boldsymbol{\phi}^*(\boldsymbol{\psi}^*\boldsymbol{\phi}^* - \boldsymbol{I}_d) = 0$$

a.e. in $[-\pi, \pi]$. Consequently, the difference

$$\boldsymbol{\Delta}_t := \int_{-\pi}^{\pi} e^{it\omega}\boldsymbol{\phi}(\omega)\boldsymbol{\psi}(\omega)d\mathbf{Z}_\omega - \int_{-\pi}^{\pi} e^{it\omega}d\mathbf{Z}_\omega$$

is orthogonal to itself in $H(\mathbf{X})$, so it is a zero vector. Thus we have

$$\mathbf{X}_t = \int_{-\pi}^{\pi} e^{it\omega}d\mathbf{Z}_\omega = \int_{-\pi}^{\pi} e^{it\omega}\boldsymbol{\phi}(\omega)\boldsymbol{\psi}(\omega)d\mathbf{Z}_\omega = \int_{-\pi}^{\pi} e^{it\omega}\boldsymbol{\phi}(\omega)d\mathbf{V}_\omega. \tag{4.18}$$

It is clear by (4.12) that each entry $\phi_{k\ell}(\omega)$ is square integrable, so has a Fourier expansion converging in $L^2([-\pi, \pi], \mathcal{B}, d\omega)$:

$$\boldsymbol{\phi}(\omega) = \sum_{j=-\infty}^{\infty} \boldsymbol{b}(j)e^{-ij\omega}, \quad \boldsymbol{b}(j) = \frac{1}{2\pi}\int_{-\pi}^{\pi} \boldsymbol{\phi}(\omega)d\omega, \quad \sum_{j=-\infty}^{\infty} \|\boldsymbol{b}(j)\|_F^2 < \infty. \tag{4.19}$$

Finally, (4.17), (4.18), and (4.19) imply that

$$\mathbf{X}_t = \sum_{j=-\infty}^{\infty} \boldsymbol{b}(j)\boldsymbol{\xi}_{t-j}, \quad \boldsymbol{b}(j) = [b_{k\ell}(j)]_{d\times r},$$

which completes the proof of statement (a) of the theorem.

Conversely, assume that $\{\mathbf{X}_t\}_{t\in\mathbb{Z}}$ has an MA representation (4.10). Let $\boldsymbol{\phi} = [\phi_{j\ell}]_{d\times r}$ be the Fourier series with coefficients $\boldsymbol{b}(j)$ as defined by (4.19) and $\mathbf{V}_\omega = (V_1(\omega), \dots, V_r(\omega))$ be the random Fourier spectrum of the orthonormal time series $\{\boldsymbol{\xi}_t\}_{t\in\mathbb{Z}}$ as defined by (4.17).

The Hilbert space $H(\boldsymbol{\xi}) \subset L^2(\Omega, \mathcal{F}, \mathbb{P})$ generated by the orthonormal time series $\{\boldsymbol{\xi}_t\}_{t\in\mathbb{Z}}$ coincides with the Hilbert space $H(\mathbf{V})$ generated by random variables of the form

$$\zeta = \int_{-\pi}^{\pi} \mathbf{h}^*(\omega)\, d\mathbf{V}(\omega), \quad \mathbf{h}(\omega) = (h_1(\omega), \dots, h_r(\omega)),$$

where $h_\ell \in L^2([-\pi, \pi], \mathcal{B}, d\omega)$ $(1 \le \ell \le r)$. By (4.10), it is also clear that $H(\mathbf{X}) = H(\boldsymbol{\xi}) = H(\mathbf{V})$.

Since the orthonormal time series $\{\boldsymbol{\xi}_t\}_{t\in\mathbb{Z}}$ has $\frac{1}{2\pi}\boldsymbol{I}_r$ as its spectral density matrix, (4.10) implies that the spectral measure of $\{\mathbf{X}_t\}_{t\in\mathbb{Z}}$ is also absolutely continuous w.r.t. Lebesgue measure with spectral density matrix

$$\boldsymbol{f} = \frac{1}{2\pi}\boldsymbol{\phi}\boldsymbol{\phi}^* \text{ and } \operatorname{rank}(\boldsymbol{f}(\omega)) \le r \quad \text{for a.e.} \quad \omega \in [-\pi, \pi].$$

Indeed,

$$\boldsymbol{C}(h) = \mathbb{E}(\mathbf{X}_{t+h}\mathbf{X}_t^*) = \sum_{j,k=-\infty}^{\infty} \boldsymbol{b}(j)\mathbb{E}(\boldsymbol{\xi}_{t+h-j}\boldsymbol{\xi}_{t-k}^*)\boldsymbol{b}^*(k) = \sum_{k=-\infty}^{\infty} \boldsymbol{b}(k+h)\boldsymbol{b}^*(k),$$

while the Fourier series $\boldsymbol{\phi}(\omega) = \sum_{k=-\infty}^{\infty} \boldsymbol{b}(k)e^{-ik\omega}$ converges in L^2, so $\frac{1}{2\pi}\boldsymbol{\phi}\boldsymbol{\phi}^* \in L^1$, thus

$$\frac{1}{2\pi}\int_{-\pi}^{\pi} e^{ih\omega}\boldsymbol{\phi}(\omega)\boldsymbol{\phi}^*(\omega)d\omega = \frac{1}{2\pi}\int_{-\pi}^{\pi} e^{ih\omega} \sum_{j,k=-\infty}^{\infty} \boldsymbol{b}(j)e^{-ij\omega}\boldsymbol{b}^*(k)e^{ik\omega}\, d\omega$$

$$= \sum_{k=-\infty}^{\infty} \boldsymbol{b}(k+h)\boldsymbol{b}^*(k) = \boldsymbol{C}(h) = \int_{-\pi}^{\pi} e^{ih\omega} d\boldsymbol{F}(\omega). \tag{4.20}$$

Now, indirectly suppose that on a set of positive Lebesgue measure, $\operatorname{rank}(\boldsymbol{f}(\omega)) < r$. Then there exists $\mathbf{h}(\omega) = (h_1(\omega), \dots, h_r(\omega))$, $h_\ell \in L^2([-\pi,\pi], \mathcal{B}, d\omega)$, $\int_{-\pi}^{\pi} |\mathbf{h}(\omega)|^2 d\omega \neq 0$, such that $\boldsymbol{\phi}(\omega)\mathbf{h}(\omega) = 0$ for a.e. $\omega \in [-\pi, \pi]$. This means that the random variable

$$\zeta = \int_{-\pi}^{\pi} \mathbf{h}^*(\omega)\, d\mathbf{V}(\omega) \in H(\mathbf{V}) = H(\mathbf{X}), \quad \|\zeta\| \neq 0,$$

is orthogonal to $H(\mathbf{X})$, since

$$\langle X_t^j, \zeta\rangle = \mathbb{E}(X_t^j\overline{\zeta}) = \int_{-\pi}^{\pi} e^{it\omega}\sum_{\ell=1}^{r} \phi_{j\ell}(\omega)\overline{h_\ell(\omega)}d\omega = 0 \quad (j = 1, \dots, d;\ t \in \mathbb{Z}).$$

This contradiction completes the proof of statement (b) of the theorem. □

Remark 4.1. Observe that the proof of statement (a) of the previous theorem depended only on the fact that one has a rank $r \leq d$ factorization of the spectral density of the form

$$\boldsymbol{f}(\omega) = \frac{1}{2\pi}\boldsymbol{\phi}(\omega)\boldsymbol{\phi}^*(\omega), \quad \boldsymbol{\phi}(\omega) \in \mathbb{C}^{d\times r}, \quad \text{for a.e.} \quad \omega \in [-\pi, \pi].$$

Remark 4.2. Similarly as in 1D in Section 2.3, here we can also introduce the impulse response function $\{\boldsymbol{b}(j)\}_{j\in\mathbb{Z}}$ and the transfer function $\boldsymbol{H}(z) = \sum_{j\in\mathbb{Z}} \boldsymbol{b}(j)z^j$, which belongs to $L^2(T)$ and with which we can write that $\mathbf{X}_t = \boldsymbol{H}(z)\boldsymbol{\xi}_t$, $t \in \mathbb{Z}$. Clearly, for the factor $\boldsymbol{\phi}$ we have $\boldsymbol{\phi}(\omega) = \boldsymbol{H}(e^{-i\omega})$. This and (4.12) imply that

$$\boldsymbol{f}(\omega) = \frac{1}{2\pi}\boldsymbol{H}(e^{-i\omega})\boldsymbol{H}(e^{-i\omega})^*, \quad \omega \in [-\pi, \pi].$$

By (4.10), the covariance function of $\{\mathbf{X}_t\}$ can be written in terms of $\{\boldsymbol{b}(j)\}$ as in 1D:

$$\boldsymbol{C}(h) = \mathbb{E}(\mathbf{X}_{t+h}\mathbf{X}_t^*) = \sum_{k=-\infty}^{\infty} \boldsymbol{b}(k+h)\boldsymbol{b}^*(k), \quad h \in \mathbb{Z}. \tag{4.21}$$

4.4 Multidimensional Wold decomposition

The multidimensional version of Wold decomposition goes similarly as its 1D case in Section 2.6. The notations, definitions, and a part of the material of Section 2.6 about *singular and regular processes* can be extended to the multidimensional case with no essential change, therefore will not be repeated here.

4.4.1 Decomposition with an orthonormal process

Theorem 4.2. *(See e.g. Rozanov's book [47, Sections II.2–3].)*

Assume that $\{\mathbf{X}_t\}_{t\in\mathbb{Z}}$ is an d-dimensional non-singular stationary time series. Then it can be represented in the form

$$\mathbf{X}_t = \mathbf{R}_t + \mathbf{Y}_t = \sum_{j=0}^{\infty} \boldsymbol{b}(j)\boldsymbol{\xi}_{t-j} + \mathbf{Y}_t \quad (t \in \mathbb{Z}), \tag{4.22}$$

where

(1) $\{\mathbf{R}_t\}$ is a d-dimensional regular time series causally subordinated to $\{\mathbf{X}_t\}$;

(2) $\{\mathbf{Y}_t\}$ is an d-dimensional singular time series causally subordinated to $\{\mathbf{X}_t\}$;

(3) $\{\boldsymbol{\xi}_t\}$ is an r-dimensional ($r \le d$) WN($\boldsymbol{I}_r$) (orthonormal) sequence causally subordinated to $\{\mathbf{X}_t\}$;

(4) $\{\mathbf{R}_t\}$ and $\{\mathbf{Y}_t\}$ are orthogonal to each other: $\mathbb{E}(\mathbf{R}_t\mathbf{Y}_s^*) = \boldsymbol{O}$ for $t, s \in \mathbb{Z}$;

(5) $\boldsymbol{b}(j) = [b_{k\ell}(j)] \in \mathbb{C}^{d\times r}$ for $j \ge 0$ and $\sum_{j=0}^{\infty} \|\boldsymbol{b}(j)\|_F^2 < \infty$.

Proof. Let $H_k^- := \overline{\text{span}}\{\mathbf{X}_j : j \le k\}$ denote the *past of* $\{\mathbf{X}_t\}$ *until time* k and $H_{-\infty} := \bigcap_{k\in\mathbb{Z}} H_k^-$ is the *remote past of* $\{\mathbf{X}_t\}$.

Since $\{\mathbf{X}_t\}$ is assumed to be non-singular, $H_{-1}^- \ne H_0^- = \overline{\text{span}}\{H_{-1}^-, \mathbf{X}_0\}$. (One may choose an arbitrary initial time.) Let D_0 be the orthogonal complement of H_{-1}^- in H_0^-:

$$H_0^- = H_{-1}^- \oplus D_0, \tag{4.23}$$

where $\oplus$ denotes orthogonal direct sum. Choose an orthonormal basis $\{\xi_0^1, \dots, \xi_0^r\}$ in the subspace D_0. Then clearly, $r \le d$. Set $\boldsymbol{\xi}_0 := (\xi_0^1, \dots, \xi_0^r)$.

Let S denote the unitary operator of right time-shift in $H(\mathbf{X})$. Define $\boldsymbol{\xi}_t := S^t\boldsymbol{\xi}_0$ for $t \in \mathbb{Z}$. Clearly, $\boldsymbol{\xi}_t \in D_t := S^t D_0$, so $\boldsymbol{\xi}_t \in H_t^-$, $\boldsymbol{\xi}_t \perp H_{t-1}^-$ for each t, and $H_t^- = \overline{\text{span}}\{H_{t-1}^-, \boldsymbol{\xi}_t\}$. Thus $\{\boldsymbol{\xi}_t\}_{t\in\mathbb{Z}}$ is an r-dimensional orthonormal sequence:

$$\mathbb{E}(\boldsymbol{\xi}_t\boldsymbol{\xi}_s^*) = \delta_{ts}\boldsymbol{I}_r \quad (t, s \in \mathbb{Z})$$

and $\boldsymbol{\xi}_t \perp H_{-\infty}$ for each t.

It is important that

$$H_k^- = H_{-\infty} \oplus \bigoplus_{j=-\infty}^{k} D_j, \quad H(\mathbf{X}) = H_{-\infty} \oplus \bigoplus_{j=-\infty}^{\infty} D_j,$$

and each $\{\xi_k^1, \dots, \xi_k^r\}$ is an orthonormal basis in D_k ($k \in \mathbb{Z}$). Let us expand $\mathbf{X}_0$ into its orthogonal series w.r.t. $\{\boldsymbol{\xi}_t\}$:

$$X_0^k = Y_0^k + \sum_{j=0}^{\infty} \sum_{\ell=1}^{r} b_{k\ell}(j) \xi_{-j}^{\ell}, \quad b_{k\ell}(j) := \langle X_0^k, \xi_{-j}^{\ell} \rangle, \quad k = 1, \dots, d;$$

that is,

$$\mathbf{X}_0 = \mathbf{Y}_0 + \sum_{j=0}^{\infty} \boldsymbol{b}(j) \boldsymbol{\xi}_{-j}. \tag{4.24}$$

The vector $\mathbf{Y}_0$ is simply the remainder term, which of course is $\mathbf{0}$ if the bases $\{\xi_t^1, \dots, \xi_t^r\}_{t \in \mathbb{Z}}$ span $H(\mathbf{X})$, but not in general. It is not hard to see that $\mathbf{Y}_0 \in H_{-\infty}$.

Now apply the operator S^t to (4.24):

$$\mathbf{X}_t = \sum_{j=0}^{\infty} \boldsymbol{b}(j) \boldsymbol{\xi}_{t-j} + \mathbf{Y}_t =: \mathbf{R}_t + \mathbf{Y}_t \quad (t \in \mathbb{Z}),$$

where we have defined $\mathbf{Y}_t := S^t \mathbf{Y}_0$ ($t \in \mathbb{Z}$). Thus $\{\mathbf{Y}_t\}_{t \in \mathbb{Z}}$ is a stationary process, and $\mathbf{Y}_t \in H_{-\infty}$ for all t, so it is singular. $\{\mathbf{R}_t\}_{t \in \mathbb{Z}}$ is a stationary causal MA(∞) process, that is, a regular process. $\{\mathbf{R}_t\}$ and $\{\mathbf{Y}_t\}$ are orthogonal to each other. Formula (4.7) shows that $\{\mathbf{R}_t\}$ is a linear transform of $\{\mathbf{X}_t\}$, and since $\mathbf{Y}_t = \mathbf{X}_t - \mathbf{R}_t$ for all t, it follows for $\{\mathbf{Y}_t\}$ as well. The relationships

$$H_k^-(\mathbf{R}) \subset H_k^-(\mathbf{X}), \quad H_k^-(\mathbf{Y}) \subset H_k^-(\mathbf{X}) \quad (k \in \mathbb{Z})$$

clearly holds: thus $\{\mathbf{R}_t\}$ and $\{\mathbf{Y}_t\}$ are subordinated to $\{\mathbf{X}_t\}$.

As we are going to see in the next section, the orthonormal series $\{\boldsymbol{\xi}_t\}$ can be chosen the same as the orthonormal series in Theorem 4.1, which was obtained by linear transformation $\boldsymbol{\psi}(\omega)$ from $\{\mathbf{X}_t\}$ according to formula (4.17). Also, by the definition $\{\boldsymbol{\xi}_t\}$ in the present theorem, $H_k^-(\boldsymbol{\xi}) \subset H_k^-(\mathbf{X})$, so $\{\boldsymbol{\xi}_t\}$ is also subordinated to $\{\mathbf{X}_t\}$. This completes the proof of the theorem. □

Remark 4.3. It is clear from the proof that $\boldsymbol{\xi}_0$ is unique up to pre-multiplication by an arbitrary $r \times r$ unitary matrix $\boldsymbol{U}$. Indeed, we can take any other initial orthonormal basis in the subspace D_0 in the form

$$\tilde{\boldsymbol{\xi}}_0 = \boldsymbol{U} \boldsymbol{\xi}_0,$$

but otherwise the subspace D_0 is fixed. Also, the Wold decomposition of the theorem is uniquely defined up to this rotation of the initial basis.

The transfer function of the regular part $\{\mathbf{R}_t\}$ is

$$\boldsymbol{H}_{\mathbf{R}}(z) = \sum_{j=0}^{\infty} \boldsymbol{b}(j) z^j \quad (z \in T), \quad \mathbf{R}_t = \boldsymbol{H}_{\mathbf{R}}(z)\boldsymbol{\xi}_t \quad (t \in \mathbb{Z}). \tag{4.25}$$

For simplicity, let us fix the present time as time 0. Then the *best linear h-step ahead prediction* $\hat{\mathbf{X}}_h$ of $\mathbf{X}_h$ is by definition the projection of $\mathbf{X}_h$ onto the past until 0, that is, to H_0^-. By the Wold decomposition (4.22), the best linear h-step ahead prediction is

$$\hat{\mathbf{X}}_h = \sum_{j=h}^{\infty} \boldsymbol{b}(j)\boldsymbol{\xi}_{h-j} + \mathbf{Y}_t = \sum_{k=-\infty}^{0} \boldsymbol{b}(h-k)\boldsymbol{\xi}_k + \mathbf{Y}_t \quad (h \geq 1), \tag{4.26}$$

since the right hand side of (4.26) is in H_0^- and the error

$$\mathbf{X}_h - \hat{\mathbf{X}}_h = \sum_{j=0}^{h-1} \boldsymbol{b}(j)\boldsymbol{\xi}_{h-j}$$

is orthogonal to H_0^-. Hence the mean square error of the prediction is given by

$$\sigma_h^2 := \|\mathbf{X}_h - \hat{\mathbf{X}}_h\|^2 = \sum_{j=0}^{h-1} \|\boldsymbol{b}(j)\|_F^2 > 0. \tag{4.27}$$

4.4.2 Decomposition with innovations

(See e.g. the approach introduced by Wold [62] and used by Wiener and Masani [59]. Now we briefly describe that approach too, since we need it later.)

Let H_k^- denote again the Hilbert space spanned by the past of $\{\mathbf{X}_t\}$ until time k and let $\mathrm{Proj}_{H_{t-1}^-}\mathbf{X}_t$ denote the orthogonal projection of $\mathbf{X}_t$ to H_{t-1}^-, which exists uniquely by the projection theorem and Lemma 1.1. Then

$$\|\mathbf{X}_t - \mathrm{Proj}_{H_{t-1}^-}\mathbf{X}_t\| \leq \|\mathbf{X}_t - \mathbf{Z}\| \quad \text{for any} \quad \mathbf{Z} \in H_{t-1}^-.$$

Define the *process of innovations*

$$\boldsymbol{\eta}_t := \mathbf{X}_t - \mathrm{Proj}_{H_{t-1}^-}\mathbf{X}_t, \quad t \in \mathbb{Z}. \tag{4.28}$$

If the covariance matrix of $\{\boldsymbol{\eta}_t\}$ is

$$\boldsymbol{\Sigma} := \mathbb{E}(\boldsymbol{\eta}_0\boldsymbol{\eta}_0^*) \in \mathbb{C}^{d\times d}, \tag{4.29}$$

then $\{\boldsymbol{\eta}_t\}$ is a WN($\boldsymbol{\Sigma}$) process, $\mathbb{E}(\boldsymbol{\eta}_t\boldsymbol{\eta}_s^*) = \delta_{ts}\boldsymbol{\Sigma}$ for any $t, s \in \mathbb{Z}$. The one-step ahead prediction of $\mathbf{X}_t$ based on H_{t-1}^- is exactly $\mathrm{Proj}_{H_{t-1}^-}\mathbf{X}_t$, the prediction error is $\boldsymbol{\eta}_t$, and the covariance matrix of this error is $\boldsymbol{\Sigma}$. Further, by (4.28),

$$\mathrm{Cov}(\mathbf{X}_t, \boldsymbol{\eta}_t) = \mathbb{E}(\mathbf{X}_t \boldsymbol{\eta}_t^*) = \mathbb{E}(\boldsymbol{\eta}_t \boldsymbol{\eta}_t^*) = \boldsymbol{\Sigma}. \tag{4.30}$$

The rank $r \le d$ of $\boldsymbol{\Sigma}$ is the *rank of the process* $\{\mathbf{X}_t\}$. It is clear from the definition (4.23) of D_0 that $\mathrm{rank}(\boldsymbol{\Sigma}) \le \dim(D_0)$. On the other hand, $\mathrm{rank}(\boldsymbol{\Sigma}) < \dim(D_0)$ would lead to a contradiction. So it follows that

$$r = \mathrm{rank}(\boldsymbol{\Sigma}) = \dim(D_0), \tag{4.31}$$

and so each D_t, $t \in \mathbb{Z}$, can be called *innovation subspace.* If $\{\mathbf{X}_t\}$ is non-singular, then clearly we must have $r \ge 1$. In the special case when $r = d$ the process is called *full rank.*

In general, there is the following relationship between the approach described in Theorem 4.2 and the present one. The components of $\boldsymbol{\xi}_0$ are an orthonormal basis in the innovation subspace D_0, Thus the components of the innovation $\boldsymbol{\eta}_0$ can be expressed as linear combinations of them. It means that there exists a $d \times r$ matrix $\boldsymbol{A}$ such that $\boldsymbol{\eta}_0 = \boldsymbol{A}\boldsymbol{\xi}_0$. By the stationarity of the time series $\{\mathbf{X}_t\}$, the operator S of right (forward) time shift takes this relationship into

$$\boldsymbol{\eta}_t = S^t \boldsymbol{\eta}_0 = \boldsymbol{A} S^t \boldsymbol{\xi}_0 = \boldsymbol{A}\boldsymbol{\xi}_t \quad (t \in \mathbb{Z}), \quad \boldsymbol{\Sigma} = \mathbb{E}(\boldsymbol{A}\boldsymbol{\xi}_t \boldsymbol{\xi}_t^* \boldsymbol{A}^*) = \boldsymbol{A}\boldsymbol{A}^*. \tag{4.32}$$

With the $\mathrm{WN}(\boldsymbol{\Sigma})$ innovation process $\{\boldsymbol{\eta}_t\}$ the Wold decomposition described in the proof of Theorem 4.2 can go similarly as with the orthonormal process $\{\boldsymbol{\xi}_t\}$ there, resulting

$$\mathbf{X}_t = \mathbf{R}_t + \mathbf{Y}_t = \sum_{j=0}^{\infty} \boldsymbol{a}(j)\boldsymbol{\eta}_{t-j} + \mathbf{Y}_t \quad (t \in \mathbb{Z}). \tag{4.33}$$

Here $\{\mathbf{Y}_t\}$ is a singular process, $\mathbf{Y}_t \in H_{-\infty}$, $\mathbf{R}_s \perp \mathbf{Y}_t$ for any $s, t \in \mathbb{Z}$. Further, $\{\mathbf{R}_t\}$ is a regular process, $\boldsymbol{a}(j) \in \mathbb{C}^{d \times d}$ $(j \ge 0)$, and by (4.32) and (4.22),

$$\mathbf{R}_t = \sum_{j=0}^{\infty} \boldsymbol{a}(j)\boldsymbol{\eta}_{t-j} = \sum_{j=0}^{\infty} \boldsymbol{a}(j)\boldsymbol{A}\boldsymbol{\xi}_{t-j} = \sum_{j=0}^{\infty} \boldsymbol{b}(j)\boldsymbol{\xi}_{t-j}, \quad t \in \mathbb{Z}.$$

This implies that $\boldsymbol{b}(j) = \boldsymbol{a}(j)\boldsymbol{A}$ for $j \ge 0$. Moreover, by (4.33) and (4.30),

$$\begin{aligned} &\mathrm{Cov}(\mathbf{X}_0, \boldsymbol{\eta}_{-j}) = \mathbb{E}(\mathbf{X}_0 \boldsymbol{\eta}_{-j}^*) = \boldsymbol{a}(j)\boldsymbol{\Sigma}, \\ &\mathrm{Cov}(\mathbf{X}_0, \boldsymbol{\eta}_0) = \mathrm{Cov}(\boldsymbol{\eta}_0, \boldsymbol{\eta}_0) \quad \Rightarrow \quad \boldsymbol{a}(0)\boldsymbol{\Sigma} = \boldsymbol{\Sigma}. \end{aligned} \tag{4.34}$$

The covariance matrix function of $\{\mathbf{R}_t\}$ is

$$\boldsymbol{C}^R(h) = \mathbb{E}\left(\sum_{j,k=0}^{\infty} \boldsymbol{a}(j)\boldsymbol{\eta}_{t+h-j}\boldsymbol{\eta}_{t-k}^* \boldsymbol{a}^*(k) \right) = \sum_{k=0}^{\infty} \boldsymbol{a}(k+h)\,\boldsymbol{\Sigma}\,\boldsymbol{a}^*(k), \tag{4.35}$$

and its squared norm is

$$\|\mathbf{R}_t\|^2 = \operatorname{tr}(\boldsymbol{C}^R(0)) = \sum_{k=0}^{\infty} \operatorname{tr}\left(\boldsymbol{a}(k)\,\boldsymbol{\Sigma}^{\frac{1}{2}}\,\boldsymbol{\Sigma}^{\frac{1}{2}}\,\boldsymbol{a}^*(k)\right) = \sum_{k=0}^{\infty} \|\boldsymbol{a}(k)\boldsymbol{\Sigma}^{\frac{1}{2}}\|_F^2 < \infty. \tag{4.36}$$

In the present setting, the transfer function of the regular part $\{\mathbf{R}_t\}$ is

$$\boldsymbol{H}_{\mathbf{R}}(z) = \sum_{j=0}^{\infty} \boldsymbol{a}(j) z^j \quad (z \in T), \quad \mathbf{R}_t = \boldsymbol{H}_{\mathbf{R}}(z)\boldsymbol{\eta}_t \quad (t \in \mathbb{Z}),$$

compare with (4.25).

4.5 Regular and singular time series

Assume that $\{\mathbf{X}_t\}_{t\in\mathbb{Z}}$ is a regular stationary time series. By Wold's decomposition (4.22),

$$\mathbf{X}_t = \sum_{j=0}^{\infty} \boldsymbol{b}(j)\boldsymbol{\xi}_{t-j} \quad (t \in \mathbb{Z}), \quad \boldsymbol{b}(j) = [b_{k\ell}(j)]_{d\times r}, \tag{4.37}$$

$$\sum_{j=0}^{\infty} \|\boldsymbol{b}(j)\|_F^2 < \infty, \qquad \{\boldsymbol{\xi}_t\}_{t\in\mathbb{Z}} \sim \mathrm{WN}(\boldsymbol{I}_r).$$

By Theorem 4.1, this representation of $\{\mathbf{X}_t\}$ implies that $\{\mathbf{X}_t\}$ has an absolutely continuous spectral measure with density matrix $\boldsymbol{f}(\omega)$ and with constant rank $\tilde{r}$ for a.e. $\omega \in [-\pi, \pi]$,

$$\boldsymbol{f}(\omega) = \frac{1}{2\pi}\boldsymbol{\phi}(\omega)\boldsymbol{\phi}^*(\omega), \quad \boldsymbol{\phi}(\omega) = [\phi_{k\ell}(\omega)]_{d\times\tilde{r}}, \quad \text{for a.e. } \omega \in [-\pi, \pi] \tag{4.38}$$

and

$$\mathbf{X}_t = \sum_{j=-\infty}^{\infty} \tilde{\boldsymbol{b}}(j)\tilde{\boldsymbol{\xi}}_{t-j} \quad (t \in \mathbb{Z}), \quad \tilde{\boldsymbol{b}}(j) = [\tilde{b}_{k\ell}(j)]_{d\times\tilde{r}}, \tag{4.39}$$

$$\sum_{j=-\infty}^{\infty} \|\tilde{\boldsymbol{b}}(j)\|_F^2 < \infty, \qquad \{\tilde{\boldsymbol{\xi}}_t\}_{t\in\mathbb{Z}} \sim \mathrm{WN}(\boldsymbol{I}_r).$$

Compare now (4.37) and (4.39). Remark 4.3 says that the orthonormal process $\{\boldsymbol{\xi}_t\}$ in (4.37) is unique up to pre-multiplication by an arbitrary $r \times r$ unitary matrix $\boldsymbol{U}$. Thus it follows that

1. $\tilde{r} = r$,

2. $\tilde{\boldsymbol{b}}(j) = 0$ if $j < 0$,
3. $\tilde{\boldsymbol{\xi}}_0 = \boldsymbol{U}\boldsymbol{\xi}_0$, $\tilde{\boldsymbol{\xi}}_k = S^k \boldsymbol{U}\boldsymbol{\xi}_0$ $(k \in \mathbb{Z})$.

Corollary 4.1. *The dimension r of the WN($\boldsymbol{I}_r$) (orthonormal) innovation process $\{\boldsymbol{\xi}_t\}$ in (4.37) and of the rank of the WN($\boldsymbol{\Sigma}$) innovation process $\{\boldsymbol{\eta}_t\}$ in (4.28) and (4.31) are equal to the a.e. constant rank of the spectral density matrix $\boldsymbol{f}$ of the regular time series $\{\mathbf{X}_t\}$.*

From now on we assume that $\boldsymbol{\xi}_0 = \tilde{\boldsymbol{\xi}}_0$ is chosen as in (4.37), so we may omit all 'tildes' in (4.39). Let us write the Fourier expansion (4.19) for the present regular time series $\{\mathbf{X}_t\}$:

$$\boldsymbol{\phi}(\omega) = \sum_{j=0}^{\infty} \boldsymbol{b}(j) e^{-ij\omega}, \quad \|\boldsymbol{\phi}\|_2^2 = \sum_{j=0}^{\infty} \|\boldsymbol{b}(j)\|_F^2 < \infty.$$

Corollary 4.2. *A stationary time series $\{\mathbf{X}_t\}_{t\in\mathbb{Z}}$ is regular if and only it has an absolutely continuous spectral measure with spectral density of rank $r \le d$ that can be factored in the form*

$$\boldsymbol{f}(\omega) = \frac{1}{2\pi}\boldsymbol{\phi}(\omega)\boldsymbol{\phi}^*(\omega), \quad \boldsymbol{\phi}(\omega) = [\phi_{k\ell}(\omega)]_{d\times r}, \quad \textit{for a.e. } \omega \in [-\pi, \pi],$$

where

$$\boldsymbol{\phi}(\omega) = \sum_{j=0}^{\infty} \boldsymbol{b}(j) e^{-ij\omega}, \quad \|\boldsymbol{\phi}\|_2^2 = \sum_{j=0}^{\infty} \|\boldsymbol{b}(j)\|_F^2 < \infty.$$

That is,

$$\boldsymbol{\phi}(\omega) = \boldsymbol{\Phi}(e^{-i\omega}), \quad \boldsymbol{\Phi}(z) = \sum_{j=0}^{\infty} \boldsymbol{b}(j) z^j. \quad z \in D, \tag{4.40}$$

so the entries of the spectral factor $\boldsymbol{\Phi}(z) = [\Phi_{jk}(z)]_{d\times r}$ *are analytic functions in the open unit disc D and belong to the class $L^2(T)$ on the unit circle, consequently, they belong to H^2, see Section A.3. Briefly, we write that $\boldsymbol{\Phi} \in H^2$.*

Proof. If $\{\mathbf{X}_t\}$ is regular, then it has constant rank and so its spectral density $\boldsymbol{f}$ can be factored as (4.38) and $\boldsymbol{\phi}$ can be expressed as an L^2-convergent Fourier sum with coefficients $\boldsymbol{b}(j)$ by (4.11), but now (4.22) implies that $\boldsymbol{b}(j) = 0$ when $j < 0$. Conversely, if $\boldsymbol{f}$ can be factored as stated, then Theorem 4.1, Remark 4.1 and Theorem 4.2 imply that it is regular. □

Take the spectral factor $\boldsymbol{\Phi}(z) = [\Phi_{k\ell}(z)]_{d\times r} \in H^2$ defined in (4.40). Not only does it have the boundary value $\boldsymbol{\phi}(\omega)$ on the unit circle T, but by Theorem A.10 it has radial limits:

$$\lim_{\rho\to 1} \|\boldsymbol{\phi}(\omega) - \boldsymbol{\Phi}(\rho e^{-i\omega})\|_2 = 0.$$

The spectral factor $\mathbf{\Phi}(z)$ contains all information needed for finding the orthonormal innovation process $\{\boldsymbol{\xi}_t\}$ since by (4.17) in the proof of Theorem 4.1, we may compute $\{\boldsymbol{\xi}_t\}$ by linear transformation of $\{\mathbf{X}_t\}$ using $\boldsymbol{\psi}$, and by (4.13) $\boldsymbol{\psi}$ can be obtained from the system of linear equations

$$\boldsymbol{\psi}(\omega)\boldsymbol{\phi}(\omega) = \boldsymbol{I}_r,$$

where $\boldsymbol{\phi}$ is the boundary value of $\mathbf{\Phi}$. Also, the coefficients $\boldsymbol{b}(j)$ can be obtained from $\mathbf{\Phi}$ by power series expansion. As soon as we have these information, we may get the optimal linear prediction and its mean square error by formulas (4.26) and (4.27). Moreover, it follows from Remark 4.2, that $\mathbf{\Phi}(z)$ is equal to the transfer function $\boldsymbol{H}(z)$ on the unit circle T, so we can write that $\mathbf{X}_t = \boldsymbol{H}(z)\boldsymbol{\xi}_t = \mathbf{\Phi}(z)\boldsymbol{\xi}_t$, $t \in \mathbb{Z}$.

In the case of a regular process $\{\mathbf{X}_t\}$, (4.21) takes the following form:

$$\boldsymbol{C}(h) = \sum_{k=0}^{\infty} \boldsymbol{b}(k+h)\boldsymbol{b}^*(k) \quad (h \in \mathbb{Z}), \quad \boldsymbol{C}(-h) = \boldsymbol{C}^*(h).$$

4.5.1 Full rank processes

Assume that $\{\mathbf{X}_t\}$ is a d-dimensional stationary time series, with spectral measure matrix

$$d\boldsymbol{F} = d\boldsymbol{F}_a + d\boldsymbol{F}_s, \quad d\boldsymbol{F}_a \ll d\omega, \quad d\boldsymbol{F}_s \perp d\omega,$$

where $d\omega$ denotes Lebesgue measure, $d\boldsymbol{F}_a(\omega) = \boldsymbol{f}(\omega)d\omega$, $\boldsymbol{f}$ is the spectral density matrix of $\{\mathbf{X}_t\}$, $d\boldsymbol{F}_s$ is supported on a zero Lebesgue measure subset of $[-\pi, \pi]$. We say that $\{\mathbf{X}_t\}$ has *full rank* if $\operatorname{rank}(\boldsymbol{f}(\omega)) = d$ for a.e. $\omega \in [-\pi, \pi]$. It means that $\boldsymbol{f}(\omega)$ is a.e. non-singular, positive definite matrix.

The next theorem is an extension of Theorem 2.4 and the Kolmogorov–Szegő formula, Corollary 2.3, to the multidimensional full rank case.

Theorem 4.3. *A d-dimensional stationary time series $\{\mathbf{X}_t\}$ is of full rank non-singular process if and only if* $\log \det \boldsymbol{f} \in L^1$, *that is,*

$$\int_{-\pi}^{\pi} \log \det \boldsymbol{f}(\omega) d\omega > -\infty.$$

In this case if $\mathbf{\Sigma}$ denotes the covariance matrix of the innovation process $\{\boldsymbol{\eta}_t\}$, that is, of the one-step ahead prediction error process defined in (4.28) *and* (4.29), *then*

$$\det \mathbf{\Sigma} = (2\pi)^d \exp \int_{-\pi}^{\pi} \log \det \boldsymbol{f}(\omega) \frac{d\omega}{2\pi}. \tag{4.41}$$

The proof of this theorem requires some technical tools that we are going to discuss first. The arguments will essentially follow the lines of Wiener and Masani [59].

Let $\phi : [-\pi, \pi] \to \mathbb{C}$ be a measurable function. When $p > 0$, we say that $\phi \in L^p$ if

$$\int_{-\pi}^{\pi} |\phi(\omega)|^p \, d\omega < \infty.$$

When $\boldsymbol{f} = [f_{jk}]_{d\times d}$, we say that $\boldsymbol{f} \in L^p$ if each $f_{jk} \in L^p$. Recall that each L^p is a vector space, however it is a Banach space only when $p \geq 1$.

Lemma 4.1. *If $\boldsymbol{f} = [f_{jk}]_{d\times d} \in L^p$, $p > 0$, then $\det \boldsymbol{f} \in L^{p/d}$.*

Proof. The function $\det \boldsymbol{f}$ is the sum of $d!$ terms of the form

$$g(\omega) := \pm f_{1k_1}(\omega) \cdots f_{dk_d}(\omega).$$

Then by the inequality between the geometric and arithmetic means,

$$|g(\omega)|^{p/d} = (|f_{1k_1}(\omega)|^p \cdots |f_{dk_d}(\omega)|^p)^{1/d} \leq \frac{1}{d} \sum_{j=1}^{d} |f_{jk_j}(\omega)|^p,$$

thus $g \in L^{p/d}$. It implies that $\det \boldsymbol{f} \in L^{p/d}$ as well. □

Lemma 4.2. *Let $\boldsymbol{A}$ and $\boldsymbol{B}$ be $d \times d$ self-adjoint, non-negative definite matrices. Then*

(1) $(\det(\boldsymbol{A}+\boldsymbol{B}))^{1/d} \geq (\det \boldsymbol{A})^{1/d} + (\det \boldsymbol{B})^{1/d}$,

(2) $\det(\boldsymbol{A}+\boldsymbol{B}) \operatorname{tr}(\boldsymbol{A}) \geq \det(\boldsymbol{A}) \operatorname{tr}(\boldsymbol{A}+\boldsymbol{B})$,

(3) $\det(\boldsymbol{A}+\boldsymbol{B}) \geq \det(\boldsymbol{A})$.

Proof. (1) is Minkowski's inequality, see [30, p. 35].

(2) We may suppose that $\boldsymbol{A}$ is a diagonal matrix, since we may diagonalise it by means of a unitary transformation that does not affect the determinants and traces. If $\boldsymbol{A} = \operatorname{diag}[a_1, \dots, a_d]$, then

$$\det(\boldsymbol{A}+\boldsymbol{B}) = \begin{vmatrix} a_1 + b_{11} & b_{12} & \cdots & b_{1d} \\ b_{21} & a_2 + b_{22} & \cdots & b_{2d} \\ \vdots & \vdots & \ddots & \vdots \\ b_{d1} & b_{d2} & \cdots & a_d + b_{dd} \end{vmatrix}$$
$$= \det \boldsymbol{B} + \sum_{1\leq j\leq d} a_j \det \boldsymbol{B}_j + \sum_{1\leq j<k\leq d} a_j a_k \det \boldsymbol{B}_{jk} + \cdots + a_1 \cdots a_d,$$

where $\det \boldsymbol{B}_{jk\dots}$, is the principal minor obtained by deleting the jth, kth, ... rows and columns of $\boldsymbol{B}$. Since $\boldsymbol{A}$ and $\boldsymbol{B}$ are non-negative definite, each term in the last expansion is non-negative. Retaining only the last two terms, we obtain

$$\det(\boldsymbol{A}+\boldsymbol{B}) \geq \sum_{1\leq j\leq d} a_1 \cdots a_{j-1} a_{j+1} \cdots a_d \, b_{jj} + \det \boldsymbol{A}.$$

Thus

$$\begin{aligned}\det(\boldsymbol{A}+\boldsymbol{B})\operatorname{tr}(\boldsymbol{A}) &\geq \sum_{1\leq j\leq d} \operatorname{tr}(\boldsymbol{A})\, a_1\cdots a_{j-1}a_{j+1}\cdots a_d\, b_{jj} + \operatorname{tr}(\boldsymbol{A})\det \boldsymbol{A}\\ &\geq \sum_{1\leq j\leq d} (a_1\cdots a_d)b_{jj} + \operatorname{tr}(\boldsymbol{A})\det \boldsymbol{A}\\ &= \det(\boldsymbol{A})(\operatorname{tr}(\boldsymbol{B})+\operatorname{tr}(\boldsymbol{A})) = \det(\boldsymbol{A})\operatorname{tr}(\boldsymbol{A}+\boldsymbol{B}).\end{aligned}$$

(3) follows from (2), and also from (1). □

Lemma 4.3. *If $\boldsymbol{f} = [f_{jk}]_{d\times d} \in L^1$ and $\boldsymbol{f}(\omega)$ is a self-adjoint, non-negative definite matrix for a.e. $\omega \in [-\pi, \pi]$, then*

$$\log\det\left\{\frac{1}{2\pi}\int_{-\pi}^{\pi} \boldsymbol{f}(\omega)d\omega\right\} \geq \frac{1}{2\pi}\int_{-\pi}^{\pi} \log\det \boldsymbol{f}(\omega)d\omega \geq -\infty.$$

Proof. Apply the inequality in Lemma 4.2 (1) to a finite number of self-adjoint non-negative definite matrices $\boldsymbol{A}_j$, $j = 1, \dots, k$:

$$\left\{\det\left(\sum_{j=1}^{k} c_j \boldsymbol{A}_j\right)\right\}^{1/d} \geq \sum_{j=1}^{k} c_j\,(\det A_j)^{1/d}, \tag{4.42}$$

where each $c_j \geq 0$.

Since $\boldsymbol{f}$ is integrable on $[-\pi, \pi]$, there exists a sequence of partitions $\{B_j^n\}_{j=1}^{k(n)}$ of $[-\pi, \pi]$ and points $\omega_j^n \in B_j^n$, $n = 1, 2, \dots$, such that

$$\lim_{n\to\infty} \sum_{j=1}^{k(n)} \boldsymbol{f}(\omega_j^n)\, d\omega(B_j^n) = \int_{-\pi}^{\pi} \boldsymbol{f}(\omega)d\omega, \tag{4.43}$$

where $d\omega$ denotes Lebesgue measure. Then by Lemma 4.1,

$$\lim_{n\to\infty} \sum_{j=1}^{k(n)} \left(\det \boldsymbol{f}(\omega_j^n)\right)^{1/d} d\omega(B_j^n) = \int_{-\pi}^{\pi} \left(\det \boldsymbol{f}(\omega)\right)^{1/d} d\omega, \tag{4.44}$$

Formulas (4.42), (4.43), and (4.44) imply that

$$\left\{\det\left(\frac{1}{2\pi}\int_{-\pi}^{\pi} \boldsymbol{f}(\omega)d\omega\right)\right\}^{1/d} \geq \frac{1}{2\pi}\int_{-\pi}^{\pi} (\det \boldsymbol{f}(\omega)^{1/d})d\omega.$$

Taking logarithms on both sides and applying Jensen's inequality to the right hand side, we get the statement of the lemma. □

Lemma 4.4. *Suppose that $\boldsymbol{\phi} = [\phi_{k\ell}]_{d\times d} \in L^p$, $p \geq 1$, and the Fourier coefficients of $\boldsymbol{\phi}$:*

$$\boldsymbol{b}(j) := \frac{1}{2\pi}\int_{-\pi}^{\pi} e^{ij\omega}\boldsymbol{\phi}(\omega)d\omega \in \mathbb{C}^{d\times d},$$

are zero for $j < 0$. Define the $d \times d$ matrix valued function

$$\mathbf{\Phi}(z) := \sum_{j=0}^{\infty} \boldsymbol{b}(j) z^j, \quad z \in D = \{z \in \mathbb{C} : |z| < 1\}.$$

Then

(a) $\mathbf{\Phi} = [\Phi_{k\ell}]_{d \times d}$ *is analytic in D, moreover,* $\mathbf{\Phi} \in H^p$*, that is, each entry* $\Phi_{k\ell}$ *is in the Hardy space* H^p*;* $\boldsymbol{\phi}$ *is the boundary value of* $\mathbf{\Phi}$*;*

(b) $\det \mathbf{\Phi} \in H^{p/d}$, $\det \boldsymbol{\phi} \in L^{p/d}$*;*

(c) *either* $\det \mathbf{\Phi}$ *vanishes identically in D, or* $\log|\det \boldsymbol{\phi}| \in L^1$ *and*

$$\log|\det \mathbf{\Phi}(0)| \leq \frac{1}{2\pi} \int_{-\pi}^{\pi} \log|\det \boldsymbol{\phi}(\omega)| d\omega. \tag{4.45}$$

Proof. (a) $\mathbf{\Phi} \in H^p$ follows from Theorem A.15. Theorem A.10 implies that $\boldsymbol{\phi}$ is the boundary value of $\mathbf{\Phi}$:

$$\lim_{r \to 1} \mathbf{\Phi}(re^{i\omega}) = \boldsymbol{\phi}(\omega), \quad \text{a.e.} \quad \omega \in [-\pi, \pi]. \tag{4.46}$$

(b) By Lemma 4.1, $\det \boldsymbol{\phi} \in L^{p/d}$. Since each entry $\phi_{k\ell} \in L^p$ and its jth Fourier coefficient is zero if $j < 0$, it follows that $\Phi_{k\ell} \in H^p$.

At this point the proof is similar to the one of Lemma 4.1. The function $\det \mathbf{\Phi}(z)$ is the sum of $d!$ terms of the form

$$g(z) := \pm \Phi_{1\ell_1}(z) \cdots \Phi_{d\ell_d}(z).$$

Then by the inequality between the geometric and arithmetic means,

$$|g(z)|^{p/d} = \left(|\Phi_{1\ell_1}(z)|^p \cdots |\Phi_{d\ell_d}(z)|^p\right)^{1/d} \leq \frac{1}{d} \sum_{k=1}^{d} |\Phi_{k\ell_k}(z)|^p, \quad z \in D.$$

By substituting $z = re^{i\omega}$, $0 \leq r < 1$, and integrating over $[-\pi, \pi]$, we see that $g \in H^{p/d}$. It implies that $\det \mathbf{\Phi} \in H^{p/d}$ as well. Therefore by Theorem A.10, the radial limit $\det \boldsymbol{\phi} \in L^{p/d}$.

(c) Since $\det \mathbf{\Phi}$ is analytic in D, if it is not identically 0, then by [50, Theorem 17.3] $\log|\det \mathbf{\Phi}|$ is subharmonic in D. This, (4.46), and Theorem A.13 imply that $\log|\det \boldsymbol{\phi}| \in L^1$ and (4.45) holds. □

Lemma 4.5. *Assume that* $\{\mathbf{X}_t\}_{t \in \mathbb{Z}}$ *is a non-singular d-dimensional stationary time series with Wold decomposition given by*

$$\mathbf{X}_t = \mathbf{R}_t + \mathbf{Y}_t = \sum_{j=0}^{\infty} \boldsymbol{a}(j) \boldsymbol{\eta}_{t-j} + \mathbf{Y}_t, \quad t \in \mathbb{Z},$$

as described by equations (4.28)–(4.34). Let $d\boldsymbol{F}$*,* $d\boldsymbol{F}^R$ *and* $d\boldsymbol{F}^Y$ *denote the spectral measure matrices of* $\{\mathbf{X}_t\}$*,* $\{\boldsymbol{R}_t\}$*, and* $\{\mathbf{Y}_t\}$*, respectively. Then* *(a)* $d\boldsymbol{F} = d\boldsymbol{F}^R + d\boldsymbol{F}^Y$*;*

(b) $d\boldsymbol{F}^R$ *is absolutely continuous with density* $\boldsymbol{f}^R$*,*

$$\boldsymbol{f}^R(\omega) = \frac{1}{2\pi}\boldsymbol{\phi}(\omega)\boldsymbol{\phi}^*(\omega), \quad \boldsymbol{\phi}(\omega) = \sum_{j=0}^{\infty} \boldsymbol{a}(j)\boldsymbol{\Sigma}^{\frac{1}{2}}e^{-ij\omega}, \quad \boldsymbol{a}(j)\boldsymbol{\Sigma} = \mathbb{E}(\mathbf{X}_0\boldsymbol{\eta}_{-j}^*);$$

(c) if $\{\mathbf{X}_t\}$ *has full rank, then* $\log\det \boldsymbol{f}^R \in L^1$*, and*

$$\det \boldsymbol{\Sigma} \le (2\pi)^d \exp \int_{-\pi}^{\pi} \log\det \boldsymbol{f}^R(\omega)\frac{d\omega}{2\pi}.$$

Observe that here $\boldsymbol{\phi}(\omega) \in \mathbb{C}^{d\times d}$*, in contrast to the case of Corollary 4.2, which depended on the Wold decomposition of Theorem 4.2.*

Proof. (a) By the Wold decomposition, $\boldsymbol{R}_s \perp \mathbf{Y}_t$ for any $s, t \in \mathbb{Z}$. Hence

$$\boldsymbol{C}(h) = \mathbb{E}(\mathbf{X}_{t+h}\mathbf{X}_t^*) = \boldsymbol{C}^R(h) + \boldsymbol{C}^Y(h),$$

where $\boldsymbol{C}^R$ and $\boldsymbol{C}^Y$ denote the covariance functions of $\{\boldsymbol{R}_t\}$ and $\{\mathbf{Y}_t\}$, respectively. This implies (a).

(b) This proof is essentially the same as (4.20). $\{\boldsymbol{R}_t\}$ is a regular process and by (4.35), (4.36), and (4.34),

$$\boldsymbol{C}^R(h) = \sum_{k=0}^{\infty} \boldsymbol{a}(k+h)\,\boldsymbol{\Sigma}\,\boldsymbol{a}^*(k), \quad \sum_{k=0}^{\infty} \|\boldsymbol{a}(k)\boldsymbol{\Sigma}^{\frac{1}{2}}\|_F^2 < \infty, \quad \boldsymbol{a}(j)\boldsymbol{\Sigma} = \mathbb{E}(\mathbf{X}_0\boldsymbol{\eta}_{-j}^*).$$

Hence the Fourier series

$$\boldsymbol{\phi}(\omega) := \sum_{j=0}^{\infty} \boldsymbol{a}(j)\boldsymbol{\Sigma}^{\frac{1}{2}}e^{-ij\omega}$$

converges in L^2, $\frac{1}{2\pi}\boldsymbol{\phi}\boldsymbol{\phi}^* \in L^1$, and

$$\begin{aligned}
&\frac{1}{2\pi}\int_{-\pi}^{\pi} e^{ih\omega}\boldsymbol{\phi}(\omega)\boldsymbol{\phi}^*(\omega)d\omega \\
&= \frac{1}{2\pi}\int_{-\pi}^{\pi} e^{ih\omega}\sum_{j=0}^{\infty} e^{-ij\omega}\left\{\sum_{k=0}^{\infty} \boldsymbol{a}(k+j)\boldsymbol{\Sigma}^{\frac{1}{2}}\boldsymbol{\Sigma}^{\frac{1}{2}}\boldsymbol{a}^*(k)\right\}d\omega \\
&= \sum_{k=0}^{\infty} \boldsymbol{a}(k+h)\,\boldsymbol{\Sigma}\,\boldsymbol{a}^*(k) = \boldsymbol{C}^R(h) = \int_{-\pi}^{\pi} e^{ih\omega}d\boldsymbol{F}^R(\omega).
\end{aligned}$$

This shows that $\boldsymbol{F}^R$ is absolutely continuous with density $\boldsymbol{f}^R = \frac{1}{2\pi}\boldsymbol{\phi}\boldsymbol{\phi}^*$. This proves (b).

(c) If $\{\mathbf{X}_t\}$ has full rank, then $\mathbf{\Sigma}$ is invertible. By (4.34), $\boldsymbol{a}(0)\mathbf{\Sigma} = \mathbf{\Sigma}$, which now shows that $\boldsymbol{a}(0) = \boldsymbol{I}_d$. Defining

$$\mathbf{\Phi}(z) := \sum_{j=0}^{\infty} \boldsymbol{a}(j)\mathbf{\Sigma}^{\frac{1}{2}} z^j, \quad z \in D,$$

it follows that $\det \mathbf{\Phi}(0) = \det(\mathbf{\Sigma}^{\frac{1}{2}}) > 0$. Then the integrability of $\log\det \boldsymbol{f}^R$ and the inequality in (c) follows from Lemma 4.4(c), since $\log\det \mathbf{\Sigma} = 2\log\det \mathbf{\Phi}(0)$ and

$$\frac{1}{2\pi}\int_{-\pi}^{\pi} \log\det \boldsymbol{f}^R(\omega)d\omega = \frac{1}{2\pi}\int_{-\pi}^{\pi} \log\det\left(\frac{1}{2\pi}\boldsymbol{\phi}(\omega)\boldsymbol{\phi}^*(\omega)\right) d\omega$$
$$= -d\log(2\pi) + 2\cdot\frac{1}{2\pi}\int_{-\pi}^{\pi} \log|\det \boldsymbol{\phi}(\omega)|d\omega.$$

□

Let

$$\boldsymbol{P}(z) := \sum_{j=0}^{N} \boldsymbol{A}_j z^j, \quad z \in \mathbb{C},$$

be a matrix polynomial with coefficients $\boldsymbol{A}_j \in \mathbb{C}^{d\times d}$. We consider z as the left shift (backward shift) operator and define

$$\boldsymbol{P}(\mathbf{X}) := \boldsymbol{P}(z)\mathbf{X}_0 := \sum_{j=0}^{N} \boldsymbol{A}_j \mathbf{X}_{-j}. \tag{4.47}$$

Lemma 4.6. *Let $\{\mathbf{X}_t\}$ be a d-dimensional stationary time series with spectral measure matrix $d\boldsymbol{F}$. Then*

$$\mathbb{E}(\boldsymbol{P}(\mathbf{X})\boldsymbol{P}(\mathbf{X})^*) = \int_{-\pi}^{\pi} \boldsymbol{P}(e^{-i\omega})\, d\boldsymbol{F}(\omega)\, \boldsymbol{P}(e^{-i\omega})^*;$$

Proof.

$$\mathbb{E}\left\{\sum_{j=0}^{N} \boldsymbol{A}_j\mathbf{X}_{-j} \sum_{k=0}^{N} \mathbf{X}_{-k}^*\boldsymbol{A}_k^*\right) = \sum_{j,k=0}^{N} \boldsymbol{A}_j\mathbb{E}(\mathbf{X}_{-j}\mathbf{X}_{-k}^*)\boldsymbol{A}_k^*$$
$$= \sum_{j,k=0}^{N} \boldsymbol{A}_j \cdot \int_{-\pi}^{\pi} e^{i(k-j)\omega} d\boldsymbol{F}(\omega)\cdot \boldsymbol{A}_k^*$$
$$= \int_{-\pi}^{\pi} \left(\sum_{j=0}^{N} \boldsymbol{A}_j e^{-ij\omega}\right) d\boldsymbol{F}(\omega) \left(\sum_{k=0}^{N} \boldsymbol{A}_k e^{-ik\omega}\right)^*,$$

which proves the statement. □

Let $\{\mathbf{X}_t\}$ be a d-dimensional stationary time series with spectral measure matrix $d\boldsymbol{F}$. Then there exists a unique Lebesgue decomposition

$$d\boldsymbol{F} = d\boldsymbol{F}_a + d\boldsymbol{F}_s, \quad d\boldsymbol{F}_a \ll d\omega, \quad d\boldsymbol{F}_s \perp d\omega,$$

where $d\omega$ denotes Lebesgue measure, $d\boldsymbol{F}_a(\omega) = \boldsymbol{f}(\omega)d\omega$, $\boldsymbol{f}$ is the spectral density matrix of $\{\mathbf{X}_t\}$, $d\boldsymbol{F}_s$ is supported on a zero Lebesgue measure subset of $[-\pi, \pi]$.

Lemma 4.7. *Let $\{\mathbf{X}_t\}$ be a d-dimensional stationary time series with spectral measure matrix $d\boldsymbol{F} = d\boldsymbol{F}_a + d\boldsymbol{F}_s$ and spectral density matrix $\boldsymbol{f}$, as defined above. Let $\boldsymbol{P}(\mathbf{X})$ be defined by (4.47). Then*

$$\log\det\left\{\frac{1}{2\pi}\mathbb{E}(\boldsymbol{P}(\mathbf{X})\boldsymbol{P}(\mathbf{X})^*)\right\} \geq \frac{1}{2\pi}\int_{-\pi}^{\pi}\log\det \boldsymbol{f}(\omega)d\omega + 2\log|\det \boldsymbol{A}_0|.$$

Proof. It is not difficult to see that

$$\int_{-\pi}^{\pi}\boldsymbol{P}(e^{-i\omega})\,d\boldsymbol{F}(\omega)\,\boldsymbol{P}(e^{-i\omega})^* - \int_{-\pi}^{\pi}\boldsymbol{P}(e^{-i\omega})\,\boldsymbol{f}(\omega)\,\boldsymbol{P}(e^{-i\omega})^*d\omega$$

is a self-adjoint, non-negative definite matrix. Thus by Lemmas 4.2 (3) and 4.6,

$$\begin{aligned}
&\det\left\{\frac{1}{2\pi}\mathbb{E}(\boldsymbol{P}(\mathbf{X})\boldsymbol{P}(\mathbf{X})^*)\right\} = \det\left\{\frac{1}{2\pi}\int_{-\pi}^{\pi}\boldsymbol{P}(e^{-i\omega})\,d\boldsymbol{F}(\omega)\,\boldsymbol{P}(e^{-i\omega})^*\right\}\\
&\geq \det\left\{\frac{1}{2\pi}\int_{-\pi}^{\pi}\boldsymbol{P}(e^{-i\omega})\,\boldsymbol{f}(\omega)\,\boldsymbol{P}(e^{-i\omega})^*d\omega\right\}. \qquad (4.48)
\end{aligned}$$

The values of $\boldsymbol{P}(e^{-i\omega})\,\boldsymbol{f}(\omega)\,\boldsymbol{P}(e^{-i\omega})^*$ are also self-adjoint, non-negative definite matrices for a.e. ω, so by Lemma 4.3 we have

$$\begin{aligned}
&\log\det\left\{\frac{1}{2\pi}\int_{-\pi}^{\pi}\boldsymbol{P}(e^{-i\omega})\,\boldsymbol{f}(\omega)\,\boldsymbol{P}(e^{-i\omega})^*\right\}d\omega\\
&\geq \frac{1}{2\pi}\int_{-\pi}^{\pi}\log\det\left(\boldsymbol{P}(e^{-i\omega})\,\boldsymbol{f}(\omega)\,\boldsymbol{P}(e^{-i\omega})^*\right)d\omega\\
&= \frac{1}{2\pi}\int_{-\pi}^{\pi}2\log|\det \boldsymbol{P}(e^{-i\omega})|d\omega + \frac{1}{2\pi}\int_{-\pi}^{\pi}\log\det \boldsymbol{f}(\omega)d\omega. \qquad (4.49)
\end{aligned}$$

Now $\det \boldsymbol{P}(z)$ is a polynomial in z, therefore $\log|\det \boldsymbol{P}(z)|$ is a subharmonic function, see e.g. [50, Theorem 17.3]:

$$\frac{1}{2\pi}\int_{-\pi}^{\pi}\log|\det \boldsymbol{P}(e^{-i\omega})|d\omega \geq \log|\det \boldsymbol{P}(0)| = \log|\det \boldsymbol{A}_0|. \qquad (4.50)$$

Inequalities (4.48), (4.49), and (4.50) prove the statement of the lemma. □

Proof of Theorem 4.3. Assume first that $\{\mathbf{X}_t\}$ is a full rank non-singular process. By the Wold decomposition described in (4.33)–(4.34) and by Lemma 4.5, $\mathbf{X}_t = \mathbf{R}_t + \mathbf{Y}_t$ with spectral measures $d\boldsymbol{F} = d\boldsymbol{F}^R + d\boldsymbol{F}^Y$, $d\boldsymbol{F}^R(\omega) = \boldsymbol{f}^R(\omega)\,d\omega$. Also, $\log\det \boldsymbol{f}^R \in L^1$ and

$$\det \boldsymbol{\Sigma} \leq (2\pi)^d \exp \int_{-\pi}^{\pi} \log\det \boldsymbol{f}^R(\omega) \frac{d\omega}{2\pi}. \tag{4.51}$$

We may take the Lebesgue decompositions $d\boldsymbol{F} = d\boldsymbol{F}_a + d\boldsymbol{F}_s$, $d\boldsymbol{F}_a(\omega) = \boldsymbol{f}(\omega)d\omega$ and $d\boldsymbol{F}^Y = d\boldsymbol{F}_a^Y + d\boldsymbol{F}_s^Y$, $d\boldsymbol{F}_a^Y(\omega) = \boldsymbol{f}^Y(\omega)d\omega$ as well. Then $\boldsymbol{f} = \boldsymbol{f}^R + \boldsymbol{f}^Y$. Since $\boldsymbol{f}^Y$ is self-adjoint, non-negative definite, by Lemma 4.2 (3) we see that

$$\det \boldsymbol{f}^R(\omega) \leq \det \boldsymbol{f}(\omega). \tag{4.52}$$

Inequalities (4.51) and (4.52) imply that

$$\det \boldsymbol{\Sigma} \leq (2\pi)^d \exp \int_{-\pi}^{\pi} \log\det \boldsymbol{f}(\omega) \frac{d\omega}{2\pi}. \tag{4.53}$$

Since we assumed that $\{\mathbf{X}_t\}$ has full rank, $\det \boldsymbol{\Sigma} \neq 0$. Thus the integral in (4.53) cannot be $-\infty$. It cannot be $+\infty$ either, since by Lemma 4.3 it is dominated by

$$\log\det \left\{ \frac{1}{2\pi} \int_{-\pi}^{\pi} \boldsymbol{f}(\omega) d\omega \right\} < \infty.$$

This proves that $\log\det \boldsymbol{f} \in L^1$.

Conversely, assume that $\log\det \boldsymbol{f} \in L^1$. By definition (4.28), take the innovation at time 0:

$$\boldsymbol{\eta}_0 = \mathbf{X}_0 - \mathrm{Proj}_{H_{-1}^-} \mathbf{X}_0.$$

Consider an approximation of $\boldsymbol{\eta}_0$ by a finite past:

$$\boldsymbol{\eta}_0^N := \mathbf{X}_0 - \sum_{j=1}^{N} \boldsymbol{A}_j^N \mathbf{X}_{-j}, \quad \lim_{N\to\infty} \boldsymbol{\eta}_0^N = \boldsymbol{\eta}_0 \quad \text{in} \quad H_0^- \subset L^2(\Omega, \mathcal{F}, \mathbb{P}).$$

Then $\boldsymbol{\eta}_0^N$ is a matrix polynomial $\boldsymbol{P}(\mathbf{X})$ as defined by (4.47), with $\boldsymbol{A}_0 = \boldsymbol{I}_d$. Hence by Lemma 4.7,

$$\log\det \left\{ \frac{1}{2\pi} \mathbb{E}(\boldsymbol{\eta}_0^N \boldsymbol{\eta}_0^{N*}) \right\} \geq \frac{1}{2\pi} \int_{-\pi}^{\pi} \log\det \boldsymbol{f}(\omega) d\omega.$$

Now if $N \to \infty$, $\mathbb{E}(\boldsymbol{\eta}_0^N \boldsymbol{\eta}_0^{N*}) \to \mathbb{E}(\boldsymbol{\eta}_0 \boldsymbol{\eta}_0^*) = \boldsymbol{\Sigma}$. Thus it follows that

$$\log\det \left(\frac{1}{2\pi} \boldsymbol{\Sigma} \right) \geq \int_{-\pi}^{\pi} \log\det \boldsymbol{f}(\omega) \frac{d\omega}{2\pi}. \tag{4.54}$$

By our assumption, the integral is finite, so $\det \boldsymbol{\Sigma} \neq 0$. This proves that $\{\mathbf{X}_t\}$ is non-singular and has full rank.

Thus we have proved the equivalence of the conditions that $\log\det \boldsymbol{f} \in L^1$ and that $\{\mathbf{X}_t\}$ is non-singular and of full rank. Under these conditions both (4.53) and (4.54) hold, which prove the multidimensional Kolmogorov–Szegő formula (4.41). □

Corollary 4.3. *If $\{\mathbf{X}_t\}$ is of full rank non-singular time series, then the singular process $\{\mathbf{Y}_t\}$ in its Wold decomposition $\mathbf{X}_t = \mathbf{R}_t + \mathbf{Y}_t$ has singular spectral measure. More exactly, using the notations introduced above,*

$$d\boldsymbol{F}^R = d\boldsymbol{F}_a, \quad d\boldsymbol{F}^Y = d\boldsymbol{F}_s.$$

Proof. We know that in the Wold decomposition $d\boldsymbol{F} = d\boldsymbol{F}^R + d\boldsymbol{F}^Y$, $d\boldsymbol{F}^R$ is an absolutely continuous measure, see Lemma 4.5 (b). Hence it is enough to show that for the spectral density of $\mathbf{Y}$ we have $\boldsymbol{f}^Y = 0$ a.e. on $[-\pi, \pi]$.

Also, by Lemma 4.5 (b),

$$\boldsymbol{f} = \boldsymbol{f}^R + \boldsymbol{f}^Y = \frac{1}{2\pi}\boldsymbol{\phi}\boldsymbol{\phi}^* + \boldsymbol{f}^Y, \quad \boldsymbol{\phi} \in L^2.$$

Since $\boldsymbol{f}^R$ and $\boldsymbol{f}^Y$ are self-adjoint and non-negative definite, Lemma 4.2 (2) gives that

$$\det(\boldsymbol{f})\operatorname{tr}(\boldsymbol{f}^R) \geq \det(\boldsymbol{f}^R)\operatorname{tr}(\boldsymbol{f}^R + \boldsymbol{f}^Y), \quad \operatorname{tr}(\boldsymbol{f}^R) = \frac{1}{2\pi}\|\boldsymbol{\phi}\|_F^2,$$
$$\det(\boldsymbol{f}) \geq \det(\boldsymbol{f}^R)\left(1 + \frac{\operatorname{tr}(\boldsymbol{f}^Y)}{\frac{1}{2\pi}\|\boldsymbol{\phi}\|_F^2}\right).$$

Take logarithm and integrate over $[-\pi, \pi]$:

$$\int_{-\pi}^{\pi} \log\det \boldsymbol{f}(\omega)d\omega \geq \int_{-\pi}^{\pi} \log\det \boldsymbol{f}^R(\omega)d\omega + \int_{-\pi}^{\pi} \log\left(1 + \frac{\operatorname{tr}(\boldsymbol{f}^Y(\omega))}{\frac{1}{2\pi}\|\boldsymbol{\phi}(\omega)\|_F^2}\right)d\omega$$
$$\geq 2\pi \log\det\left(\frac{1}{2\pi}\boldsymbol{\Sigma}\right) + \int_{-\pi}^{\pi} \log\left(1 + \frac{\operatorname{tr}(\boldsymbol{f}^Y(\omega))}{\frac{1}{2\pi}\|\boldsymbol{\phi}(\omega)\|_F^2}\right)d\omega, \tag{4.55}$$

where we were allowed to use Lemma 4.5 (c) since $\{\mathbf{X}_t\}$ has full rank. By (4.41), the integral on the left hand side of (4.55) is also

$$\int_{-\pi}^{\pi} \log\det \boldsymbol{f}(\omega)d\omega = 2\pi \log\det\left(\frac{1}{2\pi}\boldsymbol{\Sigma}\right).$$

Since the integrand on the right hand side of (4.55) is non-negative, it follows that

$$\log\left(1 + \frac{\operatorname{tr}(\boldsymbol{f}^Y(\omega))}{\frac{1}{2\pi}\|\boldsymbol{\phi}(\omega)\|_F^2}\right) = 0 \quad \text{a.e.} \quad \Rightarrow \quad \operatorname{tr}(\boldsymbol{f}^Y(\omega)) = 0 \quad \text{a.e.},$$

because the denominator here must be finite a.e., since $\boldsymbol{\phi} \in L^2$. We know that $\boldsymbol{f}^Y(\omega)$ is self-adjoint and non-negative definite a.e., thus we conclude that $\boldsymbol{f}^Y(\omega) = 0$ a.e. This completes the proof. □

Corollary 4.4. *A stationary time series $\{\mathbf{X}_t\}$ is regular and of full rank if and only if*

(1) it has an absolutely continuous spectral measure matrix $d\boldsymbol{F}$ with density matrix $\boldsymbol{f}$;

(2) $\log\det \boldsymbol{f} \in L^1$.

Then the Kolmogorov–Szegő formula (4.41) *also holds.*

Proof. First assume that $\{\mathbf{X}_t\}$ is regular and of full rank. Then in the Wold decomposition $\mathbf{X}_t = \mathbf{R}_t + \mathbf{Y}_t$ we have $\mathbf{Y}_t = 0$, so $\boldsymbol{F}$ is absolutely continuous, see Lemma 4.5. By Theorem 4.3, $\log\det \boldsymbol{f} \in L^1$, since the process has full rank.

Conversely, assume that assumptions (1) and (2) hold. Then $\{\mathbf{X}_t\}$ has full rank by Theorem 4.3. Then by Corollary 4.3, $d\boldsymbol{F}^R = d\boldsymbol{F}_a$ and $d\boldsymbol{F}^Y = d\boldsymbol{F}_s$. Since $d\boldsymbol{F}_s = 0$ by assumption (1), it follows that $\mathbf{X}_t = \mathbf{R}_t$, a regular process. □

4.5.2 Generic regular processes

The next theorem is an extension of Corollary 4.4 to the general, not necessarily full rank, case. Let $\{\mathbf{X}_t\}$ be a d-dimensional stationary time series. Assume that its spectral measure matrix $d\boldsymbol{F}$ is absolutely continuous with density matrix $\boldsymbol{f}(\omega)$ which has rank r, $1 \le r \le d$, for a.e. $\omega \in [-\pi, \pi]$. Take the parsimonious spectral decomposition of the self-adjoint, non-negative definite matrix $\boldsymbol{f}(\omega)$ as described in Definition B.2:

$$\boldsymbol{f}(\omega) = \sum_{j=1}^{r} \lambda_j(\omega)\mathbf{u}_j(\omega)\mathbf{u}_j^*(\omega) = \tilde{\boldsymbol{U}}(\omega)\boldsymbol{\Lambda}_r(\omega)\tilde{\boldsymbol{U}}^*(\omega), \tag{4.56}$$

where

$$\boldsymbol{\Lambda}_r(\omega) = \operatorname{diag}[\lambda_1(\omega), \dots, \lambda_r(\omega)], \quad \lambda_1(\omega) \ge \dots \ge \lambda_r(\omega) > 0, \tag{4.57}$$

and $\tilde{\boldsymbol{U}}(\omega) \in \mathbb{C}^{d\times r}$ is a sub-unitary matrix.

Then, still, we have

$$\boldsymbol{\Lambda}_r(\omega) = \tilde{\boldsymbol{U}}^*(\omega)\boldsymbol{f}(\omega)\tilde{\boldsymbol{U}}(\omega). \tag{4.58}$$

Take care that here we use the word ‘spectral’ in two different meanings. On one hand, we use the spectral density of a time series in terms of a Fourier spectrum, on the other hand we take the spectral decomposition of a matrix in terms of eigenvalues and eigenvectors.

Theorem 4.4. *A d-dimensional stationary time series $\{\mathbf{X}_t\}$ is regular and of rank r, $1 \le r \le d$, if and only if each of the following conditions holds:*

(1) It has an absolutely continuous spectral measure matrix $d\boldsymbol{F}$ *with density matrix* $\boldsymbol{f}(\omega)$ *which has rank* r *for a.e.* $\omega \in [-\pi, \pi]$.

(2) For $\boldsymbol{\Lambda}_r(\omega)$ *defined by* (4.57) *one has* $\log\det \boldsymbol{\Lambda}_r \in L^1 = L^1([-\pi,\pi], \mathcal{B}, d\omega)$, *equivalently,*

$$\int_{-\pi}^{\pi} \log \lambda_r(\omega)\, d\omega > -\infty. \tag{4.59}$$

(3) The sub-unitary matrix function $\tilde{\boldsymbol{U}}(\omega)$ *appearing in the spectral decomposition of* $\boldsymbol{f}(\omega)$ *in* (4.56) *belongs to the Hardy space* $H^\infty \subset H^2$, *so*

$$\tilde{\boldsymbol{U}}(\omega) = \sum_{j=0}^{\infty} \boldsymbol{\psi}(j) e^{-ij\omega}, \quad \boldsymbol{\psi}(j) \in \mathbb{C}^{d\times r}, \quad \sum_{j=0}^{\infty} \|\boldsymbol{\psi}(j)\|_F^2 < \infty.$$

Proof. Assume first that $\{\mathbf{X}_t\}$ is regular and of rank r, $1 \le r \le d$. Here we are going to use the Wold decomposition given in Theorem 4.2. By Corollary 4.2 it follows that $\{\mathbf{X}_t\}$ has an absolutely continuous spectral measure with density matrix $\boldsymbol{f}(\omega)$, which has constant rank r for a.e. ω. This shows that condition (1) of the theorem holds.

In order to check that $\log\det \boldsymbol{\Lambda}_r \in L^1$, it is enough to show that (4.59) holds. First,

$$\int_{-\pi}^{\pi} \log\det \boldsymbol{\Lambda}_r(\omega) d\omega < \infty$$

is always true. Namely, using the inequality $\log x < x$ if $x > 0$, we see that

$$\log\det \boldsymbol{\Lambda}_r(\omega) = \sum_{j=1}^{r} \log \lambda_j(\omega) < \operatorname{tr}(\boldsymbol{\Lambda}_r(\omega)) = \operatorname{tr}(\boldsymbol{f}(\omega)) \in L^1,$$

since the spectral density $\boldsymbol{f}$ is an integrable function. Second, by (4.57),

$$\int_{-\pi}^{\pi} \sum_{j=1}^{r} \log \lambda_j(\omega) d\omega > -\infty \quad \Leftrightarrow \quad \int_{-\pi}^{\pi} \log \lambda_r(\omega)\, d\omega > -\infty. \tag{4.60}$$

By Corollary 4.2 it follows that

$$\boldsymbol{f}(\omega) = \frac{1}{2\pi} \boldsymbol{\Phi}(e^{-i\omega}) \boldsymbol{\Phi}^*(e^{-i\omega}), \tag{4.61}$$

where the spectral factor $\boldsymbol{\Phi}(z) = [\Phi_{jk}(z)]_{d\times r}$ is an analytic function in the open unit disc D and $\boldsymbol{\Phi} \in H^2$. Equality (4.61) implies that every principal minor $M(\omega) = \det[f_{j_p j_q}]_{p,q=1}^r$ of $\boldsymbol{f}$ is a constant times the product of a minor $M_{\boldsymbol{\Phi}}(e^{-i\omega})$ of $\boldsymbol{\Phi}(e^{-i\omega})$ and its conjugate:

$$\begin{aligned} M(\omega) &= (2\pi)^{-r} \det[\Phi_{j_p k}(e^{-i\omega})]_{p,k=1}^r \det[\overline{\Phi_{j_p k}(e^{-i\omega})}]_{p,k=1}^r \\ &= (2\pi)^{-r} \left|\det[\Phi_{j_p k}(e^{-i\omega})]_{p,k=1}^r\right|^2 =: (2\pi)^{-r} |M_{\boldsymbol{\Phi}}(e^{-i\omega})|^2. \end{aligned} \tag{4.62}$$

The row indices of the minor $M_{\mathbf{\Phi}}(z)$ in the matrix $\mathbf{\Phi}(z)$ are the same indices j_p, $p = 1, \dots, r$, that define the principal minor $M(\omega)$ in the matrix $\boldsymbol{f}(\omega)$. Since the function $M_{\mathbf{\Phi}}(z) = \det[\Phi_{j_p k}(z)]_{p,k=1}^r$ is analytic in D, it is either identically zero or is different from zero a.e. The rank of $\boldsymbol{f}$ is r a.e., therefore the sum of all its principal minors of order r (which are non-negative since $\boldsymbol{f}$ is non-negative definite) must be different from zero a.e. The last two sentences imply that there exists a principal minor $M(\omega)$ of order r which is different from zero a.e. We are using this principal minor $M(\omega)$ from now on.

The entries of $\mathbf{\Phi}(z)$ are in H^2, so by Lemma 4.4 (b) it follows that the determinant $M_{\mathbf{\Phi}}(z) \in H^{2/r}$. Then Theorem A.13 in the Appendix implies that $\log |M_{\mathbf{\Phi}}(e^{-i\omega})| \in L^1$, which in turn with (4.62) imply that

$$\log M(\omega) = \log \left\{ (2\pi)^{-r} \left| M_{\mathbf{\Phi}}(e^{-i\omega}) \right|^2 \right\} = -r \log 2\pi + 2 \log \left| M_{\mathbf{\Phi}}(e^{-i\omega}) \right| \in L^1. \tag{4.63}$$

Further, let us define the corresponding minor of the matrix $\tilde{\boldsymbol{U}}(\omega)$ by $M_{\tilde{\boldsymbol{U}}}(\omega) := \det[\tilde{U}_{j_p k}]_{p,k=1}^r$. Since $\tilde{\boldsymbol{U}}(\omega)$ is a sub-unitary matrix, its each entry has absolute value less than or equal to 1. Consequently, $|M_{\tilde{\boldsymbol{U}}}(\omega)| \le r!$. By (4.56),

$$M(\omega) = M_{\tilde{\boldsymbol{U}}}(\omega) \det \mathbf{\Lambda}_r(\omega) \overline{M_{\tilde{\boldsymbol{U}}}(\omega)} = \det \mathbf{\Lambda}_r(\omega) \, |M_{\tilde{\boldsymbol{U}}}(\omega)|^2.$$

It follows that

$$\log \det \mathbf{\Lambda}_r(\omega) = \log M(\omega) - 2 \log |M_{\tilde{\boldsymbol{U}}}(\omega)| \ge \log M(\omega) - 2 \log(r!),$$

which with (4.63) and (4.60) shows (4.59), and this proves condition (2) of the theorem.

The matrix function $\mathbf{\Lambda}_r(\omega)$ is a self-adjoint, positive definite function, $\mathrm{tr}(\mathbf{\Lambda}_r(\omega)) = \mathrm{tr}(\boldsymbol{f}(\omega))$, where $\boldsymbol{f}(\omega)$ is the density function of a finite spectral measure. This shows that the integral of $\mathrm{tr}(\mathbf{\Lambda}_r(\omega))$ over $[-\pi, \pi]$ is finite. Hence $\mathbf{\Lambda}_r(\omega)$ can be considered as the spectral density function of an r-dimensional stationary time series $\{\mathbf{V}_t\}_{t \in \mathbf{Z}}$ of full rank r. In fact, by linear filtering we define

$$\mathbf{V}_t := \int_{-\pi}^{\pi} e^{it\omega} \tilde{\boldsymbol{U}}^*(\omega) d\mathbf{Z}_\omega, \quad t \in \mathbb{Z}. \tag{4.64}$$

Then by (4.4) its auto-covariance function and spectral density are really given by

$$\boldsymbol{C}_{\mathbf{V}}(h) = \int_{-\pi}^{\pi} e^{ih\omega} \tilde{\boldsymbol{U}}^*(\omega) \boldsymbol{f}(\omega) \tilde{\boldsymbol{U}}(\omega) d\omega = \int_{-\pi}^{\pi} e^{ih\omega} \mathbf{\Lambda}_r(\omega) d\omega, \quad h \in \mathbb{Z}. \tag{4.65}$$

Then Corollary 4.4 and (4.63) show that $\{\mathbf{V}_t\}$ is a regular time series.

By the characterization of regular time series in Corollary 4.2, it follows that $\mathbf{\Lambda}_r$ can be factored:

$$\mathbf{\Lambda}_r(\omega) = \frac{1}{2\pi} \boldsymbol{D}_r(\omega) \boldsymbol{D}_r(\omega), \tag{4.66}$$

where $\boldsymbol{D}_r(\omega) = \sqrt{2\pi}\,\mathrm{diag}[\sqrt{\lambda_1(\omega)}, \dots, \sqrt{\lambda_r(\omega)}]$. Then

$$\boldsymbol{D}_r(\omega) = \sum_{j=0}^{\infty} \boldsymbol{\delta}(j) e^{-ij\omega}, \quad \sum_{j=0}^{\infty} \|\boldsymbol{\delta}(j)\|_F^2 < \infty,$$

and

$$\begin{aligned} \boldsymbol{\Delta}(z) &:= \sum_{j=0}^{\infty} \boldsymbol{\delta}(j) z^j, \quad \boldsymbol{\Delta}(z) = [\Delta_{k\ell}(z)]_{r\times r}, \quad z \in D, \\ \boldsymbol{\Delta}(e^{-i\omega}) &= \boldsymbol{D}_r(\omega), \end{aligned} \tag{4.67}$$

$\boldsymbol{\Delta}(z)$ is an H^2 analytic function. Since the non-diagonal entries of the boundary value $\boldsymbol{D}_r(\omega)$ are zero, it follows that the non-diagonal entries of $\boldsymbol{\Delta}(z)$ are also zero, see the integral formulas in Theorem A.10.

The factorizations (4.12), (4.40), and (4.56) show that

$$\boldsymbol{\Phi}(e^{-i\omega}) = \tilde{\boldsymbol{U}}(\omega)\boldsymbol{\Delta}(e^{-i\omega}).$$

Since $\boldsymbol{D}_r(\omega) = \boldsymbol{\Delta}(e^{-i\omega})$ is a diagonal matrix and each diagonal entry is positive, we may write that

$$\tilde{\boldsymbol{U}}(\omega) = \boldsymbol{\Phi}(e^{-i\omega}) \begin{bmatrix} \Delta_{11}^{-1}(e^{-i\omega}) & \cdots & 0 \\ \vdots & \ddots & \vdots \\ 0 & \cdots & \Delta_{rr}^{-1}(e^{-i\omega}) \end{bmatrix}.$$

The components of the process $\{\mathbf{V}_t\}$ are regular 1D time series, so by Theorem 2.3, the H^2-functions $\Delta_{kk}(z)$ $(k = 1, \dots, r)$ have no zeros in the open unit disc D. Consequently, the entries on the right hand side are boundary values of the ratio of two H^2-functions, and the denominator has no zeros in D. Hence $\tilde{\boldsymbol{U}}(\omega)$ is the boundary value of an analytic function $\boldsymbol{W}(z) = [w_{k\ell}(z)]_{d\times r}$ defined in D:

$$\tilde{\boldsymbol{U}}(\omega) = \boldsymbol{W}(e^{-i\omega}). \tag{4.68}$$

Moreover, since the boundary value $\boldsymbol{U}(\omega)$ is unitary, its entries are bounded functions. It implies that

$$\begin{aligned} \boldsymbol{W}(z) &= \sum_{j=0}^{\infty} \boldsymbol{\psi}(j) z^j \in H^\infty \subset H^2, \\ \tilde{\boldsymbol{U}}(\omega) &= \sum_{j=0}^{\infty} \boldsymbol{\psi}(j) e^{-ij\omega}, \quad \sum_{j=0}^{\infty} \|\boldsymbol{\psi}(j)\|_F^2 < \infty. \end{aligned} \tag{4.69}$$

This proves condition (3) of the theorem.

Conversely, assume that conditions (1), (2), and (3) of the theorem hold. Conditions (1) and (2) give that $\boldsymbol{\Lambda}_r(\omega)$ is the spectral density of an r-dimensional regular stationary time series $\{\mathbf{V}_t\}_{t\in\mathbb{Z}}$ of full rank r, just like in the first part of the proof. Then $\mathbf{\Lambda}_r(\omega)$ has the factorization (4.66), (4.67). Condition (3) implies that $\tilde{\boldsymbol{U}}(\omega) \in H^\infty \subset H^2$, with properties (4.68) and (4.69). The spectral decomposition (4.56) can be written as

$$\boldsymbol{f}(\omega) = \tilde{\boldsymbol{U}}(\omega)\mathbf{\Lambda}_r(\omega)\tilde{\boldsymbol{U}}^*(\omega) = \frac{1}{2\pi}\tilde{\boldsymbol{U}}(\omega)\boldsymbol{D}_r(\omega)\boldsymbol{D}_r(\omega)\tilde{\boldsymbol{U}}^*(\omega) = \frac{1}{2\pi}\boldsymbol{\phi}(\omega)\boldsymbol{\phi}^*(\omega).$$

So

$$\boldsymbol{\phi}(\omega) = \tilde{\boldsymbol{U}}(\omega)\boldsymbol{D}_r(\omega) = \sum_{j=0}^{\infty}\boldsymbol{\psi}(j)e^{-ij\omega}\sum_{k=0}^{\infty}\boldsymbol{\delta}(j)e^{-ik\omega} = \sum_{\ell=0}^{\infty}\boldsymbol{b}(\ell)e^{-i\ell\omega},$$

$$\boldsymbol{b}(\ell) = \sum_{j=0}^{\ell}\boldsymbol{\psi}(j)\boldsymbol{\delta}(\ell-j).$$

These imply that

$$\mathbf{\Phi}(z) := \sum_{\ell=0}^{\infty}\boldsymbol{b}(\ell)z^\ell = \boldsymbol{W}(z)\mathbf{\Delta}(z), \quad z \in D,$$

where $\boldsymbol{W} \in H^\infty$ and $\mathbf{\Delta} \in H^2$, thus $\mathbf{\Phi}(z) \in H^2$. By Corollary 4.2 it means that the time series $\{\mathbf{X}_t\}$ is regular. This completes the proof of the theorem. □

Remark 4.4. Assume that $\{\mathbf{X}_t\}$ is a d-dimensional regular time series of rank r. Assume as well that its spectral density matrix $\boldsymbol{f}$ has the spectral decomposition (4.56). Then the r-dimensional time series $\{\mathbf{V}_t\}$ can be given by (4.64), whose spectral density by (4.65) is $\boldsymbol{f}_{\mathbf{V}}(\omega) = \mathbf{\Lambda}_r(\omega)$.

By (4.69) and (4.64) we obtain that $\tilde{\boldsymbol{U}}^*(\omega) = \sum_{j=0}^{\infty}\boldsymbol{\psi}^*(j)e^{ij\omega}$, $\sum_{j=0}^{\infty}\|\boldsymbol{\psi}^*(j)\|_F^2 < \infty$, and

$$\begin{aligned}\mathbf{V}_t &= \int_{-\pi}^{\pi} e^{it\omega}\sum_{j=0}^{\infty}\boldsymbol{\psi}^*(j)e^{ij\omega}d\mathbf{Z}_\omega = \sum_{j=0}^{\infty}\boldsymbol{\psi}^*(j)\int_{-\pi}^{\pi} e^{i(t+j)\omega}d\mathbf{Z}_\omega \\ &= \sum_{j=0}^{\infty}\boldsymbol{\psi}^*(j)\mathbf{X}_{t+j}.\end{aligned} \tag{4.70}$$

It is interesting that $\{\mathbf{V}_t\}$ is *not* causally subordinated to $\{\mathbf{X}_t\}$, see (4.8) and (4.9).

Remark 4.5. Comparing Corollary 4.4 and Theorem 4.4 shows that in the full rank case, condition (3) in Theorem 4.4 follows from conditions (1) and (2).

Corollary 4.5. *Assume that $\{\mathbf{X}_t\}$ is a d-dimensional regular stationary time series of rank r, $1 \leq r \leq d$. Then a Kolmogorov–Szegő formula holds:*

$$\det \mathbf{\Sigma}_r = (2\pi)^r \exp\int_{-\pi}^{\pi}\log\det\mathbf{\Lambda}_r(\omega)\frac{d\omega}{2\pi} = (2\pi)^r \exp\int_{-\pi}^{\pi}\sum_{j=1}^{r}\log\lambda_j(\omega)\frac{d\omega}{2\pi},$$

where $\mathbf{\Lambda}_r$ is defined by (4.57) and $\mathbf{\Sigma}_r$ is the covariance matrix of the innovation process of an r-dimensional subprocess $\{\mathbf{X}_t^{(r)}\}$ of rank r, as defined below in the proof.

Proof. Let $\boldsymbol{f}_r(\omega)$ be the submatrix of $\boldsymbol{f}(\omega)$ whose determinant $\boldsymbol{M}(\omega)$ was defined in the first part of the proof of Theorem 4.4:

$$\boldsymbol{f}_r(\omega) = [f_{j_p j_q}(\omega)]_{p,q=1}^r, \quad \det \boldsymbol{f}_r(\omega) = M(\omega) \neq 0 \quad \text{for a.e.} \quad \omega \in [-\pi, \pi].$$

The indices j_p, $p = 1, \dots, r$, define a subprocess $\mathbf{X}_t^{(r)} = [X_t^{j_1}, \dots, X_t^{j_r}]^T$ of the original time series $\{\mathbf{X}_t\}$. Then $\{\mathbf{X}_t^{(r)}\}$ has an absolutely continuous spectral measure with density $\boldsymbol{f}_r(\omega)$, and by (4.63),

$$\log \det \boldsymbol{f}_r = \log M \in L^1.$$

Hence by Corollary 4.4, $\{\mathbf{X}_t^{(r)}\}$ is a regular process of full rank r and

$$\det \mathbf{\Sigma}_r = (2\pi)^r \exp \int_{-\pi}^{\pi} \log \det \boldsymbol{f}_r(\omega) \frac{d\omega}{2\pi},$$

where $\mathbf{\Sigma}_r$ is the covariance matrix of the innovation process of $\{\mathbf{X}_t^{(r)}\}$ as defined by (4.28) and (4.29):

$$\boldsymbol{\eta}_t^{(r)} := \mathbf{X}_t^{(r)} - \text{Proj}_{H_{t-1}^-} \mathbf{X}_t^{(r)} \quad (t \in \mathbb{Z}), \quad \mathbf{\Sigma}_r := \mathbb{E}\left(\boldsymbol{\eta}_0^{(r)} \boldsymbol{\eta}_0^{(r)*}\right).$$

Here we used that the past until $(t-1)$ of the subprocess $\{\mathbf{X}_t^{(r)}\}$ is the same as the past H_{t-1}^- of $\{\mathbf{X}_t\}$. Really, by (4.33), for the regular process $\{\mathbf{X}_t\}$ of rank r we have a causal MA(∞) form:

$$\mathbf{X}_t = \sum_{j=0}^{\infty} \boldsymbol{a}(j) \boldsymbol{\eta}_{t-j}, \quad t \in \mathbb{Z},$$

and for each t, $\boldsymbol{\eta}_t$ is linearly dependent on $\boldsymbol{\eta}_t^{(r)}$. Thus

$$\overline{\text{span}}\{\mathbf{X}_{t-1}^{(r)}, \mathbf{X}_{t-2}^{(r)}, \mathbf{X}_{t-3}^{(r)}, \dots\} = \overline{\text{span}}\{\mathbf{X}_{t-1}, \mathbf{X}_{t-2}, \mathbf{X}_{t-3}, \dots\} = H_{t-1}^-.$$

□

4.5.3 Classification of non-regular multidimensional time series

The classification of multidimensional time series is — not surprisingly — more complex than the one-dimensional ones, see the 1D classes in Section 2.10. We call a time series *non-regular* if either it is singular or its Wold decomposition contains two orthogonal, non-vanishing processes: a regular and a singular one. The classification below follows from Theorem 4.4.

In dimension $d > 1$ a non-singular process beyond its regular part may have a singular part with non-vanishing spectral density. For example, if $d = 3$ and the components $\{(X_t^1, X_t^2, X_t^3)\}$ are orthogonal to each other, it is possible that $\{X_t^1\}$ is regular of rank 1, $\{X_t^2\}$ is Type (1) singular, and $\{X_t^3\}$ is Type (2) singular.

Below we are considering a d-dimensional stationary time series $\{\mathbf{X}_t\}$ with spectral measure $d\boldsymbol{F}$.

- *Type (0) non-regular processes.* In this case the spectral measure $d\boldsymbol{F}$ of the time series $\{\mathbf{X}_t\}$ is singular w.r.t. the Lebesgue measure in $[-\pi, \pi]$. Clearly, type (0) non-regular processes are simply singular ones. Like in the 1D case, we may further divide this class into processes with a discrete spectrum or processes with a continuous singular spectrum or processes with both.

- *Type (1) non-regular processes.* The time series has an absolutely continuous spectral measure with density $\boldsymbol{f}$, but rank$(\boldsymbol{f})$ is not constant. It means that there exist measurable subsets $A, B \subset [-\pi, \pi]$ such that $d\omega(A) > 0$ and $d\omega(B) > 0$, rank$(\boldsymbol{f}(\omega)) = r_1$ if $\omega \in A$, rank$(\boldsymbol{f}(\omega)) = r_2$ if $\omega \in B$, and $r_1 \neq r_2$. Here $d\omega$ denotes Lebesgue measure in $[-\pi, \pi]$.

- *Type (2) non-regular processes.* The time series has an absolutely continuous spectral measure with density $\boldsymbol{f}$ which has constant rank r a.e., $1 \leq r \leq d$, but

  $$\int_{-\pi}^{\pi} \log \det \boldsymbol{\Lambda}_r(\omega) d\omega = \int_{-\pi}^{\pi} \sum_{j=1}^{r} \log \lambda_j(\omega)\, d\omega = -\infty,$$

  where $\boldsymbol{\Lambda}_r$ is defined by (4.57).

- *Type (3) non-regular processes.* The time series has an absolutely continuous spectral measure with density $\boldsymbol{f}$ which has constant rank r a.e., $1 \leq r < d$,

  $$\int_{-\pi}^{\pi} \log \det \boldsymbol{\Lambda}_r(\omega) d\omega = \int_{-\pi}^{\pi} \sum_{j=1}^{r} \log \lambda_j(\omega)\, d\omega > -\infty,$$

  but the unitary matrix function $\tilde{\boldsymbol{U}}(\omega)$ appearing in the spectral decomposition of $\boldsymbol{f}(\omega)$ in (4.58) does not belong to the Hardy space H^2.

By Corollary 4.3, if $\{\mathbf{X}_t\}$ has full rank $r = d$ and it is non-singular, then it may have only a Type (0) singular part.

4.6 Low rank approximation

The aim of this section is to approximate a time series of constant rank r with one of smaller rank k. This problem was treated by Brillinger in [8] and also in [9, Chapter 9] where it was called Principal Component Analysis (PCA) in the Frequency Domain. We show the important fact that when the process is regular, the low rank approximation can also be chosen regular.

4.6.1 Approximation of time series of constant rank

Assume that $\{\mathbf{X}_t\}$ is a d-dimensional stationary time series of constant rank r, $1 \le r \le d$. By Theorem 4.1, it is equivalent to the assumption that $\{\mathbf{X}_t\}$ can be written as a sliding summation of form (4.10). The spectral density $\boldsymbol{f}$ of the process has rank r a.e., and so we may write its eigenvalues as

$$\lambda_1(\omega) \ge \cdots \ge \lambda_r(\omega) > 0, \quad \lambda_{r+1}(\omega) = \cdots = \lambda_d(\omega) = 0. \tag{4.71}$$

Also, the spectral decomposition of $\boldsymbol{f}$ is

$$\boldsymbol{f}(\omega) = \sum_{j=1}^{r} \lambda_j(\omega)\mathbf{u}_j(\omega)\mathbf{u}_j^*(\omega) = \tilde{\boldsymbol{U}}_r(\omega)\tilde{\boldsymbol{\Lambda}}_r(\omega)\tilde{\boldsymbol{U}}_r^*(\omega), \qquad \text{a.e. } \omega \in [-\pi, \pi], \tag{4.72}$$

where $\tilde{\boldsymbol{\Lambda}}_r(\omega) := \operatorname{diag}[\lambda_1(\omega), \dots, \lambda_r(\omega)]$, $\mathbf{u}_j(\omega) \in \mathbb{C}^d$ $(j = 1, \dots, r)$ are the corresponding orthonormal eigenvectors, and $\tilde{\boldsymbol{U}}_r(\omega) \in \mathbb{C}^{d \times r}$ is the matrix of these column vectors.

Now the problem we are treating can be described as follows. Given an integer k, $1 \le k \le r$, find a process $\{\mathbf{X}_t^{(k)}\}$ of constant rank k which is a linear transform of $\{\mathbf{X}_t\}$ and which minimizes the distance

$$\begin{aligned}\|\mathbf{X}_t - \mathbf{X}_t^{(k)}\|^2 &= \mathbb{E}\left\{(\mathbf{X}_t - \mathbf{X}_t^{(k)})^* \, (\mathbf{X}_t - \mathbf{X}_t^{(k)})\right\} \\ &= \operatorname{tr} \operatorname{Cov}\left\{(\mathbf{X}_t - \mathbf{X}_t^{(k)}), (\mathbf{X}_t - \mathbf{X}_t^{(k)})\right\}, \quad \forall t \in \mathbb{Z}. \end{aligned} \tag{4.73}$$

In Brillinger's book [9] this is called *Principal Component Analysis (PCA) in the Frequency Domain.*

Consider the spectral representations of $\{\mathbf{X}_t\}$ and $\{\mathbf{X}_t^{(k)}\}$, see (4.3):

$$\mathbf{X}_t = \int_{-\pi}^{\pi} e^{it\omega} d\mathbf{Z}_\omega, \quad \mathbf{X}_t^{(k)} = \int_{-\pi}^{\pi} e^{it\omega}\boldsymbol{T}(\omega) d\mathbf{Z}_\omega, \quad t \in \mathbb{Z}.$$

Then by (4.4) and (4.72) we can rewrite (4.73) as

$$\begin{aligned}\|\mathbf{X}_t - \mathbf{X}_t^{(k)}\|^2 &= \operatorname{tr} \int_{-\pi}^{\pi} (\boldsymbol{I}_d - \boldsymbol{T}(\omega))\boldsymbol{f}(\omega)(\boldsymbol{I}_d - \boldsymbol{T}^*(\omega))\, d\omega \\ &= \operatorname{tr} \int_{-\pi}^{\pi} (\boldsymbol{I}_d - \boldsymbol{T}(\omega))\tilde{\boldsymbol{U}}_r(\omega)\tilde{\boldsymbol{\Lambda}}_r(\omega)\tilde{\boldsymbol{U}}_r^*(\omega)(\boldsymbol{I}_d - \boldsymbol{T}^*(\omega))\, d\omega, \end{aligned} \tag{4.74}$$

which clearly does not depend on $t \in \mathbb{Z}$.

To find the minimizing linear transformation $\boldsymbol{T}(\omega)$, we have to study the non-negative definite quadratic form

$$\mathbf{v}^*\boldsymbol{f}(\omega)\mathbf{v} = \mathbf{v}^*\tilde{\boldsymbol{U}}_r(\omega)\tilde{\boldsymbol{\Lambda}}_r(\omega)\tilde{\boldsymbol{U}}_r^*(\omega)\mathbf{v}, \quad \mathbf{v} \in \mathbb{C}^d, \quad |\mathbf{v}| = 1.$$

By (4.71), there is a monotonicity: taking the orthogonal projections $\mathbf{u}_j\mathbf{u}_j^*$ $(j = 1, \dots, r)$ in the space $\mathbb{C}^d$ one-by-one, the sequence

$$\mathbf{v}_j^* \boldsymbol{f}(\omega)\mathbf{v}_j, \qquad \mathbf{v}_j \in \mathbf{u}_j(\omega)\mathbf{u}_j^*(\omega)\,\mathbb{C}^d \qquad (j = 1, \dots, r),$$

is non-increasing. Since $\boldsymbol{I}_d = \sum_{j=1}^d \mathbf{u}_j(\omega)\mathbf{u}_j^*(\omega)$ and $\boldsymbol{T}(\omega)$ must have rank k a.e., (4.74) implies that the minimizing linear transformation must be the orthogonal projection

$$\boldsymbol{T}(\omega) = \sum_{j=1}^{k} \mathbf{u}_j(\omega)\mathbf{u}_j^*(\omega) = \tilde{\boldsymbol{U}}_k(\omega)\tilde{\boldsymbol{U}}_k^*(\omega); \tag{4.75}$$

see Corollary B.1 as well. Thus we have proved that

$$\mathbf{X}_t^{(k)} = \int_{-\pi}^{\pi} e^{it\omega} \tilde{\boldsymbol{U}}_k(\omega)\tilde{\boldsymbol{U}}_k^*(\omega) d\mathbf{Z}_\omega, \quad t \in \mathbb{Z}. \tag{4.76}$$

Then by (4.5), the spectral density of $\{\mathbf{X}_t^{(k)}\}$ is

$$\begin{aligned}
\boldsymbol{f}_k(\omega) &= \tilde{\boldsymbol{U}}_k(\omega)\tilde{\boldsymbol{U}}_k^*(\omega)\tilde{\boldsymbol{U}}_r(\omega)\tilde{\boldsymbol{\Lambda}}_r(\omega)\tilde{\boldsymbol{U}}_r^*(\omega)\tilde{\boldsymbol{U}}_k(\omega)\tilde{\boldsymbol{U}}_k^*(\omega) \\
&= \tilde{\boldsymbol{U}}_k(\omega) \begin{bmatrix} I_k & 0_{k\times(r-k)} \end{bmatrix} \tilde{\boldsymbol{\Lambda}}_r(\omega) \begin{bmatrix} I_k \\ 0_{(r-k)\times k} \end{bmatrix} \tilde{\boldsymbol{U}}_k^*(\omega) \\
&= \tilde{\boldsymbol{U}}_k(\omega)\tilde{\boldsymbol{\Lambda}}_k(\omega)\tilde{\boldsymbol{U}}_k^*(\omega), \quad \omega \in [-\pi, \pi].
\end{aligned} \tag{4.77}$$

Further, the covariance function of $\{\mathbf{X}_t^{(k)}\}$ is

$$\boldsymbol{C}_k(h) := \int_{-\pi}^{\pi} e^{ih\omega} \boldsymbol{f}_k(\omega) d\omega, \qquad h \in \mathbb{Z}. \tag{4.78}$$

The next theorem summarizes the results above.

Theorem 4.5. *Assume that $\{\mathbf{X}_t\}$ is a d-dimensional stationary time series of constant rank r, $1 \le r \le d$, with spectral density $\boldsymbol{f}$. Let* (4.71) *and* (4.72) *be the spectral decomposition of $\boldsymbol{f}$.*

(a) Then

$$\mathbf{X}_t^{(k)} = \int_{-\pi}^{\pi} e^{it\omega} \tilde{\boldsymbol{U}}_k(\omega)\tilde{\boldsymbol{U}}_k^*(\omega) d\mathbf{Z}_\omega, \quad t \in \mathbb{Z},$$

is the approximating process of rank k, $1 \le k \le r$, which minimizes the mean square error of the approximation.

(b) For the mean square error we have

$$\|\mathbf{X}_t - \mathbf{X}_t^{(k)}\|^2 = \int_{-\pi}^{\pi} \sum_{j=k+1}^{r} \lambda_j(\omega)\, d\omega, \qquad t \in \mathbb{Z}, \tag{4.79}$$

and

$$\frac{\|\mathbf{X}_t - \mathbf{X}_t^{(k)}\|^2}{\|\mathbf{X}_t\|^2} = \frac{\int_{-\pi}^{\pi} \sum_{j=k+1}^{r} \lambda_j(\omega)\, d\omega}{\int_{-\pi}^{\pi} \sum_{j=1}^{r} \lambda_j(\omega)\, d\omega}, \qquad t \in \mathbb{Z}. \tag{4.80}$$

(c) If condition

$$\lambda_k(\omega) \geq \Delta > \epsilon \geq \lambda_{k+1}(\omega) \qquad \forall \omega \in [-\pi, \pi], \tag{4.81}$$

holds then we also have

$$\|\mathbf{X}_t - \mathbf{X}_t^{(k)}\| \leq (2\pi(r-k)\epsilon)^{1/2}, \quad t \in \mathbb{Z}$$

and

$$\frac{\|\mathbf{X}_t - \mathbf{X}_t^{(k)}\|}{\|\mathbf{X}_t\|} \leq \left(\frac{(r-k)\epsilon}{r\Delta}\right)^{\frac{1}{2}}, \qquad t \in \mathbb{Z}.$$

Proof. Statement (a) was shown above.

By (4.74) and (4.75),

$$\begin{aligned}
&\|\mathbf{X}_t - \mathbf{X}_t^{(k)}\|^2 \\
&= \operatorname{tr} \int_{-\pi}^{\pi} \left\{ \sum_{j=k+1}^{r} \mathbf{u}_j(\omega)\mathbf{u}_j^*(\omega) \right\} \tilde{\boldsymbol{U}}_r(\omega)\tilde{\boldsymbol{\Lambda}}_r(\omega)\tilde{\boldsymbol{U}}_r^*(\omega) \left\{ \sum_{j=k+1}^{r} \mathbf{u}_j(\omega)\mathbf{u}_j^*(\omega) \right\} d\omega \\
&= \operatorname{tr} \int_{-\pi}^{\pi} \begin{bmatrix} 0_{d\times k} & \mathbf{u}_{k+1}(\omega) & \cdots & \mathbf{u}_r(\omega) \end{bmatrix} \tilde{\boldsymbol{\Lambda}}_r(\omega) \begin{bmatrix} 0_{k\times d} \\ \mathbf{u}_{k+1}^*(\omega) \\ \vdots \\ \mathbf{u}_r^*(\omega) \end{bmatrix} d\omega \\
&= \operatorname{tr} \int_{-\pi}^{\pi} \tilde{\boldsymbol{U}}_r(\omega)\operatorname{diag}[0,\ldots,0,\lambda_{k+1},\ldots,\lambda_r]\tilde{\boldsymbol{U}}_r^*(\omega) d\omega \\
&= \int_{-\pi}^{\pi} \sum_{j=k+1}^{r} \lambda_j(\omega)\, d\omega,
\end{aligned}$$

where we finally used that the trace equals the sum of the eigenvalues of a matrix. This proves (4.79). Since (4.79) holds for $k = 0$ as well, we get (4.80).

Finally, condition (4.81) and (b) imply (c). □

Corollary 4.6. *For the difference of the covariance functions of* $\{\mathbf{X}_t\}$ *and* $\{\mathbf{X}_t^{(k)}\}$ *we have the following estimate:*

$$\|\boldsymbol{C}(h) - \boldsymbol{C}_k(h)\| \leq \int_{-\pi}^{\pi} \lambda_{k+1}(\omega) d\omega, \quad h \in \mathbb{Z},$$

where the matrix norm is the spectral norm, see Appendix B. If condition (4.81) *holds then we have the bound*

$$\|\boldsymbol{C}(h) - \boldsymbol{C}_k(h)\| \leq 2\pi\epsilon, \quad h \in \mathbb{Z}.$$

Proof. By (4.72), (4.77), and (4.78) it follows that

$$\begin{aligned}
\|\boldsymbol{C}(h) - \boldsymbol{C}_k(h)\| &= \left\| \int_{-\pi}^{\pi} e^{ih\omega}(f(\omega) - f_k(\omega))d\omega \right\| \\
&= \left\| \int_{-\pi}^{\pi} e^{ih\omega} \tilde{\boldsymbol{U}}_r(\omega) \left\{ \tilde{\boldsymbol{\Lambda}}_r(\omega) - \operatorname{diag}[\lambda_1(\omega), \ldots, \lambda_k(\omega), 0, \ldots, 0] \right\} \tilde{\boldsymbol{U}}_r^*(\omega) d\omega \right\| \\
&\leq \int_{-\pi}^{\pi} \|\tilde{\boldsymbol{U}}_r(\omega)\| \cdot \|\operatorname{diag}[0, \ldots, 0, \lambda_{k+1}(\omega), \ldots, \lambda_r(\omega)]\| \cdot \|\tilde{\boldsymbol{U}}_r^*(\omega)\| d\omega \\
&= \int_{-\pi}^{\pi} \|\operatorname{diag}[0, \ldots, 0, \lambda_{k+1}(\omega), \ldots, \lambda_r(\omega)]\| \, d\omega \\
&= \int_{-\pi}^{\pi} \lambda_{k+1}(\omega) d\omega.
\end{aligned}$$

□

Equation (4.76) can be factored. As in Theorem 4.4, one can take the Fourier series of the sub-unitary matrix function $\tilde{\boldsymbol{U}}_k(\omega) \in L^2$:

$$\tilde{\boldsymbol{U}}_k(\omega) = \sum_{j=-\infty}^{\infty} \boldsymbol{\psi}(j) e^{-ij\omega}, \quad \boldsymbol{\psi}(j) = \frac{1}{2\pi} \int_{-\pi}^{\pi} e^{ij\omega} \tilde{\boldsymbol{U}}_k(\omega) d\omega \in \mathbb{C}^{d \times k},$$

where $\sum_{j=-\infty}^{\infty} \|\boldsymbol{\psi}(j)\|_F^2 < \infty$. Consequently,

$$\tilde{\boldsymbol{U}}_k^*(\omega) = \sum_{j=-\infty}^{\infty} \boldsymbol{\psi}^*(j) e^{ij\omega}, \quad \omega \in [-\pi, \pi].$$

If the time series $\{\mathbf{V}_t\}$ is defined by the linear filter (4.64), then similarly to (4.70) it follows that $\{\mathbf{V}_t\}$ can be obtained from the original time series $\{\mathbf{X}_t\}$ by a sliding summation:

$$\mathbf{V}_t = \sum_{j=-\infty}^{\infty} \boldsymbol{\psi}^*(j) \mathbf{X}_{t+j}, \quad t \in \mathbb{Z},$$

and similarly to (4.65), its spectral density is a diagonal matrix:

$$f_{\mathbf{V}}(\omega) = \boldsymbol{\Lambda}_k(\omega) = \operatorname{diag}[\lambda_1(\omega), \ldots, \lambda_k(\omega)].$$

It means that the covariance matrix function of $\{\mathbf{V}_t\}$ is also diagonal:

$$\boldsymbol{C}_{\mathbf{V}}(h) = \operatorname{diag}[c_{11}(h), \ldots, c_{kk}(h)], \quad c_{jj}(h) = \int_{-\pi}^{\pi} e^{ih\omega} \lambda_j(\omega) d\omega, \quad h \in \mathbb{Z},$$

that is, the components of the process $\{\mathbf{V}_t\}$ are orthogonal to each other. Thus the process $\{\mathbf{V}_t\}$ can be called k-dimensional *Dynamic Principal Components (DPC)* of the d-dimensional process $\{\mathbf{X}_t\}$.

Using a second linear filtration, which is the adjoint ψ of the previous filtration ψ^*, one can obtain the k-rank approximation $\{\mathbf{X}_t^{(k)}\}$ from $\{\mathbf{V}_t\}$:

$$\mathbf{X}_t^{(k)} = \int_{-\pi}^{\pi} e^{it\omega} \tilde{\boldsymbol{U}}_k(\omega)\tilde{\boldsymbol{U}}_k^*(\omega) d\mathbf{Z}_\omega = \int_{-\pi}^{\pi} e^{it\omega} \tilde{\boldsymbol{U}}_k(\omega) d\mathbf{Z}_\omega^{\mathbf{V}}$$
$$= \sum_{j=-\infty}^{\infty} \psi(j)\mathbf{V}_{t-j}, \quad t \in \mathbb{Z}.$$

Notice the *dimension reduction* in this approximation. Dimension d of the original process $\{\mathbf{X}_t\}$ can be reduced to dimension $k < d$ with the cross-sectionally orthogonal DPC process $\{\mathbf{V}_t\}$, obtained by linear filtration, from which the low-rank approximation $\{\mathbf{X}_t^{(k)}\}$ can be reconstructed also by linear filtration. Of course, this is useful only if the error of the approximation given by Theorem 4.5 is small enough.

Since $\tilde{\boldsymbol{U}}_k \tilde{\boldsymbol{U}}_k^* \in L^2$ as well, one can take the L^2-convergent Fourier series

$$\tilde{\boldsymbol{U}}_k(\omega)\tilde{\boldsymbol{U}}_k^*(\omega) = \sum_{j,\ell=-\infty}^{\infty} \psi(j) e^{-ij\omega} \psi^*(\ell) e^{i\ell\omega} = \sum_{m=-\infty}^{\infty} \boldsymbol{w}(m) e^{-im\omega},$$

where $\omega \in [-\pi, \pi]$ and

$$\boldsymbol{w}(m) = \sum_{j=-\infty}^{\infty} \psi(j)\, \psi^*(j-m) \in \mathbb{C}^{d\times d}, \qquad \sum_{m=-\infty}^{\infty} \|\boldsymbol{w}(m)\|_F^2 < \infty.$$

By (4.7) it implies that the filtered process $\{\mathbf{X}_t^{(k)}\}$ can be obtained directly from $\{\mathbf{X}_t\}$ by a two-sided sliding summation:

$$\mathbf{X}_t^{(k)} = \sum_{m=-\infty}^{\infty} \boldsymbol{w}(m)\mathbf{X}_{t-m}, \quad t \in \mathbb{Z}.$$

4.6.2 Approximation of regular time series

Proposition 4.1. *Assume that $\{\mathbf{X}_t\}$ is a d-dimensional regular time series of rank r. Then the rank k approximation $\{\mathbf{X}_t^{(k)}\}$, $1 \le k \le r$, defined in (4.76) is also a regular time series.*

Proof. We have to check that the conditions in Theorem 4.4 hold for $\{\mathbf{X}_t^{(k)}\}$. By (4.77), $\{\mathbf{X}_t^{(k)}\}$ has an absolutely continuous spectral measure with density of constant rank k, so condition (1) holds.

If $\int_{-\pi}^{\pi} \log \lambda_k(\omega) d\omega = -\infty$, then that would contradict the regularity of the original process $\{\mathbf{X}_t\}$, and this proves condition (2).

It follows from Theorem 4.4 that $\tilde{\boldsymbol{U}}_r(\omega)$ belongs to the Hardy space H^∞, so the same holds for $\tilde{\boldsymbol{U}}_k(\omega)$ as well, since its columns are a subset of the former's. This proves condition (3). □

Theorem 4.5 and Corollary 4.6 are valid for regular processes without change. However, the factorization of the approximation discussed above is different in the regular case, because several of the summations become one-sided. Thus we have

$$\tilde{\boldsymbol{U}}_k^*(\omega) = \sum_{j=0}^{\infty} \boldsymbol{\psi}^*(j) e^{ij\omega}, \quad \omega \in [-\pi, \pi].$$

Consequently, the k-dimensional, cross-sectionally orthogonal process $\{\mathbf{V}_t\}$ becomes

$$\mathbf{V}_t = \sum_{j=0}^{\infty} \boldsymbol{\psi}^*(j) \mathbf{X}_{t+j}, \quad t \in \mathbb{Z}.$$

Further, the reconstruction of the k-rank approximation $\{\mathbf{X}_t^{(k)}\}$ from $\{\mathbf{V}_t\}$ is

$$\mathbf{X}_t^{(k)} = \sum_{j=0}^{\infty} \boldsymbol{\psi}(j) \mathbf{V}_{t-j}, \quad t \in \mathbb{Z}.$$

The direct evaluation of $\{\mathbf{X}_t^{(k)}\}$ from $\{\mathbf{X}_t\}$ takes now the following form:

$$\tilde{\boldsymbol{U}}_k(\omega)\tilde{\boldsymbol{U}}_k^*(\omega) = \sum_{j,\ell=0}^{\infty} \boldsymbol{\psi}(j) e^{-ij\omega} \boldsymbol{\psi}^*(\ell) e^{i\ell\omega} = \sum_{m=-\infty}^{\infty} \boldsymbol{w}(m) e^{-im\omega}, \quad \omega \in [-\pi, \pi],$$

where $\boldsymbol{w}(m) = \sum_{j=\max(0,m)}^{\infty} \boldsymbol{\psi}(j)\, \boldsymbol{\psi}^*(j-m)$. It implies that the filtered process $\{\mathbf{X}_t^{(k)}\}$ is *not* causally subordinated to the original regular process $\{\mathbf{X}_t\}$ in general, since it can be obtained from $\{\mathbf{X}_t\}$ by a two-sided sliding summation:

$$\mathbf{X}_t^{(k)} = \sum_{m=-\infty}^{\infty} \boldsymbol{w}(m) \mathbf{X}_{t-m}, \quad t \in \mathbb{Z},$$

see (4.7). On the other hand, it is clear that if $\|\boldsymbol{\psi}(j)\|_F$ goes to 0 fast enough as $j \to \infty$, one does not have to use too many 'future' terms of $\{\mathbf{X}_t\}$ to get a good enough approximation of $\{\mathbf{X}_t^{(k)}\}$. In practice one can also replace the future values of $\{\mathbf{X}_t\}$ by $\mathbf{0}$ to get a causal approximation of $\{\mathbf{X}_t^{(k)}\}$.

4.7 Rational spectral densities

An important subclass of the class of regular stationary time series, which in turn is a subclass of time series with constant rank, is such that each entry $f^{k\ell}(\omega)$ in the spectral density matrix $\boldsymbol{f}$ is a rational complex function in $z = e^{-i\omega}$. As we are going to see in Theorem 4.8, this subclass is the same as that of the stable VARMA(p, q) processes. Also, this is the subclass of stable stochastic linear systems with finite dimensional state space representation. Moreover, this is the subclass of stable stochastic linear systems with rational transfer function. In sum, this is the subclass of weakly stationary time series that can be described by finitely many complex valued parameters.

Remark 4.6. Every minor (that is, the determinant of a sub-matrix) of a rational matrix is a rational function which is either identically zero or has zeros and poles at only finitely many points. Hence a weakly stationary time series with a spectral density $\boldsymbol{f}$ which is a rational matrix in $z = e^{-i\omega}$ must be of constant rank r.

Remark 4.7. Suppose that the spectral density $\boldsymbol{f}$ of a weakly stationary time series is a rational matrix in $z = e^{-i\omega}$. Since the entries of the spectral measure $d\boldsymbol{F}$ are finite measures on $[-\pi, \pi]$ by formula (1.23), the entries of $\boldsymbol{f}$ can have no poles on the unit circle T. See Remark 2.3 as well.

4.7.1 Smith–McMillan form

The Smith–McMillan form is a useful tool by which one can diagonalize a non-negative definite rational matrix so that both the obtained diagonal matrix and the transformation matrix used for the diagonalization are rational matrices. The usual technique of linear algebra which uses eigenvalues and eigenvectors does not have this important property, since the eigenvalues and the entries of eigenvectors of a rational matrix are not rational functions in general.

Lemma 4.8. *Let $\boldsymbol{A}(z) = [a_{jk}(z)]_{d\times d}$ be a rational matrix which is self-adjoint and non-negative definite for $z \in T$, having no poles on T, and whose rank is r, $1 \le r \le d$, for $z \in T \setminus Z$, where Z is a finite set. Then we can write $\boldsymbol{A}(z)$ in Smith–McMillan form:*

$$\boldsymbol{A}(z) = \boldsymbol{E}(z)\boldsymbol{\Lambda}(z)\boldsymbol{E}^*(z) \quad (z \in T \setminus Z), \tag{4.82}$$

where $\boldsymbol{\Lambda}(z)$ and $\boldsymbol{E}(z)$ are rational matrices,

$$\boldsymbol{\Lambda}(z) = \operatorname{diag}[\lambda_1(z), \dots, \lambda_r(z)], \quad \lambda_j(z) > 0 \quad (j = 1, \dots, r;\ z \in T \setminus Z).$$

Here $\operatorname{diag}[\lambda_1, \dots, \lambda_r]$ *denotes an $r \times r$ diagonal matrix with entries λ_j $(j = 1, \dots, r)$ in its main diagonal. Also,*

$$\boldsymbol{E}(z) = [e_{jk}(z)]_{d\times r} \quad (z \in T \setminus Z),$$

is a lower unit trapezoidal matrix:

1. $e_{jk}(z) = 0$ *if* $k > j$,
2. $e_{jk}(z) = 1$ *if* $k = j$.

Proof. Since $\boldsymbol{A}(z)$ is a self-adjoint rational matrix, has no poles on T, and has rank r for $z \in T \setminus Z$, it has a principal minor $M(z)$ of order r which is a rational function, has no poles for $z \in T$ and no zeros for $z \in T \setminus Z$, while any minor of order larger than r is identically zero. We may assume that the sub-matrix corresponding to $M(z)$ stands in the upper left corner of $\boldsymbol{A}(z)$, since by rearranging rows and columns (that is, using elementary row and column operations) we may achieve this. Since $\boldsymbol{A}(z)$ is non-negative definite, any upper left corner minor $M_j(z)$ of size j $(j = 1, \dots, r)$ obtained from $M(z) = M_r(z)$ must be positive if $z \in T \setminus Z$.

In particular, $M_1(z) = a_{11}(z) > 0$ if $z \in T \setminus Z$. Using elementary row and column operations we may transform the rational matrix $\boldsymbol{A}^{(1)}(z) = \boldsymbol{A}(z)$ into another rational matrix of the form

$$\boldsymbol{A}^{(2)}(z) = [a_{jk}^{(2)}(z)]_{d \times d} = \begin{bmatrix} a_{11}(z) & 0_{1\times(n-1)} \\ 0_{(n-1)\times 1} & \boldsymbol{B}^{(2)}(z) \end{bmatrix},$$

$$\boldsymbol{B}^{(2)}(z) = [b_{jk}^{(2)}(z)]_{j,k=2,\dots,n}, \quad b_{jk}^{(2)}(z) = a_{jk}(z) - \frac{a_{j1}(z)a_{1k}(z)}{a_{11}(z)}.$$

If $r \geq 2$, then $b_{22}^{(2)}(z) = M_2(z)/M_1(z) > 0$ if $z \in T \setminus Z$, and we can apply the same transformation for $\boldsymbol{B}^{(2)}(z)$ as for $\boldsymbol{A}^{(1)}(z)$, and so on. Finally, after the rth step we obtain

$$\begin{aligned} &\boldsymbol{A}^{(r)}(z) = \operatorname{diag}[\lambda_1(z), \dots, \lambda_r(z), 0, \dots, 0]_{d \times d}, \quad \lambda_1(z) = M_1(z) > 0, \\ &\lambda_j(z) = M_j(z)/M_{j-1}(z) > 0 \qquad (j = 2, \dots, r;\ z \in T \setminus Z). \end{aligned} \tag{4.83}$$

Define

$$\boldsymbol{\Lambda}(z) := \operatorname{diag}[\lambda_1(z), \dots, \lambda_r(z)] \quad (z \in T \setminus Z). \tag{4.84}$$

The way one obtains $\boldsymbol{A}^{(\ell+1)}(z)$ from $\boldsymbol{A}^{(\ell)}(z)$ is an application of a sequence of elementary row and column operations. Such an elementary row operation is adding $\alpha(z)$ times the ith row to the jth row, where $\alpha(z)$ is a rational complex function. This operation is equivalent to left multiplication by the elementary matrix $\boldsymbol{E}_{i,j}(\alpha(z))$, see (D.1).

Similarly, the elementary column operation of adding $\overline{\alpha(z)}$ times the ith column to the jth column is equivalent to right multiplication by $\boldsymbol{E}_{i,j}(\alpha(z))^*$. It is important that $\boldsymbol{E}_{i,j}(\alpha(z))^{-1} = \boldsymbol{E}_{i,j}(-\alpha(z))$. Using this notation, we can write that

$$\boldsymbol{A}^{(\ell+1)}(z) = \boldsymbol{E}^{(\ell)}(z)\boldsymbol{A}^{(\ell)}(z)\boldsymbol{E}^{(\ell)}(z)^*,$$

where

$$\boldsymbol{E}^{(\ell)}(z) := \prod_{j=0}^{d-\ell-1} \boldsymbol{E}_{\ell,d-j}\left(-a_{d-j,\ell}^{(\ell)}(z)/a_{\ell\ell}^{(\ell)}(z)\right).$$

(Product of a sequence of matrices is understood so that the one with the first index, in the present case $j = 0$, is the first factor from the left, and so on.) Observe that each matrix $\boldsymbol{E}^{(\ell)}(z)$ is a lower unit triangular matrix.

Thus

$$\boldsymbol{A}^{(r)}(z) = \tilde{\boldsymbol{E}}(z)\boldsymbol{A}(z)\tilde{\boldsymbol{E}}^*(z), \quad \tilde{\boldsymbol{E}}(z) := \prod_{\ell=1}^{r-1} \boldsymbol{E}^{r-\ell}(z), \text{ if } z \in T \setminus Z,$$

where $\tilde{\boldsymbol{E}}(z)$ is also a lower unit triangular matrix. Then

$$\boldsymbol{A}(z) = \tilde{\boldsymbol{E}}(z)^{-1}\boldsymbol{A}^{(r)}(z)\tilde{\boldsymbol{E}}^*(z)^{-1} \text{ if } z \in T \setminus Z.$$

Since the last $d-r$ rows and columns of $\boldsymbol{A}^{(r)}(z)$ are zero, we may omit the last $d-r$ columns of $\tilde{\boldsymbol{E}}(z)^{-1}$. The resulting matrix of size $d \times r$ will be denoted by $\boldsymbol{E}(z)$, with which we still have

$$\boldsymbol{A}(z) = \boldsymbol{E}(z)\boldsymbol{\Lambda}(z)\boldsymbol{E}^*(z) \text{ if } z \in T \setminus Z,$$

where $\boldsymbol{E}(z)$ is a lower unit trapezoidal matrix and, correspondingly, $\boldsymbol{E}(z)^*$ is an upper unit trapezoidal matrix. This with equations (4.83) and (4.84) proves the lemma. □

4.7.2 Spectral factors of a rational spectral density matrix

Theorem 4.6. *[47, Section I.10]*

Let $\boldsymbol{A}(z) = [a_{jk}(z)]_{d\times d}$ be a rational matrix which is self-adjoint and non-negative definite for $z \in T$, having no poles on T, and whose rank is r for all $z \in T \setminus Z$, where Z is a finite set. Then we can write a special Gram decomposition

$$\boldsymbol{A}(z) = \frac{1}{2\pi}\boldsymbol{\Phi}(z)\boldsymbol{\Phi}^*(z) \quad (z \in T \setminus Z),$$

where $\boldsymbol{\Phi}(z) = [\Phi_{jk}(z)]_{d\times r}$ is a rational matrix, analytic in the open unit disc $D = \{z : |z| < 1\}$, and has rank r for any $z \in T \setminus Z$.

Proof. By Lemma 4.8, we may write $\boldsymbol{A}(z) = \boldsymbol{E}(z)\boldsymbol{\Lambda}(z)\boldsymbol{E}(z)^*$, $z \in T \setminus Z$, with the properties of $\boldsymbol{\Lambda}(z) = \operatorname{diag}[\lambda_1(z), \dots, \lambda_r(z)]$ and $\boldsymbol{E}(z) = [e_{jk}(z)]_{d\times r}$ described in the statement of the lemma. Denote

$$e_{jk}(z) = \frac{q_{jk}(z)}{p_{jk}(z)} \quad (z \in T \setminus Z),$$

where q_{jk} and p_{jk} are coprime (relative prime) polynomials for $j = 1, \dots, d$ and $k = 1, \dots, r$. Fixing a column $k = 1, \dots, r$, let $\zeta_\ell^{(k)}$ $(\ell = 1, \dots, \ell_0(k))$ denote all zeros in D of the polynomials $p_{jk}(z)$, each with maximal multiplicity for the indices $j = 1, \dots, d$. Set

$$c_k(z) := \prod_{\ell=1}^{\ell_0(k)} (z - \zeta_\ell^{(k)}), \quad D_k(z) := \frac{\lambda_k(z)}{|c_k(z)|^2} \quad (z \in T \setminus Z).$$

By Lemma 4.8, $D_k(z)$ is a non-negative rational function with no poles on T, so Lemma 2.2 shows that we can write

$$D_k(z) = \left| \frac{Q_k(z)}{P_k(z)} \right|^2 ,$$

where the coprime polynomials Q_k and P_k do not have zeros in the open unit disc D, and P_k does not have zeros on the unit circle T either.

Define the rational functions

$$\Phi_{jk}(z) := \sqrt{2\pi} e_{jk}(z) c_k(z) \frac{Q_k(z)}{P_k(z)} \quad (j = 1, \ldots, d;\ k = 1, \ldots, r)$$

which are analytic in D and set $\mathbf{\Phi}(z) := [\Phi_{jk}(z)]_{d\times r}$. Then

$$\begin{aligned} \frac{1}{2\pi}\mathbf{\Phi}(z)\mathbf{\Phi}(z)^* &= \left[\sum_{\ell=1}^{r} e_{j\ell}(z) c_\ell(z) \frac{Q_\ell(z)}{P_\ell(z)} \bar{e}_{k\ell}(z)\, \bar{c}_\ell(z)\, \frac{\bar{Q}_\ell(z)}{\bar{P}_\ell(z)} \right]_{d\times d} \\ &= \left[\sum_{\ell=1}^{r} e_{j\ell}(z) \lambda_\ell(z) \bar{e}_{k\ell}(z) \right]_{d\times d} = \boldsymbol{A}(z) \quad (z \in T \setminus Z) \end{aligned}$$

by (4.82). □

Corollary 4.7. *Let* $\{\mathbf{X}_t\}_{t\in\mathbb{Z}}$ *be a d-dimensional weakly stationary time series with spectral density* $\boldsymbol{f}(\omega)$ *which is a rational function in* $z = e^{-i\omega}$. *By Remark 4.6,* $\boldsymbol{f}(\omega)$ *has constant rank r for all* $\omega \in [-\pi, \pi] \setminus Z$, *where Z is a finite set. By Remark 4.7,* $\boldsymbol{f}$ *cannot have poles on* $[-\pi, \pi]$. *Thus Theorem 4.6 applies:*

$$\boldsymbol{f}(\omega) = \frac{1}{2\pi}\mathbf{\Phi}(e^{-i\omega})\mathbf{\Phi}^*(e^{-i\omega}), \quad \omega \in [-\pi, \pi] \setminus Z,$$

where $\mathbf{\Phi}(z)$ *is a* $d \times r$ *rational matrix, analytic in the open unit disc D, and has rank r for any* $z = e^{-i\omega}$, $\omega \in [-\pi, \pi] \setminus Z$.

4.8 Multidimensional ARMA (VARMA) processes

4.8.1 Equivalence of different approaches

Definition 4.1. A *multidimensional ARMA(p, q) process* ($p \geq 0$, $q \geq 0$) or *VARMA(p, q) process* (vector ARMA process) is a causal stationary time series solution $\{\mathbf{Y}_t\}_{t\in\mathbb{Z}}$ of the stochastic difference equation

$$\sum_{j=0}^{p} \boldsymbol{\alpha}_j \mathbf{Y}_{t-j} = \sum_{\ell=0}^{q} \boldsymbol{\beta}_\ell \mathbf{U}_{t-\ell}, \tag{4.85}$$

where each $\mathbf{Y}_t$ is a $\mathbb{C}^n$-valued random vector, $\{\mathbf{U}_t\}_{t\in\mathbb{Z}}$ is the driving $\mathbb{C}^m$-valued WN($\boldsymbol{\Sigma}$) sequence,

$$\mathbb{E}\mathbf{U}_t = 0, \qquad \mathbb{E}(\mathbf{U}_t\mathbf{U}_s^*) = \delta_{ts}\boldsymbol{\Sigma}, \qquad \forall s, t \in \mathbb{Z},$$

$\boldsymbol{\Sigma} \in \mathbb{C}^{m\times m}$ is a given self-adjoint non-negative definite matrix. The coefficients $\boldsymbol{\alpha}_j \in \mathbb{C}^{n\times n}$ and $\boldsymbol{\beta}_\ell \in \mathbb{C}^{n\times m}$ are given matrices. We always assume that $\boldsymbol{\alpha}_0 = \boldsymbol{I}_n$ and $\boldsymbol{\beta}_0 \neq 0_{n\times m}$.

We also assume that $\mathbb{E}\mathbf{Y}_t = 0$ for each t and the $\{\mathbf{Y}_t\}$ process has no remote past:

$$\mathbf{Y}_t \in \overline{\text{span}}\{\mathbf{U}_t, \mathbf{U}_{t-1}, \dots\}, \quad \forall t \in \mathbb{Z},$$

that is, $\mathbf{Y}_t$ depends only on the present and past values of the driving process. It implies that each $\mathbf{U}_t$ is orthogonal to the past $(\mathbf{Y}_{t-1}, \mathbf{Y}_{t-2}, \dots)$:

$$\mathbb{E}(\mathbf{U}_t\mathbf{Y}_s^*) = 0, \quad \forall\, s < t.$$

As in the 1D case, we introduce polynomial matrices, the *VAR polynomial* and *VMA polynomial* of the VARMA process:

$$\boldsymbol{\alpha}(z) := \sum_{j=0}^{p} \boldsymbol{\alpha}_j z^j, \quad \boldsymbol{\beta}(z) := \sum_{j=0}^{q} \boldsymbol{\beta}_j z^j, \quad z \in \mathbb{C}.$$

From now on, the complex variable z corresponds to the left (backward) shift operator L; with this convention (4.85) can be written as

$$\boldsymbol{\alpha}(L)\mathbf{Y}_t = \boldsymbol{\beta}(L)\mathbf{U}_t \quad \text{or} \quad \boldsymbol{\alpha}(z)\mathbf{Y}_t = \boldsymbol{\beta}(z)\mathbf{U}_t, \quad t \in \mathbb{Z}. \tag{4.86}$$

The special case when $p = 0$ is called VMA(q) process and when $q = 0$ is a VAR(p) process.

We can realize any multidimensional ARMA process as a stochastic linear system. The correspondence is far from unique, but a simple, though highly redundant realization is as follows.

Proposition 4.2. *First introduce a random $\mathbb{C}^{(np+mq)}$-valued state vector that contains all necessary history of $\{\mathbf{Y}_t\}$ and $\{\mathbf{U}_t\}$:*

$$\mathbf{X}_t := (\mathbf{Y}_{t-p+1}, \mathbf{Y}_{t-p+2}, \dots, \mathbf{Y}_t; \mathbf{U}_{t-q}, \mathbf{U}_{t-q+1}, \dots, \mathbf{U}_{t-1}).$$

Then the system (3.35) *with the following matrices $A \in \mathbb{C}^{(np+mq)\times(np+mq)}$, $B \in \mathbb{C}^{(np+mq)\times m}$, $C \in \mathbb{C}^{n\times(np+mq)}$, $D \in \mathbb{C}^{n\times m}$ is a realization of the*

ARMA(p, q) process:

$$A := \begin{bmatrix} 0_n & \boldsymbol{I}_n & \cdots & 0_n & 0_{n\times m} & 0_{n\times m} & \cdots & 0_{n\times m} \\ \vdots & \vdots & \ddots & \vdots & \vdots & \vdots & \ddots & \vdots \\ 0_n & 0_n & \cdots & \boldsymbol{I}_n & 0_{n\times m} & 0_{n\times m} & \cdots & 0_{n\times m} \\ -\boldsymbol{\alpha}_p & -\boldsymbol{\alpha}_{p-1} & \cdots & -\boldsymbol{\alpha}_1 & \boldsymbol{\beta}_q & \boldsymbol{\beta}_{q-1} & \cdots & \boldsymbol{\beta}_1 \\ 0_{m\times n} & 0_{m\times n} & \cdots & 0_{m\times n} & 0_m & \boldsymbol{I}_m & \cdots & 0_m \\ \vdots & \vdots & \ddots & \vdots & \vdots & \vdots & \ddots & \vdots \\ 0_{m\times n} & 0_{m\times n} & \cdots & 0_{m\times n} & 0_m & 0_m & \cdots & \boldsymbol{I}_m \\ 0_{m\times n} & 0_{m\times n} & \cdots & 0_{m\times n} & 0_m & 0_m & \cdots & 0_m \end{bmatrix},$$

$$B := [0_{n\times m}, \ldots, 0_{n\times m}, \boldsymbol{\beta}_0, 0_m, \ldots, 0_m, \boldsymbol{I}_m]^T,$$
$$C := [0_n, \cdots, 0_n, \boldsymbol{I}_n, 0_{n\times m}, \cdots, 0_{n\times m}],$$
$$D := 0_{n\times m},$$

and $\mathbf{V}_t := \mathbf{U}_t$.

The proof is obvious, so omitted.

The next theorem gives the *standard observable realization of a VARMA process*. The dimension of the state space will be significantly smaller in general than in Proposition 4.2 above. The stability condition below is a direct generalization of the condition on the zeros of AR polynomial in the 1D case.

Theorem 4.7. *Consider the VARMA(p, q) process defined in* (4.85) *with* $\boldsymbol{\alpha}_0 = \boldsymbol{I}_n$ *and* $\boldsymbol{\beta}_0 \neq 0_{n\times m}$. *Assume the* stability condition

$$\det(\boldsymbol{\alpha}(z)) \neq 0 \quad \text{if} \quad |z| \leq 1. \tag{4.87}$$

Then this process can be realized by a parsimonious, stable observable stochastic linear system (3.35) *with an* (np)*-dimensional state space* X*, where* $\mathbf{V}_t = \mathbf{U}_t$ *for all* $t \in \mathbb{Z}$*,*

$$A := \begin{bmatrix} 0_n & \boldsymbol{I}_n & 0_n & \cdots & 0_n \\ 0_n & 0_n & \boldsymbol{I}_n & \cdots & 0_n \\ \vdots & \vdots & \vdots & \ddots & \vdots \\ 0_n & 0_n & 0_n & \cdots & \boldsymbol{I}_n \\ -\boldsymbol{\alpha}_p & -\boldsymbol{\alpha}_{p-1} & -\boldsymbol{\alpha}_{p-2} & \cdots & -\boldsymbol{\alpha}_1 \end{bmatrix}_{np\times np} \qquad B := \begin{bmatrix} \boldsymbol{H}_1 \\ \boldsymbol{H}_2 \\ \vdots \\ \boldsymbol{H}_p \end{bmatrix}_{np\times n}$$

$$C := [\boldsymbol{I}_n, 0_n, \cdots, 0_n]_{n\times np} \qquad D := \boldsymbol{\beta}_0. \tag{4.88}$$

Here $\boldsymbol{H}_1, \ldots, \boldsymbol{H}_p \in \mathbb{C}^{n\times m}$ *are coefficient matrices of the transfer function*

$$\boldsymbol{H}(z) = \sum_{j=0}^{\infty} \boldsymbol{H}_j z^j := \boldsymbol{\alpha}^{-1}(z)\boldsymbol{\beta}(z), \quad \boldsymbol{H}_0 := \boldsymbol{\beta}_0. \tag{4.89}$$

Consequently, the considered VARMA process has a unique causal stationary MA(∞) *solution*

$$\mathbf{Y}_t = \sum_{j=0}^{\infty} \boldsymbol{H}_j \mathbf{U}_{t-j} = D\mathbf{U}_t + \sum_{j=1}^{\infty} CA^{j-1}B\mathbf{U}_{t-j} \quad (t \in \mathbb{Z}), \tag{4.90}$$

which converges almost surely and also in mean square:

$$\lim_{N\to\infty} \left\| \mathbf{Y}_t - \sum_{j=0}^{N} \boldsymbol{H}_j \mathbf{U}_{t-j} \right\|^2 = 0.$$

Proof. Without loss of generality we may suppose that $p \geq q$, because if $p < q$, we may add new coefficients $\boldsymbol{\alpha}_{p+1} = \cdots = \boldsymbol{\alpha}_q = 0_n$.

From now on in this proof we assume that $\det(\boldsymbol{\alpha}(z)) \neq 0$ if $|z| \leq 1$. Since $\det(\boldsymbol{\alpha}(z))$ is a continuous function of z, there exists an $\epsilon > 0$ such that the rational matrix

$$\boldsymbol{\alpha}^{-1}(z) = \operatorname{adj}(\boldsymbol{\alpha}(z))/\det(\boldsymbol{\alpha}(z)) \tag{4.91}$$

is defined and analytic for $|z| < 1+\epsilon$. Thus we can define the transfer function (a rational matrix) as

$$\boldsymbol{H}(z) := \boldsymbol{\alpha}^{-1}(z)\boldsymbol{\beta}(z) = \sum_{j=0}^{\infty} \boldsymbol{H}_j z^j \quad (\boldsymbol{H}_j \in \mathbb{C}^{n\times m}), \tag{4.92}$$

which is a convergent power series for $|z| < 1 + \epsilon$.

Using (4.92), for $|z| < 1 + \epsilon$, we see that

$$\left(\sum_{k=0}^{p} \boldsymbol{\alpha}_k z^k\right)\left(\sum_{j=0}^{\infty} \boldsymbol{H}_j z^j\right) = \sum_{\ell=0}^{q} \boldsymbol{\beta}_\ell z^\ell,$$

so

$$\boldsymbol{\beta}_\ell = \sum_{k=0}^{\ell} \boldsymbol{\alpha}_k \boldsymbol{H}_{\ell-k} \quad (\ell = 0, 1, \ldots, q). \tag{4.93}$$

It is clear that $\boldsymbol{H}_0 = \boldsymbol{\beta}_0 = D$ by the definition of D in (4.88).

Consider the system of linear block equations

$$\begin{bmatrix} \boldsymbol{W}_1 \\ \vdots \\ \boldsymbol{W}_p \end{bmatrix} = zA \begin{bmatrix} \boldsymbol{W}_1 \\ \vdots \\ \boldsymbol{W}_p \end{bmatrix} + \begin{bmatrix} \boldsymbol{H}_1 \\ \vdots \\ \boldsymbol{H}_p \end{bmatrix}, \tag{4.94}$$

where $\boldsymbol{W}_1, \ldots, \boldsymbol{W}_p \in \mathbb{C}^{n\times m}$ are unknowns and the matrix A is defined by

(4.88). That is,

$$\begin{aligned}
\boldsymbol{W}_1 &= z\boldsymbol{W}_2 + \boldsymbol{H}_1 \\
\boldsymbol{W}_2 &= z\boldsymbol{W}_3 + \boldsymbol{H}_2 \\
&\vdots \\
\boldsymbol{W}_{p-1} &= z\boldsymbol{W}_p + \boldsymbol{H}_{p-1} \\
\boldsymbol{W}_p &= -\sum_{j=1}^{p} \boldsymbol{\alpha}_{p-j+1} z\boldsymbol{W}_j + \boldsymbol{H}_p. \qquad (4.95)
\end{aligned}$$

Substitute the first equation for $\boldsymbol{W}_1$ into the last equation:

$$\boldsymbol{W}_p = (-\boldsymbol{\alpha}_p z^2 - \boldsymbol{\alpha}_{p-1} z)\boldsymbol{W}_2 - \boldsymbol{\alpha}_{p-2}\boldsymbol{W}_3 - \cdots - \boldsymbol{\alpha}_1 \boldsymbol{W}_p + \boldsymbol{H}_p - (\boldsymbol{\alpha}_p z)\boldsymbol{H}_1.$$

Then do the same with the second equation for $\boldsymbol{W}_2$, the third equation for $\boldsymbol{W}_3$, and so on, to obtain

$$\begin{aligned}
\boldsymbol{W}_p = (-\boldsymbol{\alpha}_p z^p - \boldsymbol{\alpha}_{p-1} z^{p-1} - \cdots - \boldsymbol{\alpha}_1 z)\boldsymbol{W}_p + \boldsymbol{H}_p - (\boldsymbol{\alpha}_p z)\boldsymbol{H}_1 \\
- (\boldsymbol{\alpha}_p z^2 + \boldsymbol{\alpha}_{p-1} z)\boldsymbol{H}_2 - \cdots - (\boldsymbol{\alpha}_p z^{p-1} + \boldsymbol{\alpha}_{p-1} z^{p-2} + \cdots + \boldsymbol{\alpha}_2 z)\boldsymbol{H}_{p-1}.
\end{aligned}$$

Rearranging the terms, using $\boldsymbol{\alpha}_0 = \boldsymbol{I}_n$, we get

$$\begin{aligned}
\boldsymbol{\alpha}(z)\boldsymbol{W}_p &= -z^{p-1}(\boldsymbol{\alpha}_p \boldsymbol{H}_{p-1}) - z^{p-2}(\boldsymbol{\alpha}_p \boldsymbol{H}_{p-2} + \boldsymbol{\alpha}_{p-1}\boldsymbol{H}_{p-1}) - \cdots \\
&\quad - z(\boldsymbol{\alpha}_p \boldsymbol{H}_1 + \boldsymbol{\alpha}_{p-1}\boldsymbol{H}_2 + \cdots + \boldsymbol{\alpha}_2 \boldsymbol{H}_{p-1}) + \boldsymbol{H}_p.
\end{aligned}$$

Then substitute this into the penultimate equation of (4.95), and so on, to eventually obtain

$$\begin{aligned}
\boldsymbol{\alpha}(z)\boldsymbol{W}_1 &= z^{2p-2}(\boldsymbol{\alpha}_p \boldsymbol{H}_{p-1}) - z^{2p-3}(\boldsymbol{\alpha}_p \boldsymbol{H}_{p-2} + \boldsymbol{\alpha}_{p-1}\boldsymbol{H}_{p-1}) - \cdots \\
&\quad - z^p(\boldsymbol{\alpha}_p \boldsymbol{H}_1 + \boldsymbol{\alpha}_{p-1}\boldsymbol{H}_2 + \cdots + \boldsymbol{\alpha}_2 \boldsymbol{H}_{p-1}) + z^{p-1}\boldsymbol{H}_p \\
&\quad + \boldsymbol{\alpha}(z)(z^{p-2}\boldsymbol{H}_{p-1} + \cdots + z\boldsymbol{H}_2 + \boldsymbol{H}_1).
\end{aligned}$$

Fortunately, all terms with powers z^k $(k \geq p)$ cancel on the right hand side, so by (4.93), for $|z| < 1 + \epsilon$ we get that

$$\begin{aligned}
\boldsymbol{\alpha}(z)D + z\boldsymbol{\alpha}(z)\boldsymbol{W}_1 &= z^p(\boldsymbol{\alpha}_0 \boldsymbol{H}_p + \boldsymbol{\alpha}_1 \boldsymbol{H}_{p-1} + \cdots + \boldsymbol{\alpha}_p \boldsymbol{H}_0) + \cdots \\
&\quad + z(\boldsymbol{\alpha}_0 \boldsymbol{H}_1 + \boldsymbol{\alpha}_1 \boldsymbol{H}_0) + (\boldsymbol{\alpha}_0 \boldsymbol{H}_0) = \boldsymbol{\beta}(z). \qquad (4.96)
\end{aligned}$$

This completes the first part of the proof.

Now, let us continue with the definition of the matrices A, B, C, D given in (4.88). By Lemma 4.9 below, $\det(\boldsymbol{I}_n - zA) = z^{np}\det(z^{-1}\boldsymbol{I}_n - A) \neq 0$ for $|z| \leq 1$. It implies that all eigenvalues of the matrix A are in the open unit disc $\{z : |z| < 1\}$, that is, $\rho(A) < 1$. Thus by Theorem 3.5, the linear system defined in (4.88) has a unique causal stationary MA(∞) solution

$$\mathbf{Y}_t = D\mathbf{U}_t + \sum_{j=1}^{\infty} CA^{j-1}B\mathbf{U}_{t-j} \quad (t \in \mathbb{Z}), \qquad (4.97)$$

and this series converges with probability 1 and in mean square. Here $\{\mathbf{U}_t\} \sim \mathrm{WN}(\mathbf{\Sigma})$ is the same process that appears in the definition (4.85).

By induction over j it is easy to check that

$$CA^{j-1} = [0_n, \ldots, 0_n, \boldsymbol{I}_n, 0_n, \ldots, 0_n]_{n \times np} \quad (j = 2, \ldots, p), \tag{4.98}$$

where the block $\boldsymbol{I}_n$ is at the jth position; all other blocks are 0_n. Thus

$$CA^{j-1}B = \boldsymbol{H}_j \quad (j = 1, \ldots, p).$$

On the other hand, by (4.94) we obtain that

$$\begin{bmatrix} \boldsymbol{W}_1 \\ \vdots \\ \boldsymbol{W}_p \end{bmatrix} = (\boldsymbol{I}_n - Az)^{-1} \begin{bmatrix} \boldsymbol{H}_1 \\ \vdots \\ \boldsymbol{H}_p \end{bmatrix} = (\boldsymbol{I}_n - Az)^{-1}B = \sum_{j=0}^{\infty} A^j B z^j,$$

$$\boldsymbol{W}_1 = C(\boldsymbol{I}_n - Az)^{-1}B = \sum_{j=0}^{\infty} CA^j B z^j, \tag{4.99}$$

if $|z| < 1 + \epsilon$. Comparing equations (4.96) and (4.99), it follows for the linear system defined in (4.88) that its transfer function is

$$\boldsymbol{H}(z) = \boldsymbol{\alpha}^{-1}(z)\boldsymbol{\beta}(z) = D + zC(\boldsymbol{I}_n - Az)^{-1}B = D + \sum_{j=1}^{\infty} CA^{j-1}Bz^j,$$

if $|z| < 1 + \epsilon$. This and (4.97) imply that the solution of the linear system has the form (4.90). This proves that the linear system (4.88) is a realization of the given VARMA process.

Taking the output sequence $\mathbf{\Psi}_t := \sum_{j=0}^{p} \boldsymbol{\alpha}_j \mathbf{Y}_{t-j}$ $(t \in \mathbb{Z})$, it has the transfer function $\boldsymbol{\alpha}(z)\boldsymbol{H}(z) = \boldsymbol{\beta}(z)$, which means that the sequence $\{\mathbf{Y}_t\}$ defined by (4.90) is the unique solution of the considered VARMA process, see (4.86).

By (4.98),

$$\mathrm{rank}(\mathcal{O}_p) := \mathrm{rank} \begin{bmatrix} C \\ CA \\ \vdots \\ CA^{p-1} \end{bmatrix} = \mathrm{rank}(I_{np}) = np = \dim(X) = \mathrm{rank}(\mathcal{O}),$$

so the linear system in (4.88) is indeed observable, see Section 3.3. $\square$

Lemma 4.9. *If the block-matrix A is defined by* (4.88) *and the VAR polynomial $\boldsymbol{\alpha}(z)$ is defined by* (4.91), *then*

$$\det(\boldsymbol{I}_n - zA) = \det(\boldsymbol{\alpha}(z)).$$

Proof.

$$\det(\boldsymbol{I}_n - zA) = \begin{vmatrix} \boldsymbol{I}_n & -z\boldsymbol{I}_n & 0_n & \cdots & 0_n & 0_n \\ 0_n & \boldsymbol{I}_n & -z\boldsymbol{I}_n & \cdots & 0_n & 0_n \\ \vdots & \vdots & \vdots & \ddots & \vdots & \vdots \\ 0_n & 0_n & 0_n & \cdots & \boldsymbol{I}_n & -z\boldsymbol{I}_n \\ z\boldsymbol{\alpha}_p & z\boldsymbol{\alpha}_{p-1} & z\boldsymbol{\alpha}_{p-2} & \cdots & z\boldsymbol{\alpha}_2 & \boldsymbol{I}_n + z\boldsymbol{\alpha}_1 \end{vmatrix}$$

Add z times the first block-column to the second, then z times the second to the third, etc., lastly z times the $(p-1)$th block-column to the pth. Finally, eliminate the non-diagonal blocks of the last block-row by using all the other block-rows:

$$\det(\boldsymbol{I}_n - zA) = \begin{vmatrix} \boldsymbol{I}_n & 0_n & \cdots & 0_n & 0_n \\ 0_n & \boldsymbol{I}_n & \cdots & 0_n & 0_n \\ \vdots & \vdots & \ddots & \vdots & \vdots \\ 0_n & 0_n & \cdots & \boldsymbol{I}_n & 0_n \\ z\boldsymbol{\alpha}_p & z^2\boldsymbol{\alpha}_p + z\boldsymbol{\alpha}_{p-1} & \cdots & z^{p-1}\boldsymbol{\alpha}_p + \cdots + z\boldsymbol{\alpha}_2 & \boldsymbol{\alpha}(z) \end{vmatrix}$$

$$= \begin{vmatrix} \boldsymbol{I}_n & 0_n & \cdots & 0_n & 0_n \\ 0_n & \boldsymbol{I}_n & \cdots & 0_n & 0_n \\ \vdots & \vdots & \ddots & \vdots & \vdots \\ 0_n & 0_n & \cdots & \boldsymbol{I}_n & 0_n \\ 0_n & 0_n & \cdots & 0_n & \boldsymbol{\alpha}(z) \end{vmatrix} = \det(\boldsymbol{\alpha}(z)).$$

□

By (4.5), (1.30), and (4.89), we get the *spectral density function* of a stable VARMA(p, q) process:

$$\boldsymbol{f}_Y(\omega) = \frac{1}{2\pi}\boldsymbol{H}(e^{-i\omega})\boldsymbol{\Sigma}\boldsymbol{H}(e^{-i\omega})^* = \frac{1}{2\pi}\sum_{\ell=-\infty}^{\infty} e^{-i\ell\omega} \sum_{k=\max(0,-\ell)}^{\infty} \boldsymbol{H}_{\ell+k}\boldsymbol{\Sigma}\boldsymbol{H}_k^*$$
$$= \frac{1}{2\pi}\boldsymbol{\alpha}^{-1}(e^{-i\omega})\boldsymbol{\beta}(e^{-i\omega})\boldsymbol{\Sigma}\boldsymbol{\beta}(e^{-i\omega})^*\boldsymbol{\alpha}^{-1}(e^{-i\omega})^*. \tag{4.100}$$

The next theorem summarizes results above and in Appendix D, showing equivalence of different approaches that describe weakly stationary time series determined by finitely many scalar parameters.

Theorem 4.8. *Let $\boldsymbol{H}(z) = \sum_{j=0}^{\infty} \boldsymbol{H}_j z^j$, $\boldsymbol{H}_j \in \mathbb{C}^{n\times n}$, $\boldsymbol{H}_0 = \boldsymbol{I}_n$, be a rational transfer function of a linear system, analytic in the closed unit disc $\bar{D}$. By Theorem D.1 in the Appendix, there exist left coprime polynomial matrices $\boldsymbol{\alpha}(z), \boldsymbol{\beta}(z) \in \mathbb{C}^{n\times n}[z]$ such that $\boldsymbol{H}(z)$ can be represented by a matrix fraction description (MFD)*

$$\boldsymbol{H}(z) = \boldsymbol{\alpha}^{-1}(z)\boldsymbol{\beta}(z),$$

and so the linear system can be represented as a VARMA(p, q) process as in

Definition 4.1. Since $\boldsymbol{\alpha}^{-1}(z) = \mathrm{adj}\,\boldsymbol{\alpha}(z)/\det\boldsymbol{\alpha}(z)$, *and* $\boldsymbol{H}(z)$ *is analytic in* $\bar{D}$, *it follows that* $\det(\boldsymbol{\alpha}(z)) \neq 0$ *for* $|z| \leq 1$. *Thus Theorem 4.7 implies that this matrix fraction can be realized as a stable observable stochastic linear system with coefficient matrices* $\Sigma = (A, B, C, D)$ *given by* (4.88). *Also, there exists a unique causal stationary* MA(∞) *solution of the VARMA(p, q):*

$$\mathbf{Y}_t = \sum_{k=0}^{\infty} \boldsymbol{H}_k \mathbf{U}_{t-k} \qquad (t \in \mathbb{Z}), \qquad \sum_{j=0}^{\infty} \|\boldsymbol{H}_j\|_F^2 < \infty,$$

which converges with probability 1 and in mean square, where $\{\mathbf{U}_t\} \sim$ *WN(*$\boldsymbol{I}_m$*).*

It is clear from the proof of Theorem D.1 that if $\boldsymbol{H}(z) = \boldsymbol{P}(z)/\psi(z)$, *where* $\boldsymbol{P}(z)$ *is a polynomial matrix and* $\psi(z)$ *is a polynomial, then* $\boldsymbol{\alpha}(z)$ *is obtained by simplifying with the greatest common left divisor of* $\boldsymbol{P}(z)$ *and* $\psi(z)\boldsymbol{I}_n$, *so p in VARMA(p, q) can be much smaller than the degree of* $\psi(z)$. *It means that the above realization is in this sense optimal.*

By (4.100), *any stable VARMA(p, q) process has spectral density* $\boldsymbol{f}$ *which is a rational function in* $z = e^{-i\omega}$.

In turn, by Corollaries 4.2 and 4.7 and Remark 4.7, if $\{\mathbf{X}_t\}$ *is a stationary time series with absolutely continuous spectral measure of density* $\boldsymbol{f}$ *which is a rational function in* $z = e^{-i\omega}$, *then* $\boldsymbol{f}(\omega) = \frac{1}{2\pi}\boldsymbol{\Phi}(e^{-i\omega})\boldsymbol{\Phi}^*(e^{-i\omega})$, *where the spectral factor* $\boldsymbol{\Phi}(z)$ *is a rational matrix, analytic in D, belongs to* H^2, *and has no poles on T. Thus defining the transfer function* $\boldsymbol{H}(z) = \boldsymbol{\Phi}(z)$, *the first paragraph of the present theorem shows that* $\{\mathbf{X}_t\}$ *can be given as a stable VARMA(p, q) process. By* (4.100), *its spectral density is the given* $\boldsymbol{f}$, *when* $\{\mathbf{U}_t\} \sim$ *WN(*$\boldsymbol{I}_m$*).*

4.8.2 Yule–Walker equations

In this subsection we always assume that the conditions of Theorem 4.7 are fulfilled and w.l.o.g. we suppose that $q < p$. Then there exists a unique causal stationary MA(∞) solution of the VARMA(p, q):

$$\mathbf{Y}_t = \sum_{k=0}^{\infty} \boldsymbol{H}_k \mathbf{U}_{t-k}, \qquad t \in \mathbb{Z}, \tag{4.101}$$

where the sum converges almost surely and in mean square.

Then it is easy to write the covariance function in terms of the transfer function $\boldsymbol{H}(z) = \sum_{j=0}^{\infty} \boldsymbol{H}_j z^j$ (which is convergent for $|z| \leq 1$):

$$\begin{aligned} \boldsymbol{C}_Y(k) &= \mathbb{E}(\mathbf{Y}_{t+k}\mathbf{Y}_t^*) = \mathbb{E}\left\{\left(\sum_{\ell=0}^{\infty} \boldsymbol{H}_\ell \mathbf{U}_{t+k-\ell}\right)\left(\sum_{j=0}^{\infty} \mathbf{U}_{t-j}^* \boldsymbol{H}_j^*\right)\right\} \\ &= \sum_{j=0}^{\infty} \boldsymbol{H}_{j+k} \boldsymbol{\Sigma} \boldsymbol{H}_j^*, \qquad \boldsymbol{C}_Y(-k) = \boldsymbol{C}_Y^*(k), \qquad k \geq 0. \end{aligned} \tag{4.102}$$

The important *Yule–Walker equations* can be obtained from definition (4.85) by multiplying it with $\mathbf{Y}^*_{t-k}$ and taking expectation:

$$\sum_{j=0}^{p} \boldsymbol{\alpha}_j \mathbb{E}(\mathbf{Y}_{t-j}\mathbf{Y}^*_{t-k}) = \sum_{\ell=0}^{q} \boldsymbol{\beta}_\ell \mathbb{E}(\mathbf{U}_{t-\ell}\mathbf{Y}^*_{t-k}),$$

$$\sum_{j=0}^{p} \boldsymbol{\alpha}_j \boldsymbol{C}_Y(k-j) = \sum_{\ell=k}^{q} \boldsymbol{\beta}_\ell \boldsymbol{\Sigma} \boldsymbol{H}^*_{\ell-k} \quad (k \geq 0). \tag{4.103}$$

The right hand side of (4.103) is $\boldsymbol{O}$ if $k > q$.

It is easy to extend the discussion of Yule–Walker equations in the 1D case in Section 2.4 to the present multi-D case. Substitute (4.101) into definition (4.85):

$$\sum_{k=0}^{\infty}\sum_{j=0}^{p} \boldsymbol{\alpha}_j \boldsymbol{H}_k \mathbf{U}_{t-j-k} = \sum_{\ell=0}^{q} \boldsymbol{\beta}_\ell \mathbf{U}_{t-\ell}.$$

Equate the coefficients on both sides, starting with $\mathbf{U}_t$ and working toward the past ($\boldsymbol{\alpha}_0 = \boldsymbol{I}_n$):

$$\begin{aligned}
\boldsymbol{H}_0 &= \boldsymbol{\beta}_0, \\
\boldsymbol{H}_1 + \boldsymbol{\alpha}_1 \boldsymbol{H}_0 &= \boldsymbol{\beta}_1, \\
\boldsymbol{H}_2 + \boldsymbol{\alpha}_1 \boldsymbol{H}_1 + \boldsymbol{\alpha}_2 \boldsymbol{H}_0 &= \boldsymbol{\beta}_2, \\
&\vdots \\
\boldsymbol{H}_{p-1} + \boldsymbol{\alpha}_1 \boldsymbol{H}_{p-2} + \cdots + \boldsymbol{\alpha}_{p-1} \boldsymbol{H}_0 &= \boldsymbol{\beta}_{p-1}, \\
\boldsymbol{H}_{p+k} + \boldsymbol{\alpha}_1 \boldsymbol{H}_{p+k-1} + \cdots + \boldsymbol{\alpha}_p \boldsymbol{H}_k &= 0 \qquad (k \geq 0),
\end{aligned} \tag{4.104}$$

where $\boldsymbol{\beta}_\ell = 0$ if $\ell > q$. If $\{\boldsymbol{\alpha}_j : j = 1, \ldots, p\}$ and $\{\boldsymbol{\beta}_\ell : \ell = 0, \ldots, q\}$ are known, these equations uniquely determine the coefficients $\{\boldsymbol{H}_k : k \geq 0\}$ by recursion.

From now on we omit the subscript Y from the covariances, since all of the covariances here are related to the process $\{\mathbf{Y}_t\}$. From (4.103) we obtain the following system of block equations

$$\begin{aligned}
&\begin{bmatrix} \boldsymbol{I}_n & \boldsymbol{\alpha}_1 & \cdots & \boldsymbol{\alpha}_p \end{bmatrix}
\begin{bmatrix}
\boldsymbol{C}(0) & \boldsymbol{C}(1) & \cdots & \boldsymbol{C}(p) \\
\boldsymbol{C}(-1) & \boldsymbol{C}(0) & \cdots & \boldsymbol{C}(p-1) \\
\vdots & \vdots & \ddots & \vdots \\
\boldsymbol{C}(-p) & \boldsymbol{C}(-p+1) & \cdots & \boldsymbol{C}(0)
\end{bmatrix} \\
&= \begin{bmatrix} \sum_{\ell=0}^{q} \boldsymbol{\beta}_\ell \boldsymbol{\Sigma} \boldsymbol{H}^*_\ell & \sum_{\ell=1}^{q} \boldsymbol{\beta}_\ell \boldsymbol{\Sigma} \boldsymbol{H}^*_{\ell-1} & \cdots & \boldsymbol{\beta}_q \boldsymbol{\Sigma} \boldsymbol{H}^*_0 & \boldsymbol{O} & \cdots & \boldsymbol{O} \end{bmatrix}.
\end{aligned} \tag{4.105}$$

In the special case when the process is VAR(p), that is, $q = 0$, this can be rearranged into a system of linear block equations for the unknowns $\boldsymbol{\alpha}_1, \ldots, \boldsymbol{\alpha}_p$

if the covariance matrix function $\boldsymbol{C}(h)$, $(h = 0, 1, \ldots, p)$ is known:

$$\begin{bmatrix} \boldsymbol{\alpha}_1 & \cdots & \boldsymbol{\alpha}_p \end{bmatrix} \begin{bmatrix} \boldsymbol{C}(0) & \boldsymbol{C}(1) & \cdots & \boldsymbol{C}(p-1) \\ \boldsymbol{C}(-1) & \boldsymbol{C}(0) & \cdots & \boldsymbol{C}(p-2) \\ \vdots & \ddots & \vdots & \\ \boldsymbol{C}(-p+1) & \boldsymbol{C}(-p+2) & \cdots & \boldsymbol{C}(0) \end{bmatrix}$$
$$= - \begin{bmatrix} \boldsymbol{C}(1) & \cdots & \boldsymbol{C}(p) \end{bmatrix}, \tag{4.106}$$

and also,

$$\sum_{j=0}^{p} \boldsymbol{\alpha}_j \boldsymbol{C}(j) = \boldsymbol{\beta}_0 \boldsymbol{\Sigma} \boldsymbol{\beta}_0^*,$$

since $\boldsymbol{H}_0 = \boldsymbol{\beta}_0$. Unfortunately, if $q > 0$, so when one really has an ARMA process, on the right hand side of (4.105) polynomial expressions of the unknown parameters would be obtained from (4.104).

Introducing the notations $\boldsymbol{\alpha} := [\boldsymbol{\alpha}_1, \ldots, \boldsymbol{\alpha}_p]$, $\boldsymbol{c} := [\boldsymbol{C}(1), \ldots, \boldsymbol{C}(p)]$, and

$$\mathfrak{C} := \begin{bmatrix} \boldsymbol{C}(0) & \boldsymbol{C}(1) & \cdots & \boldsymbol{C}(p-1) \\ \boldsymbol{C}(-1) & \boldsymbol{C}(0) & \cdots & \boldsymbol{C}(p-2) \\ \vdots & \ddots & \vdots & \\ \boldsymbol{C}(-p+1) & \boldsymbol{C}(-p+2) & \cdots & \boldsymbol{C}(0) \end{bmatrix},$$

we may write (4.106) as

$$\boldsymbol{\alpha} \mathfrak{C} = -\boldsymbol{c}. \tag{4.107}$$

If $\mathfrak{C}$ is positive definite, then there exists a unique solution: $\boldsymbol{\alpha} = -\boldsymbol{c}\mathfrak{C}^{-1}$. Otherwise, one may use the Moore–Penrose inverse $\mathfrak{C}^+$ instead of the inverse matrix, see Definition B.4 in Appendix B. The system (4.107) is always consistent, always has a solution, because it is a *Gauss normal equation*, see Appendix C. For, taking $\mathbf{Y} := [\mathbf{Y}_{p-1}^* \mathbf{Y}_{p-2}^* \ldots \mathbf{Y}_0^*]^*$, we have

$$\mathfrak{C} = \mathbb{E}(\mathbf{Y}\mathbf{Y}^*), \quad \boldsymbol{c} = \mathbb{E}(\mathbf{Y}_p \mathbf{Y}^*),$$

and this shows that the right hand side of (4.107) is in the row space of the left hand side.

Assuming that $n = m$, $\text{rank}(\boldsymbol{H}_0) = \text{rank}(\boldsymbol{\beta}_0) = n$, and $\text{rank}(\boldsymbol{\Sigma}) = n$, introduce the $\infty \times p$ upper trapezoidal block Toeplitz matrix

$$\mathfrak{H} := \begin{bmatrix} \boldsymbol{H}_0 & \boldsymbol{H}_1 & \cdots & \boldsymbol{H}_{p-1} & \boldsymbol{H}_p & \boldsymbol{H}_{p+1} & \cdots \\ 0 & \boldsymbol{H}_0 & \cdots & \boldsymbol{H}_{p-2} & \boldsymbol{H}_{p-1} & \boldsymbol{H}_p & \cdots \\ \vdots & \vdots & \ddots & \vdots & \vdots & \vdots & \ddots \\ 0 & 0 & \cdots & \boldsymbol{H}_0 & \boldsymbol{H}_1 & \boldsymbol{H}_2 & \cdots \end{bmatrix} \tag{4.108}$$

and the infinite block row matrix $\boldsymbol{h} := [\boldsymbol{H}_1, \boldsymbol{H}_2, \boldsymbol{H}_3, \cdots]$. Then the infinite system of linear equations (4.104) can be written as

$$\boldsymbol{\alpha} \mathfrak{H} = -\boldsymbol{h},$$

and in the light of (4.102), equation (4.107) is equivalent to

$$\boldsymbol{\alpha}\mathfrak{H}(\boldsymbol{\Sigma}\mathfrak{H}^*) = -\boldsymbol{h}(\boldsymbol{\Sigma}\mathfrak{H}^*), \quad \mathfrak{C} = \mathfrak{H}(\boldsymbol{\Sigma}\mathfrak{H}^*), \quad \boldsymbol{c} = \boldsymbol{h}(\boldsymbol{\Sigma}\mathfrak{H}^*), \tag{4.109}$$

where $(\boldsymbol{\Sigma}\mathfrak{H}^*)$ denotes block multiple of a block matrix:

$$(\boldsymbol{\Sigma}\mathfrak{H}^*) := \begin{bmatrix} \boldsymbol{\Sigma H}_0 & 0 & 0 & 0 \\ \boldsymbol{\Sigma H}_1 & \boldsymbol{\Sigma H}_0 & \cdots & 0 \\ \vdots & \vdots & \ddots & \vdots \\ \boldsymbol{\Sigma H}_{p-1} & \boldsymbol{\Sigma H}_{p-2} & \cdots & \boldsymbol{\Sigma H}_0 \\ \boldsymbol{\Sigma H}_p & \boldsymbol{\Sigma H}_{p-1} & \cdots & \boldsymbol{\Sigma H}_1 \\ \vdots & \vdots & \ddots & \vdots \end{bmatrix}.$$

Equation (4.109) gives an infinite block Cholesky-type decomposition of $\mathfrak{C}$. By our assumption $\operatorname{rank}(\boldsymbol{H}_0) = n$, so (4.108) shows that the range of $\mathfrak{H}$ (when multiplying from the left) is pn-dimensional and this pn-dimensional subspace of ℓ^2 is mapped by $(\boldsymbol{\Sigma}\mathfrak{H}^*)$ onto $\mathbb{C}^{pn}$.

Corollary 4.8. *Under the above assumptions,* $\operatorname{rank}(\mathfrak{C}) = pn$*, so the system* (4.107) *has a unique solution* $\boldsymbol{\alpha} = -\boldsymbol{c}\,\mathfrak{C}^{-1}$.

Since the covariances can be easily estimated from a random sample, see Subsection 1.5.2, the resulting estimated version of (4.106) gives a practical method for estimating the coefficients $\boldsymbol{\alpha}_1, \dots, \boldsymbol{\alpha}_p$ of a VAR(p) process.

4.8.3 Prediction, miniphase condition, and approximation by VMA processes

A very useful property of stable VARMA processes

$$\mathbf{Y}_t = -\boldsymbol{\alpha}_1\mathbf{Y}_{t-1} - \cdots - \boldsymbol{\alpha}_p\mathbf{Y}_{t-p} + \boldsymbol{\beta}_0\mathbf{U}_t + \boldsymbol{\beta}_1\mathbf{U}_{t-1} + \cdots + \boldsymbol{\beta}_q\mathbf{U}_{t-q}$$

that in mean square the *best linear one-step ahead prediction* of $\mathbf{Y}_t$ based on the infinite past $H^-_{t-1} = \overline{\operatorname{span}}\{\mathbf{Y}_{t-1}, \mathbf{Y}_{t-2}, \dots\}$ can be evaluated using only *finitely many past values*:

$$\hat{\mathbf{Y}}_t = -\boldsymbol{\alpha}_1\mathbf{Y}_{t-1} - \cdots - \boldsymbol{\alpha}_p\mathbf{Y}_{t-p} + \boldsymbol{\beta}_1\mathbf{U}_{t-1} + \cdots + \boldsymbol{\beta}_q\mathbf{U}_{t-q}. \tag{4.110}$$

The proof follows from the projection theorem: the right hand side here lies in H^-_{t-1} and the error $\mathbf{Y}_t - \hat{\mathbf{Y}}_t = \boldsymbol{\beta}_0\mathbf{U}_t$ is orthogonal to H^-_{t-1}. This finiteness is in contrast to the general case of regular processes, where one needs infinitely many data in principle, see (4.26).

The problem with the prediction (4.110) is that in practice one typically cannot observe the input process $\{\mathbf{U}_t\}$. Lemma 4.9 proves that the stability condition (4.87) is equivalent to the stability condition $\rho(A) < 1$ of the corresponding linear system. Similarly, the *strict miniphase condition*

$$\det(\boldsymbol{\beta}(z)) \neq 0, \quad \forall |z| \leq 1$$

is equivalent to the stability condition $\rho(A - BC) < 1$, see (3.40) and the 'inverse' stochastic linear system discussed in Subsection 3.6.2. Here it is assumed that $m = n$ and $\boldsymbol{\beta}_0 = \boldsymbol{I}_n$ in the definition (4.85). Under the stability condition (4.87) of the original case we had

$$\mathbf{Y}_t = \boldsymbol{H}(z)\mathbf{U}_t = \boldsymbol{\alpha}^{-1}(z)\boldsymbol{\beta}(z)\mathbf{U}_t, \quad t \in \mathbb{Z},$$

so the output could be expressed in terms of the input. Similarly, under the miniphase condition, the input can be expressed in terms of the observable output:

$$\mathbf{U}_t = \boldsymbol{H}^{-1}(z)\mathbf{Y}_t := \boldsymbol{\beta}^{-1}(z)\boldsymbol{\alpha}(z)\mathbf{Y}_t, \quad t \in \mathbb{Z}.$$

The importance and wide applicability of VARMA processes is further emphasized by the next proposition.

Proposition 4.3. *The stable VARMA processes are dense among the regular time series. More exactly, for any regular stationary time series $\{\mathbf{X}_t\}$ and for any $\epsilon > 0$ there exists a positive integer N and a VMA(N) process $\{\mathbf{Y}_t\}$ such that*

$$\|\mathbf{X}_t - \mathbf{Y}_t\| < \epsilon, \quad \forall t \in \mathbb{Z}.$$

Proof. Consider a regular time series

$$\mathbf{X}_t = \sum_{j=0}^{\infty} \boldsymbol{b}(j)\boldsymbol{\xi}_{t-j} \quad (t \in \mathbb{Z}), \quad \boldsymbol{b}(j) = [b_{k\ell}(j)]_{d\times r},$$

$$\sum_{j=0}^{\infty} |b_{k\ell}(j)|^2 < \infty \quad \forall k, \ell, \quad \{\boldsymbol{\xi}_t\}_{t\in\mathbb{Z}} \sim \mathrm{WN}(\boldsymbol{I}_r).$$

Then for any $\epsilon > 0$ and for any k, ℓ fixed, there exists a positive integer $N_{k\ell}$ such that

$$\sum_{j=N_{k\ell}+1}^{\infty} |b_{k\ell}(j)|^2 < \frac{\epsilon^2}{rd}.$$

Take $N := \max\{N_{k\ell} : k = 1, \dots, d; \ell = 1, \dots, r\}$ and define

$$\mathbf{Y}_t := \sum_{j=0}^{N} \boldsymbol{b}(j)\boldsymbol{\xi}_{t-j}, \quad t \in \mathbb{Z}.$$

Then $\{\mathbf{Y}_t\}$ is a VMA(N) process and

$$\|\mathbf{X}_t - \mathbf{Y}_t\|^2 = \sum_{j=N+1}^{\infty} \|\boldsymbol{b}(j)\|^2 < \epsilon^2, \quad \forall t \in \mathbb{Z}.$$

This proves the proposition. □

Example 4.1. Figure 4.1 shows typical trajectories of a 3D VAR(2) process, together with its impulse response functions and covariance functions. Figure 4.2 shows the spectral densities. The parameters are $\boldsymbol{\beta} = \boldsymbol{I}_3$, $\boldsymbol{\alpha}_0 = \boldsymbol{I}_3$,

$$\boldsymbol{\alpha}_1 = \begin{bmatrix} -1 & 0 & 0 \\ 1/2 & -4/5 & 0 \\ -1/2 & 1/3 & -1/6 \end{bmatrix}, \qquad \boldsymbol{\alpha}_2 = \begin{bmatrix} 1/2 & 0 & 0 \\ 1/3 & 1/5 & 0 \\ -1/2 & 1 & -1/6 \end{bmatrix}.$$

The third panel of Figure 4.1 shows that the off-diagonal entries of the matrix $\boldsymbol{H}_k$ are 0. The first panel of Figure 4.2 shows that the spectral densities of the diagonal entries are even functions, similarly the real parts of the off-diagonal entries, while the imaginary parts of the off-diagonal entries are odd functions, see Remark 1.7.

4.9 Summary

The d-dimensional, weakly stationary time series $\{\mathbf{X}_t\}$ has a spectral density matrix $\boldsymbol{f}$ of constant rank $r \le d$ (almost everywhere on $[-\pi, \pi]$) if and only if it can be factorized as

$$\boldsymbol{f}(\omega) = \frac{1}{2\pi}\boldsymbol{\phi}(\omega)\boldsymbol{\phi}^*(\omega) \quad \text{for a.e. } \omega \in [-\pi, \pi],$$

where $\boldsymbol{\phi}(\omega) \in \mathbb{C}^{d\times r}$ and

$$\boldsymbol{\phi}(\omega) = \sum_{j=-\infty}^{\infty} \boldsymbol{b}(j)e^{-ij\omega}, \qquad \sum_{j=-\infty}^{\infty} |b_{qs}(j)|^2 < \infty \quad (q = 1, \dots, d;\ s = 1, \dots, r).$$

Equivalently, $\{\mathbf{X}_t\}$ can be represented as a two-sided MA process:

$$\mathbf{X}_t = \sum_{j=-\infty}^{\infty} \boldsymbol{b}(j)\boldsymbol{\xi}_{t-j},$$

where $\boldsymbol{b}(j) \in \mathbb{C}^{d\times r}$ $(j \in \mathbb{Z})$ is the above non-random matrix-valued sequence; it consists of the Fourier coefficients of the function $\boldsymbol{\phi}(\omega)$ and called *impulse response function*; further, $\{\boldsymbol{\xi}_t\} \sim WN(\boldsymbol{I}_r)$ is the orthonormal white noise sequence of random shocks.

Akin to the 1D situation (see Chapter 2), here too we can introduce the *transfer function* (matrix) $\boldsymbol{H}(z) = \sum_{j\in\mathbb{Z}} \boldsymbol{b}(j)z^j$, which belongs to $L^2(T)$ and with which we can write that $\mathbf{X}_t = \boldsymbol{H}(z)\boldsymbol{\xi}_t$, $t \in \mathbb{Z}$. Clearly, for the *spectral factor* $\boldsymbol{\phi}$ we have $\boldsymbol{\phi}(\omega) = \boldsymbol{H}(e^{-i\omega})$. Therefore,

$$\boldsymbol{f}(\omega) = \frac{1}{2\pi}\boldsymbol{H}(e^{-i\omega})\boldsymbol{H}(e^{-i\omega})^*, \quad \omega \in [-\pi, \pi],$$

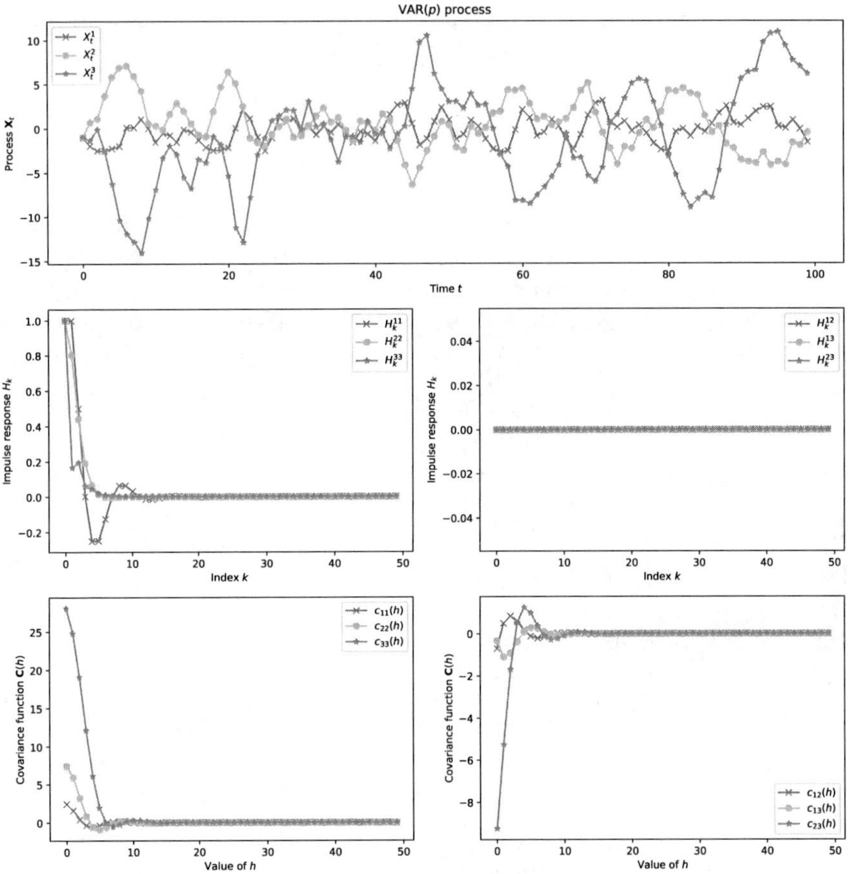

FIGURE 4.1
Typical trajectories of a 3D VAR(2) process, with its impulse response functions and covariance functions in Example 4.1.

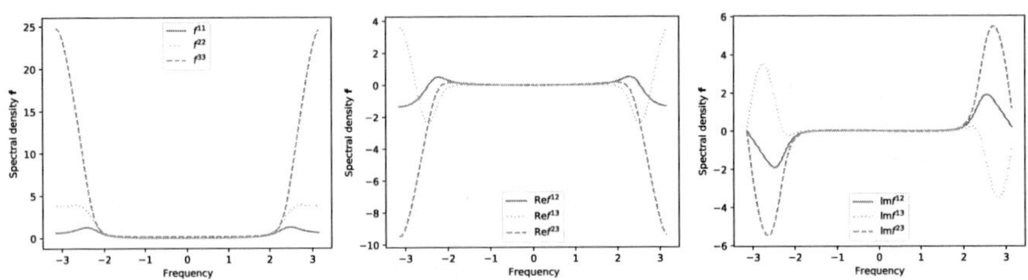

FIGURE 4.2
Spectral densities in Example 4.1.

and the covariance function of $\{\mathbf{X}_t\}$ is

$$\boldsymbol{C}(h) = \mathbb{E}(\mathbf{X}_{t+h}\mathbf{X}_t^*) = \sum_{k=-\infty}^{\infty} \boldsymbol{b}(k+h)\boldsymbol{b}^*(k), \quad h \in \mathbb{Z}.$$

Regular (purely non-deterministic) time series are subclasses of the constant rank ones in that they have a one-sided MA representation:

$$\mathbf{X}_t = \sum_{j=0}^{\infty} \boldsymbol{b}(j)\boldsymbol{\xi}_{t-j}, \quad t \in \mathbb{Z}.$$

Equivalently, their spectral density matrix factorizes as $\boldsymbol{f}(\omega) = \frac{1}{2\pi}\boldsymbol{\phi}(\omega)\boldsymbol{\phi}^*(\omega)$, where $\boldsymbol{\phi}(\omega) = \sum_{k=0}^{\infty} \boldsymbol{b}(k)e^{-ik\omega}$, so the transfer function is also one-sided.

The multi-D Wold decomposition also works as follows. Assume that $\{\mathbf{X}_t\}_{t\in\mathbb{Z}}$ is an d-dimensional non-singular stationary time series. Then it can be represented as

$$\mathbf{X}_t = \mathbf{R}_t + \mathbf{Y}_t = \sum_{j=0}^{\infty} \boldsymbol{b}(j)\boldsymbol{\xi}_{t-j} + \mathbf{Y}_t \quad (t \in \mathbb{Z}),$$

where $\{\mathbf{R}_t\}$ is a d-dimensional regular time series subordinated to $\{\mathbf{X}_t\}$; $\{\mathbf{Y}_t\}$ is an d-dimensional singular time series subordinated to $\{\mathbf{X}_t\}$; $\{\boldsymbol{\xi}_t\}$ is an r-dimensional ($r \le d$) WN($\boldsymbol{I}_r$) sequence subordinated to $\{\mathbf{X}_t\}$. Further, $\{\mathbf{R}_t\}$ and $\{\mathbf{Y}_t\}$ are orthogonal to each other; the impulse response matrices are $\boldsymbol{b}(j)$, $j \ge 0$, with square summable entries. It is important that the *orthonormal innovation process* $\{\boldsymbol{\xi}_t\}$ is not unique, but $\boldsymbol{\xi}_t$s are within the pairwise orthogonal *innovation subspaces* that are unique and their dimension is equal to the constant rank r of the spectral density matrix $\boldsymbol{f}$ of the process.

More special regular processes are the ones that can be finitely parametrized. Those are, in fact, the causal VARMA (vector ARMA) processes that also have an MFD or state space representation.

The d-dimensional VARMA(p,q) process of $\mathbf{0}$ mean is defined by

$$\boldsymbol{\alpha}(L)\,\mathbf{X}_t = \boldsymbol{\beta}(L)\,\mathbf{U}_t,$$

where $\boldsymbol{\alpha}(z) = \boldsymbol{I} + \boldsymbol{\alpha}_1 z + \cdots + \boldsymbol{\alpha}_p z^p$ and $\boldsymbol{\beta}(z) = \boldsymbol{I} + \boldsymbol{\beta}_1 z + \cdots + \boldsymbol{\beta}_q z^q$ are matrix-valued complex polynomials, namely, the AR and MA polynomials, L is the backward shift operator, and $\{\mathbf{U}_t\} \sim$ WN ($\boldsymbol{\Sigma}$) is d-dimensional white noise. The coefficient matrices $\boldsymbol{\alpha}_1, \ldots, \boldsymbol{\alpha}_p$ and $\boldsymbol{\beta}_1, \ldots, \boldsymbol{\beta}_q$ are $d \times d$ complex matrices; $\boldsymbol{\alpha}_0 = \boldsymbol{I}_d$ and $\boldsymbol{\beta}_0 = \boldsymbol{I}_d$ can be assumed. In particular, in the $q = 0$ case we have a VAR(p), whereas, in the $p = 0$ case we have a VMA(q) process.

If the condition $|\boldsymbol{\alpha}(z)| \ne 0$ for $|z| \le 1$ for the VAR polynomial is satisfied (it is called *stability*), then we have a causal representation of the process:

$$\mathbf{X}_t = \sum_{j=0}^{\infty} \boldsymbol{H}_j \mathbf{U}_{t-j},$$

with $\{\mathbf{U}_t\} \sim \mathrm{WN}(\mathbf{0}, \mathbf{\Sigma})$ and the coefficient matrices $\boldsymbol{H}_j$s come from the power series expansion of the *transfer function*:

$$\boldsymbol{H}(z) = \sum_{j=0}^{\infty} \boldsymbol{H}_j z^j = \boldsymbol{\alpha}^{-1}(z)\, \boldsymbol{\beta}(z).$$

So we can write the original process as $\mathbf{X}_t = \boldsymbol{H}(z)\, \mathbf{U}_t$.

Note that $\{\mathbf{U}_t\}$ can be transformed into an orthonormal process. Indeed, if the white noise covariance matrix $\mathbf{\Sigma}$ has rank $r \le d$, it has the Gram-decomposition (see Appendix B) like $\mathbf{\Sigma} = \boldsymbol{B}\boldsymbol{B}^*$ with the $d \times r$ matrix $\boldsymbol{B}$ of full rank, and

$$\mathbf{X}_t = \sum_{j=0}^{\infty} \boldsymbol{H}_j \mathbf{U}_{t-j} = \sum_{j=0}^{\infty} (\boldsymbol{H}_j \boldsymbol{B}) \boldsymbol{\xi}_{t-j},$$

where $\{\boldsymbol{\xi}_t\} \sim \mathrm{WN}(\boldsymbol{I}_r)$ is an orthonormal process (both longitudinally and cross-sectionally). This is the multidimensional Wold decomposition (there is no singular part).

If, in addition to the stability condition, the *inverse stability or strict miniphase condition*, i.e., $|\boldsymbol{\beta}(z)| \ne 0$ for $|z| \le 1$ also holds (concerning the MA polynomial), then $\mathbf{U}_t$ can also be expanded in terms of $\mathbf{X}_k$s $(k \le t)$.

It is important that a VAR(p) process makes rise of a finite prediction of

$$\mathbf{X}_t = \mathbf{a}_1 \mathbf{X}_{t-1} + \cdots + \mathbf{a}_p \mathbf{X}_{t-p} + \mathbf{U}_t$$

with its p-length long past:

$$\hat{\mathbf{X}}_t = \mathbf{a}_1 \mathbf{X}_{t-1} + \cdots + \mathbf{a}_p \mathbf{X}_{t-p},$$

where $\mathbf{a}_j = -\boldsymbol{\alpha}_j$ for $j = 1, \dots, p$. This extends to the infinite past prediction, see Chapter 5. The multidimensional Yule–Walker equations also work in this situation. It is also discussed when and how we can find a trivial and a minimal state space representation to a VARMA process. We prove that the stable VARMA processes are dense among the multi-D regular time series.

Regular, singular, and non-singular processes are also characterized. A d-dimensional stationary time series $\{\mathbf{X}_t\}$ is of full rank non-singular process if and only if

$$\int_{-\pi}^{\pi} \log \det \boldsymbol{f}(\omega) d\omega > -\infty.$$

In this case, if $\mathbf{\Sigma}$ denotes the covariance matrix of the innovation process $\{\boldsymbol{\eta}_t\}$, that is, of the one-step ahead prediction error process based on the infinite past (see Chapter 5), then

$$\det \mathbf{\Sigma} = (2\pi)^d \exp \int_{-\pi}^{\pi} \log \det \boldsymbol{f}(\omega) \frac{d\omega}{2\pi},$$

which is the multi-D generalization of the Kolmogorov–Szegő formula.

More generally, the d-dimensional stationary time series $\{\mathbf{X}_t\}$ is *regular* and of rank r, $1 \le r \le d$, if and only if each of the following conditions holds:

1. It has an absolutely continuous spectral measure matrix $d\boldsymbol{F}$ with density matrix $\boldsymbol{f}(\omega)$ which has rank r for a.e. $\omega \in [-\pi, \pi]$.
2. Denoting by $\boldsymbol{f}(\omega) = \tilde{\boldsymbol{U}}(\omega)\boldsymbol{\Lambda}_r(\omega)\tilde{\boldsymbol{U}}^*(\omega)$ the parsimonious spectral decomposition of the spectral density matrix, $\log\det \boldsymbol{\Lambda}_r(\omega) \in L^1$, or equivalently,
$$\int_{-\pi}^{\pi} \log \lambda_r(\omega)\, d\omega > -\infty,$$
where $\lambda_r(\omega)$ is the smallest diagonal entry of $\boldsymbol{\Lambda}_r(\omega)$.
3. The sub-unitary matrix function $\tilde{\boldsymbol{U}}(\omega)$ appearing in the spectral decomposition of $\boldsymbol{f}(\omega)$ in belongs to the Hardy space $H^\infty \subset H^2$, so
$$\tilde{\boldsymbol{U}}(\omega) = \sum_{j=0}^{\infty} \boldsymbol{\psi}(j) e^{-ij\omega}, \quad \boldsymbol{\psi}(j) \in \mathbb{C}^{d\times r}, \quad \sum_{j=0}^{\infty} \|\boldsymbol{\psi}(j)\|_F^2 < \infty.$$

In the full rank case, the first two conditions imply the third one, so it need not be stated separately.

In the possession of the above, the classification of non-regular multi-D time series is more complex than that of the 1D ones. We call a time series *non-regular* if either it is singular or its Wold decomposition contains two orthogonal, non-vanishing processes: a regular and a singular one. In dimension $d > 1$ a non-singular process beyond its regular part may have a singular part with non-vanishing spectral density. We consider a d-dimensional stationary time series $\{\mathbf{X}_t\}$ with spectral measure $d\boldsymbol{F}$. We distinguish between the following cases.

- *Type (0) non-regular processes.* In this case, $d\boldsymbol{F}$ is singular w.r.t. the Lebesgue measure in $[-\pi, \pi]$. Clearly, type (0) non-regular processes are simply singular ones. Like in the 1D case, we may further divide this class into processes with a discrete spectrum or processes with a continuous singular spectrum or processes with both.

- *Type (1) non-regular processes.* The time series has an absolutely continuous spectral measure with density $\boldsymbol{f}$, but rank($\boldsymbol{f}$) is not constant on $[-\pi, \pi]$.

- *Type (2) non-regular processes.* The time series has an absolutely continuous spectral measure with density $\boldsymbol{f}$ which has constant rank r a.e., $1 \le r \le d$, but
$$\int_{-\pi}^{\pi} \log\det \boldsymbol{\Lambda}_r(\omega) d\omega = \int_{-\pi}^{\pi} \sum_{j=1}^{r} \log \lambda_j(\omega)\, d\omega = -\infty.$$

- *Type (3) non-regular processes.* The time series has an absolutely continuous spectral measure with density $\boldsymbol{f}$ which has constant rank r a.e., $1 \le r < d$,
$$\int_{-\pi}^{\pi} \log\det \boldsymbol{\Lambda}_r(\omega) d\omega = \int_{-\pi}^{\pi} \sum_{j=1}^{r} \log \lambda_j(\omega)\, d\omega > -\infty,$$

but the unitary matrix function $\tilde{\boldsymbol{U}}(\omega)$ appearing in the spectral decomposition of $\boldsymbol{f}(\omega)$ does not belong to the Hardy space H^2.

If $\{\mathbf{X}_t\}$ has full rank $r = d$ and it is non-singular, then it may have only a Type (0) singular part.

Assume that $\{\mathbf{X}_t\}$ is a d-dimensional stationary time series of constant rank r, $1 \le r \le d$, so it can be written as a sliding summation with a WN($\boldsymbol{I}_r$) process. In Brillinger's book [9], the *Principal Component Analysis (PCA) in the Frequency Domain* aims at approximating the process $\{\mathbf{X}_t\}$ with a rank k process $\{\mathbf{X}_t^{(k)}\}$ such that the mean square error $\|\mathbf{X}_t - \mathbf{X}_t^{(k)}\|^2$ is minimized, where $k \le r$ is a fixed positive integer. The solution is as follows.

Consider the spectral decomposition $\boldsymbol{f}(\omega) = \tilde{\boldsymbol{U}}(\omega)\boldsymbol{\Lambda}_r(\omega)\tilde{\boldsymbol{U}}^*(\omega)$. Then

$$\mathbf{X}_t^{(k)} = \int_{-\pi}^{\pi} e^{it\omega}\tilde{\boldsymbol{U}}_k(\omega)\tilde{\boldsymbol{U}}_k^*(\omega)d\mathbf{Z}_\omega, \quad t \in \mathbb{Z},$$

is the best approximating process of rank k. For the mean square error we have

$$\|\mathbf{X}_t - \mathbf{X}_t^{(k)}\|^2 = \int_{-\pi}^{\pi} \sum_{j=k+1}^{r} \lambda_j(\omega)\, d\omega, \qquad t \in \mathbb{Z},$$

and

$$\frac{\|\mathbf{X}_t - \mathbf{X}_t^{(k)}\|^2}{\|\mathbf{X}_t\|^2} = \frac{\int_{-\pi}^{\pi}\sum_{j=k+1}^{r}\lambda_j(\omega)\, d\omega}{\int_{-\pi}^{\pi}\sum_{j=1}^{r}\lambda_j(\omega)\, d\omega}, \quad t \in \mathbb{Z}.$$

If condition

$$\lambda_k(\omega) \ge \Delta > \epsilon \ge \lambda_{k+1}(\omega) \qquad \forall \omega \in [-\pi, \pi],$$

holds, then we also have

$$\|\mathbf{X}_t - \mathbf{X}_t^{(k)}\| \le (2\pi(r-k)\epsilon)^{1/2}, \quad t \in \mathbb{Z}$$

and

$$\frac{\|\mathbf{X}_t - \mathbf{X}_t^{(k)}\|}{\|\mathbf{X}_t\|} \le \left(\frac{(r-k)\epsilon}{r\Delta}\right)^{\frac{1}{2}}, \qquad t \in \mathbb{Z}.$$

In particular, when $\{\mathbf{X}_t\}$ is a d-dimensional regular time series of rank r, then its best rank k approximation is also a regular time series $(1 \le k \le r)$. Application of the above for dynamic PCA is further discussed in Chapter 5.

5

Dimension reduction and prediction in the time and frequency domain

5.1 Introduction

In this chapter, we deal with the prediction of stochastic processes in general and in the weakly stationary case. We consider one-step and more-step ahead predictions based on finitely many past values or on the infinite past. Actually, the original paper of H. Wold [62] is about 1D, weakly stationary time series, and constructs the famous decomposition via one-step ahead predictions based on the n-length long past with usual multivariate regression techniques, while making use of stationarity as well. Then, at a passage to infinity ($n \to \infty$) he gets the formula for the one-step ahead prediction based on the infinite past. In this way, he decomposes the regular part of a weakly stationary 1D time series as the infinite sum of the innovations that also form a weakly stationary process, namely, a white noise process with the smallest obtainable variance of the linear prediction error. Orthogonality (uncorrelatedness) of the innovation η_t and the past $X_{t-1}, X_{t-2}, \dots$ of X_t is the consequence of the projection principle used in multivariate regression.

The generalization to a multivariate process $\{\mathbf{X}_t\}$ is straightforward with the observation that here simultaneous multivariate linear regressions are used for the components of $\mathbf{X}_t$ based on all the components of $\mathbf{X}_{t-1}, \dots, \mathbf{X}_{t-n}$. The error terms, $\boldsymbol{\eta}_t$s and their covariance matrices are obtainable by the block Cholesky decomposition of the block Toeplitz matrix $\mathfrak{C}_n$ already used in Chapter 1. At a passage to infinity, we get the multidimensional Wold decomposition that is more complicated than the 1D one in that here only the so-called innovation subspaces are unique, the dimension of which is the same as the rank of the spectral density matrix of the process $\{\mathbf{X}_t\}$. If this rank r is less than the dimension d of the process, then the error covariance matrix of $\boldsymbol{\eta}_t$ is singular (of rank r), but usually not the zero matrix. In this case, within the innovation subspaces, with usual factor analysis techniques, a $\{\boldsymbol{\xi}_t\} \sim \mathrm{WN}(\boldsymbol{I}_r)$ process can be constructed (up to orthogonal rotation) such that it appears in the multidimensional Wold decomposition instead of the innovation subspaces. Actually, this is the task of the dynamic factor analysis when the low-dimensional approximation is not always straightforward but is obtainable with spectral approximations under the conditions of the GDFM (Generalized Dynamic Factor Model) of [3, 15, 18, 31].

We also establish asymptotic relations between the spectrum of $\mathfrak{C}_n$ and the spectra of spectral density matrices at the Fourier frequencies, for "large" n. In this way, the spectra of spectra, that is the spectral decomposition of these matrices plays a crucial rule in dimension reduction and dynamic PCA (see Chapter 4), and gives rise to computationally more tractable algorithms as the above block Cholesky decomposition.

The technique of the Kálmán's filtering [1, 32, 33, 58] is also introduced together with a recursion to obtain the innovations and the newer and newer predictions for the state variable of a state space system, while using only the newcoming observed variable and the preceding estimate of the state variable. In the heart of the recursion there lies the propagation of the error covariance matrices.

5.2 1D prediction of weakly stationary processes in the time domain

5.2.1 One-step ahead prediction based on finitely many past values

We have a 1D time series $\{X_t\}$ which is not necessarily stationary, for now; we just assume the existence of the second moments (cross-autocovariances). For simplicity, the state space is $\mathbb{R}$, but the time is discrete ($t \in \mathbb{Z}$).

Assume that $\mathbb{E}(X_t) = 0$ ($t \in \mathbb{Z}$). Select a starting observation X_1 and $H_n := \mathrm{Span}\{X_1, \dots, X_n\}$ (precisely, it should be denoted by $H_n(X)$, but as the process is fixed, it is briefly denoted by H_n).

We want to linearly predict X_{n+1} based on random past values $X_1, \dots, X_n$. Let $\hat{X}_1 := 0$, and denote by $\hat{X}_{n+1}$ the best linear prediction that minimizes the mean square error $\mathbb{E}(X_{n+1} - \hat{X}_{n+1})^2$, $n = 1, 2, \dots$. If we consider the Hilbert space of the random variables with 0 expectation and finite variance, where the inner product is the covariance (see Appendix C), then we have $\mathbb{E}(X_{n+1} - \hat{X}_{n+1})^2 = \|X_{n+1} - \hat{X}_{n+1}\|^2$. By the general theory of Hilbert spaces, $\hat{X}_{n+1} = \mathrm{Proj}_{H_n} X_{n+1}$, i.e. the projection of X_{n+1} onto the linear subspace H_n. In the Gaussian case, the solution is $\hat{X}_{n+1} = \mathbb{E}(X_{n+1} \,|\, X_1, \dots, X_n)$, which is the regression plane, but the coefficients of the optimal linear predictor

$$\hat{X}_{n+1} = a_{n1} X_n + \cdots + a_{nn} X_1$$

can be obtained in the non-Gaussian case too, by solving a system of linear equations that contains the second moments and the second cross-moments of the involved random variables as follows by the theory of multivariate linear regression (see also Appendix C).

With the notations $\mathbf{a}_n = (a_{n1}, \dots, a_{nn})^T$, $\boldsymbol{C}_n = [\mathrm{Cov}(X_i, X_j)]_{i,j=1}^n$ and $\mathbf{d}_n = (\mathrm{Cov}(X_{n+1}, X_n), \dots, \mathrm{Cov}(X_{n+1}, X_1))^T$, we have to solve the following system of linear equations (Gauss normal equations):

$$\boldsymbol{C}_n \mathbf{a}_n = \mathbf{d}_n. \tag{5.1}$$

A solution (the projection) always exists, and it is unique if $\boldsymbol{C}_n$ is positive definite. Then the unique solution is $\mathbf{a}_n = \boldsymbol{C}_n^{-1}\mathbf{d}_n$. Otherwise, there are infinitely many solutions, and we can give them similarly, with any generalized inverse of the positive semidefinite matrix $\boldsymbol{C}_n$. In this case, there are linear relations between $X_1, \dots, X_n$, and so, infinitely many linear combinations of them produce the same projection of X_{n+1} onto the subspace spanned by them. In case of a singular $\boldsymbol{C}_n$ it is customary to use the (unique) Moore–Penrose inverse (see Appendix B) that gives the particular solution $\mathbf{a}_n = \boldsymbol{C}_n^{+}\mathbf{d}_n$. We will see that this issue is immaterial in the stationary case, since then zero determinant of $\boldsymbol{C}_n$ for the smallest $n \geq 1$ indicates zero prediction error, see Remark 5.5.

In particular, if $\{X_t\}$ is stationary, then $\boldsymbol{C}_n = [c(i-j)]_{i,j=1}^n$, so $\boldsymbol{C}_n$ is a Toeplitz matrix, and $d_n(j) = c(j)$, $j = 1, \dots, n$. Therefore, the solution $\mathbf{a}_n$ does not depend on the selection of the starting time of the starting observation X_1. In this case, no double indexing for the coordinates of the vector $\mathbf{a}_n$ is necessary, they can as well be written as $a_1, \dots, a_n$. Also, when $\{X_t\}$ is stationary, then under very general conditions, there is a unique solution as discussed below. Some remarks are in order.

Remark 5.1. Namely, by Proposition 5.1.1 of [11], if $c(0) > 0$ and $\lim_{h\to\infty} c(h) = 0$, then the autocovariance matrix $\boldsymbol{C}_n = [c(i-j)]_{i,j=1}^n$ of $(X_1, \dots, X_n)^T$ is positive definite for every $n \in \mathbb{N}$.

Remark 5.2. By Proposition 4.5.2 of [11], for "large" n, the eigenvalues of $\boldsymbol{C}_n$ are asymptotically the same as the union of the values of the spectral density f at the Fourier frequencies. In Section 5.4, we will generalize this statement for multidimensional time series.

Considering the decomposition

$$X_{n+1} = \hat{X}_{n+1} + \eta_{n+1},$$

where $\hat{X}_{n+1} = \mathbf{a}_n^T \mathbf{X}_n$ and η_{n+1} is the error term, it is easy to see (Appendix C) that the two right-hand side terms are orthogonal (uncorrelated), therefore their squared norms (variances) are added together:

$$\mathrm{Var}(X_{n+1}) = \mathrm{Var}(\hat{X}_{n+1}) + \mathrm{Var}(\eta_{n+1}).$$

With our notation it yields

$$\begin{aligned} c(0) &= \mathrm{Var}(\mathbf{a}_n^T \mathbf{X}_n) + \mathrm{Var}(\eta_{n+1}) = \mathrm{Var}(\mathbf{d}_n^T \boldsymbol{C}_n^{-1} \mathbf{X}_n) + \mathrm{Var}(\eta_{n+1}) \\ &= \mathbf{d}_n^T \boldsymbol{C}_n^{-1} \boldsymbol{C}_n \boldsymbol{C}_n^{-1} \mathbf{d}_n + \mathrm{Var}(\eta_{n+1}) = \mathbf{d}_n^T \boldsymbol{C}_n^{-1} \mathbf{d}_n + \mathrm{Var}(\eta_{n+1}). \end{aligned}$$

Consequently, the prediction error, that is the variance of the error term, is

$$e_n^2 = \|\eta_{n+1}\|^2 = \mathrm{Var}(\eta_{n+1}) = c(0) - \mathbf{d}_n^T \boldsymbol{C}_n^{-1} \mathbf{d}_n, \tag{5.2}$$

by Remark C.3 in Appendix C. It will be further analyzed in Section 5.2.2.

Note that Equation (5.1) is exactly the same as the first n Yule–Walker equations for estimating the parameters of a stationary AR(n) process which is

$$X_t = a_1 X_{t-1} + a_2 X_{t-2} + \cdots + a_n X_{t-n} + \eta_t, \quad t = 0, \pm 1, \pm 2, \ldots$$

where $\{\eta_t\} \sim \mathrm{WN}(\sigma^2)$ is a white noise process, and σ^2 is also estimated. In case of second order processes, due to the projection principle, it also comes out that η_t (the orthogonal component) is uncorrelated with the regressor, and so with the past values $X_{t-1}, X_{t-2}, \ldots X_{t-n}$ too.

The Yule–Walker equations based on the first n autocovariances are:

$$c(k) = \begin{cases} a_1 c(1) + \cdots + a_n c(n) + \sigma^2, & k = 0 \\ a_1 c(k-1) + \cdots + a_n c(k-n), & k = 1, \ldots, n. \end{cases} \tag{5.3}$$

For real-valued time series, the Yule–Walker equations (5.3) for $k = 1, \ldots, n$ can be written in matrix form:

$$\begin{bmatrix} c(0) & c(1) & \ldots & c(n-1) \\ c(1) & c(0) & \ldots & c(n-2) \\ \vdots & \vdots & \vdots & \vdots \\ c(n-1) & c(n-2) & \ldots & c(0) \end{bmatrix} \cdot \begin{bmatrix} a_1 \\ a_2 \\ \vdots \\ a_n \end{bmatrix} = \begin{bmatrix} c(1) \\ c(2) \\ \vdots \\ c(n) \end{bmatrix}. \tag{5.4}$$

These are the Gauss normal equations, see Appendix C. If the coefficient matrix is strictly positive definite, then we have a unique solution. Substituting this solution in the first equation of (5.3), which is the same as equation (5.2), provides the solution for σ^2. Here $\sigma^2 = e_n^2$ if the order n of the AR process is fixed.

More precisely, between the coefficients α_is of Equations (2.21) and the coefficients a_is of Equations (5.4) the relation $a_i = -\alpha_i$ holds, $i = 1, \ldots, n$ (here n stands for the order of the AR process).

Also, in case of a stable AR(n) process, the first n Yule–Walker equations imply the next ones; while in other cases, the solution of the first n Yule–Walker equations just gives the best prediction using n past values, and they are rather called Gauss normal equations.

Remark 5.3. (see [11], p. 424). If for some $n \geq 1$ the covariance matrix $\boldsymbol{C}_n$ is positive definite, then the nth degree AR polynomial $\alpha(z)$ is causal in the sense that $\alpha(z) \neq 0$ for $z \leq 1$.

Remark 5.4. Comparing Remarks 5.1 and 5.3, we can conclude the following. If for the autocovariance function of the process $\{X_t\}$, $c(0) > 0$ and $\lim_{h \to \infty} c(h) = 0$ hold, then the autocovariance matrix $\boldsymbol{C}_n = [c(i-j)]_{i,j=1}^n$

of $(X_1, \dots, X_n)^T$ assigned to the process is positive definite for every $n \in \mathbb{N}$. Consequently, the process $\{X_t\}$ has a (unique) stable AR(n) representation such that the first n autocovariances of it are $c(0), \dots, c(n-1)$, for every $n \in \mathbb{N}$. However, if the sequence $c(h)$ tends to 0, but not exponentially fast, these AR(n) representations based on just $X_1, \dots, X_n$ do not approximate the process at all, and the process is not even necessarily regular.

On the contrary, if $\boldsymbol{C}_n$ is singular for some n (and consequently, for larger ns too), then using its the generalized inverse, we get an AR(n) solution, but it is not stable.

It is also important, that in case of a stationary process, the h-step ahead prediction, i.e. the prediction of X_{n+h} based on $X_1, \dots, X_n$ can be easily concluded from the one-step ahead prediction, for $h = 1, 2, \dots$. In view of $H_n \subset H_{n+h-1}$,

$$\text{Proj}_{H_n} X_{n+h} = \text{Proj}_{H_n} \text{Proj}_{H_{n+h-1}} X_{n+h} = \text{Proj}_{H_n} \hat{X}_{n+h},$$

we get the equation

$$\boldsymbol{C}_n \mathbf{a}_n = \mathbf{d}_n(h)$$

for the coefficients of the prediction in $\mathbf{a}_n$, where

$$\mathbf{d}_n(h) = [\text{Cov}(X_{n+h}, X_n), \dots, \text{Cov}(X_{n+h}, X_1)]^T,$$

and it is $[c(h), \dots, c(n+h-1)]^T$ in the stationary case. Equation (5.1) is the special case when $h = 1$ and $\mathbf{d}_n = \mathbf{d}_n(1)$.

5.2.2 Innovations

Observe that, by the Gram–Schmidt procedure, the prediction error terms form an orthogonal sequence, and they are called *innovations*. In this way, X_ns can as well be expressed in terms of the innovations; in other words, X_n can be written as the linear combination of the normalized error terms that form a complete orthonormal system in H_n. Moreover, this is true in each step of the Gram–Schmidt process, so in the expansion of each X_n only the same and lower index error terms appear. We do it as follows.

First, $\{X_t\}$ is not necessarily stationary. Recall that $H_n := \text{Span}\{X_1, \dots, X_n\}$. Let $\eta_{n+1} := X_{n+1} - \hat{X}_{n+1}$ be the one-step ahead prediction error term, based on the n-length long past, for $n = 0, 1, \dots$. As $\hat{X}_1 = 0$ (see Section 5.2.1), $\eta_1 = X_1 \in H_1$ and the unique orthogonal decomposition

$$X_2 = \hat{X}_2 + \eta_2$$

works, where $\hat{X}_2 \in H_1$ and $\eta_2 \perp H_1$, whenever $H_1 \subset H_2$ is a proper subspace (disregard the situation $H_1 = H_2$, when $\eta_2 = 0$). Therefore, $\eta_2 \in H_2$ and $\eta_2 \perp \eta_1$. Further,

$$X_2 = l_{21}\eta_1 + \eta_2.$$

With the same considerations,

$$X_{j+1} = \hat{X}_{j+1} + \eta_{j+1}, \quad j = 2, \dots, n$$

with $\hat{X}_{j+1} \in H_j$ and $\eta_{j+1} \perp H_j$ if $H_j \subset H_{j+1}$ is a proper subspace. So $\eta_{j+1} \in H_{j+1}$ and $\eta_{j+1} \perp \eta_j$.

In this way, we get the innovations $\eta_1, \dots, \eta_n$, the linear combination of which produces X_k as

$$X_1 = \eta_1, \quad X_k = \sum_{j=1}^{k-1} l_{kj}\eta_j + \eta_k, \quad k = 2, \dots, n,$$

where the coefficients l_{kj} are obtained recursively, together with the mean square one-step ahead prediction errors $e_{k-1}^2 = \|\eta_k\|^2$ $(k = 2, \dots, n)$, $\|\eta_1\|^2 = c(0)$.

Actually, this is the LDL (variant of the Cholesky) decomposition, see Appendix B. Indeed, with the notation $\boldsymbol{\eta}_n = (\eta_1, \dots, \eta_n)^T$ and $\mathbf{X}_n = (X_1, \dots, X_n)^T$, we have to find an $n \times n$ lower triangular matrix $\boldsymbol{L}_n$ with entries l_{kj}s and all 1s along its main diagonal such that

$$\mathbf{X}_n = \boldsymbol{L}_n \boldsymbol{\eta}_n. \tag{5.5}$$

Taking the covariance matrices on both sides, yields the LDL decomposition

$$\boldsymbol{C}_n = \boldsymbol{L}_n \boldsymbol{D}_n \boldsymbol{L}_n^T. \tag{5.6}$$

If $\boldsymbol{C}_n$ is positive definite, then $\boldsymbol{D}_n = \text{diag}(c(0), e_1^2, \dots, e_{n-1}^2)$ is positive definite too. $\boldsymbol{L}_n$ is not singular (with diagonal entries 1s), hence $\boldsymbol{\eta}_n = \boldsymbol{L}_n^{-1}\mathbf{X}_n$, where $\boldsymbol{L}_n^{-1}$ is also lower triangular; therefore, the innovations can as well be written in terms of the same or lower index X_ts. So the LDL decomposition gives the prediction errors (diagonal entries of $\boldsymbol{D}_n$), and the entries of $\boldsymbol{L}_n$ (below its main diagonal, which is constantly 1) are obtainable in a nested way; therefore, n does not play an important role here, see Appendix B. With increasing n, we just extend the rows of $\boldsymbol{L}_n$.

Summarizing,

$$X_n = \sum_{j=1}^{n-1} l_{kj}\eta_j + \eta_n = \hat{X}_n + \eta_n, \quad \text{Var}(\eta_1) = c(0), \quad \text{Var}(\eta_n) = e_{n-1}^2, \tag{5.7}$$

and it is true for any $n = 1, 2, \dots$.

The situation further simplifies in the stationary case, when $\boldsymbol{C}_n$ is a Toeplitz matrix. However, $\boldsymbol{L}_n$ will not be Toeplitz, but asymptotically, it becomes more and more like a Toeplitz one, and the entries of $\boldsymbol{D}_n$ will be more and more similar to each other (the sequence e_n^2 converges) as with $n \to \infty$ the situation extends to the prediction by the infinite past case. This is the topic of the Wold decomposition, see Section 5.2.3.

In the stationary case, the innovation η_n is non-zero if $H_{n-1} \subset H_n$ is a proper subspace. In this case, η_ns are true innovations. Recall that, by Remark 5.1, this holds true at the same time for any n whenever $c(0) > 0$ and $\lim_{h\to\infty} c(h) = 0$. In this case, we can also standardize the η_ns, and write

$$X_n = \sum_{j=1}^{n} \tilde{l}_{nj}\xi_j, \quad n = 1, 2, \ldots$$

where $\xi_j = \eta_j/e_{j-1}$ and $\tilde{l}_{nj} = l_{nj}e_{j-1}$, $e_0 = \sqrt{c(0)}$, for $j = 1, \ldots, n$. Here $\xi_1, \ldots, \xi_n$ form a complete orthonormal system in H_n. The coefficients $\tilde{l}_{nj}$s are obtainable by the Gram decomposition (see Appendix B)

$$\boldsymbol{C}_n = \boldsymbol{A}_n\boldsymbol{A}_n^T$$

where $\boldsymbol{A}_n = \boldsymbol{L}_n\boldsymbol{D}_n^{1/2}$ is a lower triangular solution, but it can be post-multiplied with any orthogonal matrix.

5.2.3 Prediction based on the infinite past

Going farther, in case of a stationary, non-singular process, we can project X_{n+1} onto the infinite past $H_n^- = \overline{\text{span}}\,\{X_t : t \le n\}$ and expand it in terms of an orthonormal system, that is called Wold decomposition. This part will be the regular (causal) part of the process, whereas, the other, singular part, is orthogonal to it. Note that this singular part is of Type (0) deterministic (see Section 2.9). Recall that a singular (deterministic) process has no future, only (remote) past; whereas a regular (purely non-deterministic) process has only future, and no past. A non-singular process has future (with or without past).

Also, by stationarity, the one-step ahead prediction error

$$\sigma^2 = \|X_{n+1} - \text{Proj}_{H_n^-} X_{n+1}\| = \mathbb{E}(X_{n+1} - \text{Proj}_{H_n^-} X_{n+1})^2$$

does not depend on n, and it is positive, since the process is non-singular. Again, the Wold decomposition gives

$$X_n = \sum_{j=0}^{\infty} b_j\eta_{n-j} + Y_n,$$

where $\{Y_n\}$ is of Type (0) singular and $\{\eta_t\}$ is white noise with variance σ^2, $b_0 = 1$. If $Y_n = 0$ for all n, the process $\{X_n\}$ is regular. The coefficients b_i/σ are the *impulse responses*. Because of the stationarity and infinite past, b has a single index. Here the coefficients b_js are the limiting values of l_{nj}s when $n \to \infty$ in (5.5). It is in accord with the earlier observation that the matrix $\boldsymbol{L}_n$ will be closer and closer to a Toeplitz one, if we disregard the first finitely many rows of it.

Note that the innovation process is a MA(∞) process, which is a causal TLF. Here η_n is not considered as an error term, but rather than positive information that is not contained in the past of X_n. This is why it is called innovation.

Wold derives his celebrated decomposition theorem for real, univariate stationary time series in the following situation: the one-step ahead prediction of X_t is based on its n-length long past and $n \to \infty$.

More precisely, let us fix X_t and consider its one-step ahead prediction, based on its n-length long past. In view of Equation (5.7) and by stationarity, the mean square prediction error does not depend on t, it only depends on n, and was denoted by e_n^2. It can be written in many equivalent forms, see the theory of multivariate regression [44] and Equation (C.3):

$$e_n^2 = c(0)(1 - r^2_{X_t,(X_{t-1},\dots,X_{t-n})}) = c(0) - \mathbf{d}_n^T \boldsymbol{C}_n^{-1} \mathbf{d}_n,$$

where $r^2_{X_t,(X_{t-1},\dots,X_{t-n})}$ is the squared multiple correlation coefficient between X_t and $(X_{t-1},\dots,X_{t-n})$; it does not depend on t either, and obviously increases (does not decrease) with n, i.e. $e_1^2 \ge e_2^2 \ge \dots$. The mean square error can as well be written with the determinants of the consecutive Toeplitz matrices $\boldsymbol{C}_n$ and $\boldsymbol{C}_{n+1}$. The next proposition is also used in [62], but here we give a simple proof by means of the determinants of block matrices.

Proposition 5.1. *If for some n, $|\boldsymbol{C}_n| \neq 0$, then*

$$e_n^2 = c(0) - \mathbf{d}_n^T \boldsymbol{C}_n^{-1} \mathbf{d}_n = \frac{|\boldsymbol{C}_{n+1}|}{|\boldsymbol{C}_n|}. \tag{5.8}$$

Proof. We use block matrix techniques for the following partitioned matrix:

$$\boldsymbol{C}_{n+1} = \begin{pmatrix} \boldsymbol{C}_n & \mathbf{d}_n \\ \mathbf{d}_n^T & c(0) \end{pmatrix}.$$

It is known that

$$\begin{aligned} |\boldsymbol{C}_{n+1}| &= |\boldsymbol{C}_n - \mathbf{d}_n c^{-1}(0) \mathbf{d}_n^T| \cdot |c(0)| \\ &= c(0) |\boldsymbol{C}_n (\boldsymbol{I}_n - \boldsymbol{C}_n^{-1} \mathbf{d}_n \mathbf{d}_n^T / c(0))| \\ &= c(0) |\boldsymbol{C}_n| \cdot |\boldsymbol{I}_n - \boldsymbol{C}_n^{-1} \mathbf{d}_n \mathbf{d}_n^T / c(0)| \\ &= c(0) |\boldsymbol{C}_n| \cdot (1 - \lambda(\boldsymbol{C}_n^{-1} \mathbf{d}_n \mathbf{d}_n^T / c(0))), \end{aligned}$$

where $\lambda(\boldsymbol{C}_n^{-1} \mathbf{d}_n \mathbf{d}_n^T / c(0))$ is the only nonzero eigenvalue of the matrix $\boldsymbol{C}_n^{-1} \mathbf{d}_n \mathbf{d}_n^T / c(0)$, which is of rank 1. Indeed, the rank of the dyad $\mathbf{d}_n \mathbf{d}_n^T$ is 1, and the multiplication with another matrix cannot increase this rank. Therefore, the eigenvalues of $\boldsymbol{I}_n - \boldsymbol{C}_n^{-1} \mathbf{d}_n \mathbf{d}_n^T / c(0)$ are $1 - \lambda(\boldsymbol{C}_n^{-1} \mathbf{d}_n \mathbf{d}_n^T / c(0))$ and 1 (with multiplicity $n-1$). So its determinant is

$$\begin{aligned} 1 - \lambda(\boldsymbol{C}_n^{-1} \mathbf{d}_n \mathbf{d}_n^T / c(0)) &= 1 - \lambda(\boldsymbol{C}_n^{-1} \mathbf{d}_n \mathbf{d}_n^T)/c(0) = 1 - \operatorname{tr}(\boldsymbol{C}_n^{-1} \mathbf{d}_n \mathbf{d}_n^T)/c(0) \\ &= 1 - \operatorname{tr}(\mathbf{d}_n^T \boldsymbol{C}_n^{-1} \mathbf{d}_n)/c(0) = 1 - \mathbf{d}_n^T \boldsymbol{C}_n^{-1} \mathbf{d}_n / c(0) = \frac{c(0) - \mathbf{d}_n^T \boldsymbol{C}_n^{-1} \mathbf{d}_n}{c(0)}, \end{aligned}$$

where we used that the only nonzero eigenvalue of a rank 1 matrix is its trace and the cyclic commutativity of the trace operator. Putting things together:

$$|\boldsymbol{C}_{n+1}| = c(0)|\boldsymbol{C}_n|\frac{c(0) - \mathbf{d}_n^T \boldsymbol{C}_n^{-1}\mathbf{d}_n}{c(0)} = |\boldsymbol{C}_n|(c(0) - \mathbf{d}_n^T \boldsymbol{C}_n^{-1}\mathbf{d}_n),$$

that proves the statement. □

Remark 5.5. If $|\boldsymbol{C}_n| = 0$ for some n, then $|\boldsymbol{C}_{n+1}| = |\boldsymbol{C}_{n+2}| = \cdots = 0$ too. The smallest index n for which this happens indicates that there is a linear relation between n consecutive X_js, but no linear relation between $n-1$ consecutive ones (by stationarity, this property is irrespective of the position of the consecutive random variables). This can happen only if some X_t linearly depends on $n-1$ preceding X_js. In this case $e_{n-1}^2 = 0$ and, of course $e_n^2 = e_{n+1}^2 = \cdots = 0$ too. In any case, $e_1^2 \geq e_2^2 \geq \ldots$ is a decreasing (non-increasing) nonnegative sequence, and in view of Equation (5.8),

$$|\boldsymbol{C}_1| = c(0), \quad |\boldsymbol{C}_n| = c(0)e_1^2 \ldots e_{n-1}^2, \quad n = 2, 3, \ldots, \tag{5.9}$$

so, provided $c(0) > 0$, $|\boldsymbol{C}_n| = 0$ holds if and only if $e_{n-1}^2 = 0$. Note that in this stationary case there is no sense of using generalized inverse if $|\boldsymbol{C}_n| = 0$, since then exact one-step ahead prediction with the $n-1$ long past can be done with zero error, and this property is manifested for longer past predictions too.

Remark 5.6. Equation (5.9) can as well be obtained from the LDL decomposition of Equation (5.6). Indeed,

$$|\boldsymbol{C}_n| = |\boldsymbol{D}_n| = c(0)e_1^2 \ldots e_{n-1}^2, \tag{5.10}$$

utilizing that the diagonal entries of $\boldsymbol{D}_n$ are the prediction errors. In this way, Equation (5.10) implies another proof of Proposition 5.1.

In the light of these, there are the following possibilities:

- $\boldsymbol{C}_k$ is positive definite up to $k \leq h$, but $|\boldsymbol{C}_h| = 0$ for some positive integer h (and so, $|\boldsymbol{C}_k| = 0$ for $k > h$ too). Wold calls such a process singular of rank h. Then, by Remark 5.5,

$$e_1^2 \geq e_2^2 \geq \cdots > e_{h-1}^2 = e_h^2 = \cdots = 0.$$

  So X_t can be exactly predicted based on its $(h-1)$-length long past. This is caused by periodicities, for instance, in case of the Type(0) singular process of Section 2.10.1. In this case, $c(h)$ cannot tend to 0, otherwise all the $\boldsymbol{C}_h$s were positive definite, in view of Remark 5.1.

- $|\boldsymbol{C}_n| \neq 0$ for any n, and so, $e_n^2 > 0$ for every n, but still, $\lim_{n\to\infty} e_n^2 = 0$ in a decreasing (non-increasing) way. Wold calls such a process singular of infinite rank. This is caused by hidden periodicities, for instance, the Type (1) and Type (2) singular process of Section 2.10.2. (Then $\lim_{h\to\infty} c(h) = 0$, but not exponentially fast.)

- In the remaining (non-singular) case, $e_n^2 \to \sigma^2$ as $n \to \infty$ in a decreasing (non-increasing) way, where $0 < \sigma^2 < c(0)$. (In case of ARMA processes $\lim_{h\to\infty} c(h) = 0$ exponentially fast.)

Wold shows that the residual process $\eta_{t,n}$ (one-step ahead prediction error term of predicting $\mathbf{X}_t$ with its n-length long past) is stationary for any fixed n. After a passage to the limit, the process $\{\eta_{t,n}\}$ converges in probability to the residual process $\{\eta_t\}$ as $n \to \infty$. We cite the exact theorem (Theorem 6 in [62]):

Theorem 5.1. *A residual process $\{\eta_t\}$ obtained from a non-singular stationary process $\{X_t\}$ is stationary and non-autocorrelated. Further, η_t is non-correlated with $X_{t-1}, X_{t-2}, \dots,$ while*

$$\mathrm{Corr}(X_t, \eta_t) = \frac{\mathrm{Cov}(X_t, \eta_t)}{\sqrt{c(0)}\sqrt{\mathrm{Var}(\eta_t)}} = \frac{\mathrm{Var}(\eta_t)}{\sqrt{c(0)}\sqrt{\mathrm{Var}(\eta_t)}} = \frac{\sqrt{\mathrm{Var}(\eta_t)}}{\sqrt{c(0)}} = \frac{\sigma}{\sqrt{c(0)}}.$$

Wold notes that the arguments used in the proof of this theorem also apply to the singular cases. As the residual variables η_t are here vanishing, their correlation properties will be indeterminate. Accordingly, these cases do not need further comment.

5.3 Multidimensional prediction

5.3.1 One-step ahead prediction based on finitely many past values

We have a d-dimensional real time series $\{\mathbf{X}_t\}$ with components $\mathbf{X}_t = [X_t^1, \dots, X_t^d]^T$. It is not necessarily stationary, we just assume the existence of the second moments and cross-moments. For simplicity, the state space is $\mathbb{R}^d$, but the time is discrete ($t \in \mathbb{Z}$).

Assume that $\mathbb{E}(\mathbf{X}_t) = \mathbf{0}$ ($t \in \mathbb{Z}$). Select a starting observation $\mathbf{X}_1$ and

$$H_n := \mathrm{Span}\{X_t^j : t = 1, \dots, n;\ j = 1, \dots, d\}.$$

(Precisely, it should be denoted by $H_n(\mathbf{X})$, but as the process is fixed, it is briefly denoted by H_n. However, this H_n is not the same as in the 1D situation.)

We want to linearly predict $\mathbf{X}_{n+1}$ based on past values $\mathbf{X}_1, \dots, \mathbf{X}_n$. Let $\hat{\mathbf{X}}_1 := 0$, and denote by $\hat{\mathbf{X}}_{n+1}$ the best one-step ahead linear prediction that minimizes the mean square error

$$\mathbb{E}(\mathbf{X}_{n+1} - \hat{\mathbf{X}}_{n+1})^2 = \|\mathbf{X}_{n+1} - \hat{\mathbf{X}}_{n+1}\|^2, \quad n = 1, 2, \dots$$

in the Hilbert-space setup of Section 5.2.1. Thus, $\hat{\mathbf{X}}_{n+1} = \mathrm{Proj}_{H_n}\mathbf{X}_{n+1}$, i.e. the projection of $\mathbf{X}_{n+1}$ onto the linear subspace H_n. In the Gaussian case, the solution is $\hat{\mathbf{X}}_{n+1} = \mathbb{E}(\mathbf{X}_{n+1} \,|\, \mathbf{X}_1, \dots, \mathbf{X}_n)$, which is the instance of *simultaneous linear regressions* for the components of $\mathbf{X}_{n+1}$ by predictors $\mathbf{X}_1, \dots, \mathbf{X}_n$. In the general case, we have to solve a system of linear equations that resembles (5.1). Indeed, the projection is looked for in the form

$$\hat{\mathbf{X}}_{n+1} = \boldsymbol{A}_{n1}\mathbf{X}_n + \cdots + \boldsymbol{A}_{nn}\mathbf{X}_1, \tag{5.11}$$

where $\boldsymbol{A}_{n1}, \dots \boldsymbol{A}_{nn}$ are $d \times d$ matrices. But $(\mathbf{X}_{n+1} - \hat{\mathbf{X}}_{n+1}) \perp \mathbf{X}_{n+1-k}$ for $k = 1, \dots, n$ in the sense that

$$\mathbb{E}[(\mathbf{X}_{n+1} - \hat{\mathbf{X}}_{n+1})\mathbf{X}_{n+1-k}^T] = \boldsymbol{O}_d, \quad k = 1, \dots n, \tag{5.12}$$

where $\boldsymbol{O}_d$ is the $d \times d$ zero matrix. Equations (5.11) and (5.12) together yield the following system of linear equations:

$$\sum_{j=1}^{n} \boldsymbol{A}_{nj}\mathrm{Cov}(\mathbf{X}_{n+1-j}, \mathbf{X}_{n+1-k}) = \mathrm{Cov}(\mathbf{X}_{n+1}, \mathbf{X}_{n+1-k}), \quad k = 1, \dots, n, \tag{5.13}$$

where Cov now denotes an $d \times d$ cross-covariance matrix. This is the extension of the Gauss normal equations for parallel linear predictions with d-dimensional target, see Lemma C.2 of Appendix C.

When $\{\mathbf{X}_t\}$ is stationary, then Equation (5.13) simplifies to

$$\sum_{j=1}^{n} \boldsymbol{A}_j \boldsymbol{C}(k - j) = \boldsymbol{C}(k), \quad k = 1, \dots, n,$$

where $\boldsymbol{C}(k)$ is the kth order $d \times d$ autocovariance matrix. This provides a system of $d^2 n$ linear equations with the same number of unknowns that always has a solution. Further, the solution does not depend on the selection of the time of the starting observation $\mathbf{X}_1$, and no double indexing of the coefficient matrices is necessary. For the block matrix version, see Appendix C. The coefficient matrix is just $\mathfrak{C}_n$, which is always positive semidefinite. If positive definite, we have a unique solution; otherwise, with block matrix techniques, reduced rank innovations are obtained.

There are recursions to solve this system (e.g. the Durbin–Levinson algorithm), see [11], which resembles the set of the first n Yule–Walker equations for a multidimensional VAR(n) processes.

Proposition 5.2. ([11], p.424). *If for some $n \geq 1$ the covariance matrix of $(\mathbf{X}_{n+1}^T, \dots, \mathbf{X}_1^T)^T$ is positive definite, then the matrix polynomial $\boldsymbol{\alpha}(z) = \boldsymbol{I} - \boldsymbol{A}_1 z - \cdots - \boldsymbol{A}_n z^n$ is causal in the sense that the determinant $|\boldsymbol{\alpha}(z)| \neq 0$ for $z \leq 1$.*

5.3.2 Multidimensional innovations

Analogously to the 1D situation, $\mathbf{X}_t$ can again be expanded in terms of the now d-dimensional innovations, i.e. the prediction error terms

$$\boldsymbol{\eta}_{n+1} := \mathbf{X}_{n+1} - \hat{\mathbf{X}}_{n+1}.$$

It can be done step by step as follows. Assume that the $nd \times nd$ covariance matrix $\mathfrak{C}_n$ of the components of $\mathbf{X}_1, \dots, \mathbf{X}_n$ is positive definite for every $n \geq 1$. Let $\hat{\mathbf{X}}_1 := \mathbf{0}$, $\boldsymbol{\eta}_1 := \mathbf{X}_1$ and consider the unique orthogonal decomposition

$$\mathbf{X}_2 = \hat{\mathbf{X}}_2 + \boldsymbol{\eta}_2,$$

where $\hat{\mathbf{X}}_2 \in H_1$ and $\boldsymbol{\eta}_2 \perp H_1$, whenever $H_1 \subset H_2$ is a proper subspace (disregard the situation $H_1 = H_2$, when $\boldsymbol{\eta}_2 = \mathbf{0}$). Therefore, $\boldsymbol{\eta}_2 \in H_2$ and $\boldsymbol{\eta}_2 \perp \boldsymbol{\eta}_1$. With the same considerations,

$$\mathbf{X}_{j+1} = \hat{\mathbf{X}}_{j+1} + \boldsymbol{\eta}_{j+1}, \quad j = 2, \dots, n$$

with $\hat{\mathbf{X}}_{j+1} \in H_j$ and $\boldsymbol{\eta}_{j+1} \perp H_j$ if $H_j \subset H_{j+1}$ is a proper subspace. So $\boldsymbol{\eta}_{j+1} \in H_{j+1}$ and $\boldsymbol{\eta}_{j+1} \perp \boldsymbol{\eta}_j$.

In this way, we get the innovations $\boldsymbol{\eta}_1, \dots, \boldsymbol{\eta}_n$ that trivially have $\mathbf{0}$ expectation and form an orthogonal system in the nd-dimensional H_n (their pairwise cross-covariance matrices are zeros). We consider the first n steps, i.e. the recursive equations

$$\mathbf{X}_k = \sum_{j=1}^{k-1} \boldsymbol{B}_{kj} \boldsymbol{\eta}_j + \boldsymbol{\eta}_k, \quad k = 1, 2, \dots, n \tag{5.14}$$

in the case when the observations $\mathbf{X}_1, \dots, \mathbf{X}_n$ are available.

If our process is stationary, the coefficient matrices are irrespective of the choice of the starting time. The $\boldsymbol{\eta}_j$s are not zeros if $H_n \subset H_{n+1}$ are proper subspaces, i.e. they are true innovations. However, it can be, that though they are not zeros, they span a lower than d-dimensional subspace, i.e. their covariance matrix $\boldsymbol{E}_j = \mathbb{E}\boldsymbol{\eta}_j \boldsymbol{\eta}_j'$ is not zero, but a positive semidefinite matrix of reduced rank. When we go to the future, then look back to the "infinite" past, and obtain the multidimensional Wold decomposition (see Section 4.4 and the forthcoming explanation at the end of this section).

Multiplying the equations in (5.14) by $\mathbf{X}_j^T$ from the right, and taking expectation, the solution for the matrices $\boldsymbol{B}_{kj}$ and $\boldsymbol{E}_j$ ($k = 1. \dots, n$; $j = 1, \dots, k-1$) can be obtained via the block Cholesky (LDL) decomposition:

$$\mathfrak{C}_n = \boldsymbol{L}_n \boldsymbol{D}_n \boldsymbol{L}_n^T, \tag{5.15}$$

where $\mathfrak{C}_n$ is $nd \times nd$ positive definite block Toeplitz matrix of general entry $\boldsymbol{C}(i-j)$, see (1.3). $\boldsymbol{D}_n$ is $nm \times nm$ block diagonal and contains the positive semidefinite prediction error matrices $\boldsymbol{E}_1, \dots, \boldsymbol{E}_n$ in its diagonal blocks, whereas $\boldsymbol{L}_n$ is $nd \times nd$ lower triangular with blocks $\boldsymbol{B}_{kj}$s below its diagonal blocks which are $d \times d$ identities, so $\boldsymbol{L}_n$ is non-singular. In matrix form,

$$\boldsymbol{L}_n = \begin{bmatrix} \boldsymbol{I} & \boldsymbol{O} & \dots & \boldsymbol{O} & \boldsymbol{O} \\ \boldsymbol{B}_{21} & \boldsymbol{I} & \dots & \boldsymbol{O} & \boldsymbol{O} \\ \vdots & \vdots & \vdots & \vdots & \vdots \\ \boldsymbol{B}_{n1} & \boldsymbol{B}_{n2} & \dots & \boldsymbol{B}_{n,n-1} & \boldsymbol{I} \end{bmatrix}, \quad \boldsymbol{D}_n = \begin{bmatrix} \boldsymbol{E}_1 & \boldsymbol{O} & \dots & \boldsymbol{O} & \boldsymbol{O} \\ \boldsymbol{O} & \boldsymbol{E}_2 & \dots & \boldsymbol{O} & \boldsymbol{O} \\ \vdots & \vdots & \vdots & \vdots & \vdots \\ \boldsymbol{O} & \boldsymbol{O} & \dots & \boldsymbol{O} & \boldsymbol{E}_n \end{bmatrix}. \tag{5.16}$$

To find the block Cholesky decomposition of (5.16), the following recursion is at our disposal: for $j = 1, \dots, n$

$$\boldsymbol{E}_j := \boldsymbol{C}(0) - \sum_{k=1}^{j-1} \boldsymbol{B}_{jk} \boldsymbol{E}_k \boldsymbol{B}_{jk}^T, \quad j = 1, \dots, n \tag{5.17}$$

and for $i = j+1, \dots, n$

$$\boldsymbol{B}_{ij} := \left(\boldsymbol{C}(i-j) - \sum_{k=1}^{j-1} \boldsymbol{B}_{ik} \boldsymbol{E}_k \boldsymbol{B}_{ik}^T \right) \boldsymbol{E}_j^+, \tag{5.18}$$

where we take the Moore–Penrose inverse if necessary.

Note that Equation (5.15) implies the following:

$$|\mathfrak{C}_n| = |\boldsymbol{D}_n| = \prod_{j=1}^{n} |\boldsymbol{E}_j|$$

that is the multi-D analogue of the 1D Equation (5.10). Note that here $\boldsymbol{E}_1 = \boldsymbol{C}(0)$, and $\boldsymbol{E}_j$ is analogous to e_{j-1}^2 there.

Also note that $\boldsymbol{E}_n$ is the error covariance matrix of the prediction of $\mathbf{X}_n$ based on its $(n-1)$-length long past. In the stationary case, if we predict based on the n-length long past, then we project on a richer subspace, therefore the prediction errors of the linear combinations of the coordinates of $\mathbf{X}_n$ are decreased (better to say, not increased). Consequently, by Remark C.2 of Appendix C, the ranks of the error covariance matrices $\boldsymbol{E}_n$s are also decreased (not increased) as $n \to \infty$.

If the prediction is based on the infinite past, then with $n \to \infty$ this procedure (which is a nested one) extends to the multidimensional Wold decomposition. We can construct a causal TLF in this way. Actually, here $n = t$, and as observations arrive, $\mathbf{X}_n$ is predicted based on past values $\mathbf{X}_1, \dots, \mathbf{X}_{n-1}$, and so, $\boldsymbol{\eta}_n$ is in fact, $\boldsymbol{\eta}_{t,n}$. By stationarity, it has the same distribution for all t, especially for $t = n$. Also, if $n \to \infty$, the matrix $\boldsymbol{L}_n$ better and better approaches a Toeplitz one, and the matrices $\boldsymbol{E}_1, \dots, \boldsymbol{E}_n$ are closer and closer to $\boldsymbol{\Sigma}$, the covariance matrix of the innovation process $\{\boldsymbol{\eta}_t\}$ that is the limit of $\{\boldsymbol{\eta}_{t,n}\}$. In this way, we get the multi-D analogue of the 1D Theorem 5.1, according to which, $\boldsymbol{\eta}_n \to \boldsymbol{\eta}$ in mean square:

$$\|\boldsymbol{E}_n - \boldsymbol{\Sigma}\| = \|\mathbb{E}(\boldsymbol{\eta}_n \boldsymbol{\eta}_n^T) - \mathbb{E}(\boldsymbol{\eta}\boldsymbol{\eta}^T)\| \to 0$$

as $n \to \infty$. Consequently, $\boldsymbol{B}_{nj} \to \boldsymbol{B}_j$ as $n \to \infty$ as it continuously depends on $\boldsymbol{E}_j$s in view of Equations (5.18).

Also, if there is a gap in the spectrum of $\boldsymbol{\Sigma}$, like

$$\lambda_1 \geq \cdots \geq \lambda_r \geq \Delta \gg \varepsilon \geq \lambda_{r+1} \geq \cdots \geq \lambda_d,$$

then there is a gap in the spectrum of $\boldsymbol{E}_n$ too. Indeed, to any $\delta > 0$ there is an N such that for $n \geq N$: $\|\boldsymbol{E}_n - \boldsymbol{\Sigma}\| < \delta$. Then for the eigenvalues of $\boldsymbol{E}_n$,

$$\lambda_1^{(n)} \geq \cdots \geq \lambda_r^{(n)} \geq \Delta - \delta \gg \varepsilon + \delta \geq \lambda_{r+1}^{(n)} \geq \cdots \geq \lambda_d^{(n)}.$$

Consequently, for the best rank r approximations (with Gram-decompositions):

$$\|\boldsymbol{\Sigma} - \boldsymbol{\Sigma}^r\| \leq \varepsilon \quad \text{and} \quad \|\boldsymbol{E}_n - \boldsymbol{E}_n^r\| \leq \delta + \varepsilon$$

holds by Theorem B.5 (Weyl perturbation theorem). Therefore,

$$\|\boldsymbol{\Sigma}^r - \boldsymbol{E}_n^r\| \leq \|\boldsymbol{\Sigma}^r - \boldsymbol{\Sigma}\| + \|\boldsymbol{\Sigma} - \boldsymbol{E}_n\| + \|\boldsymbol{E}_n - \boldsymbol{E}_n^r\| \leq \varepsilon + \delta + (\delta + \varepsilon) = 2(\delta + \varepsilon)$$

that can be arbitrarily close to 2ε. At the same time, the projections onto the subspaces spanned by the eigenvectors of the r structural eigenvalues of these matrices are close to each other, in the sense of Theorem B.6 (Davis–Kahan theorem). Let $S_1 := [\Delta - \delta, \lambda_1 + \delta]$ and $S_2 := [\lambda_d + \delta, \varepsilon + \delta]$. Then for $n > N$:

$$\begin{aligned}
\|\boldsymbol{P}_{\boldsymbol{\Sigma}}(S_1) - \boldsymbol{P}_{\boldsymbol{E}_n}(S_1)\|_F^2 &= \|\boldsymbol{P}_{\boldsymbol{\Sigma}}(S_1)\|_F^2 + \|\boldsymbol{P}_{\boldsymbol{E}_n}(S_1)\|_F^2 - 2\,\mathrm{tr}[\boldsymbol{P}_{\boldsymbol{\Sigma}}(S_1)\boldsymbol{P}_{\boldsymbol{E}_n}^T(S_1)] \\
&= 2r - 2\,\mathrm{tr}[\boldsymbol{P}_{\boldsymbol{\Sigma}}(S_1)(\boldsymbol{I}_d - \boldsymbol{P}_{\boldsymbol{E}_n}^T(S_2))] \\
&= 2r - 2\,\mathrm{tr}[\boldsymbol{P}_{\boldsymbol{\Sigma}}(S_1) - \boldsymbol{P}_{\boldsymbol{\Sigma}}(S_1)\boldsymbol{P}_{\boldsymbol{E}_n}^T(S_2)] \\
&= 2r - 2r + 2\,\mathrm{tr}[\boldsymbol{P}_{\boldsymbol{\Sigma}}(S_1)\boldsymbol{P}_{\boldsymbol{E}_n}^T(S_2)] \\
&\leq 2d\|\boldsymbol{P}_{\boldsymbol{\Sigma}}(S_1)\boldsymbol{P}_{\boldsymbol{E}_n}^T(S_2)\| \\
&\leq 2d\frac{c}{\Delta - \delta - \varepsilon}\|\boldsymbol{\Sigma} - \boldsymbol{E}_n\| \leq 2d\frac{c\delta}{\Delta - \delta - \varepsilon}
\end{aligned}$$

that can be arbitrarily small if δ is arbitrarily small. Here we also used Lemma B.3 and Theorem B.6.

Going further, when the $\boldsymbol{E}_j$s are of rank $r < d$, we can find a system $\boldsymbol{\xi}_1, \dots, \boldsymbol{\xi}_n \in \mathbb{R}^r$ in the d-dimensional innovation subspaces that span the same subspace as $\boldsymbol{\eta}_1, \dots, \boldsymbol{\eta}_n$. (Though, in this situation, the block Cholesky decomposition algorithm should be modified by taking generalized inverses.) If the rank is not exactly r (may be full), but the spectral density matrix has $r < d$ structural eigenvalues, then $\boldsymbol{\xi}_j \in \mathbb{R}^r$, $\mathbb{E}\boldsymbol{\xi}_j\boldsymbol{\xi}_j' = \boldsymbol{I}_r$ is the principal component factor of $\boldsymbol{\eta}_j$ obtained from the r-factor model

$$\boldsymbol{\eta}_j = \boldsymbol{A}_j\boldsymbol{\xi}_j + \boldsymbol{\varepsilon}_j,$$

where the columns of $d \times r$ matrix $\boldsymbol{A}_j$ are $\sqrt{\lambda_{j\ell}}\mathbf{u}_{j\ell}$ with the r largest eigenvalues and the corresponding eigenvectors of $\boldsymbol{E}_j$; the vector $\boldsymbol{\varepsilon}_j$ is the error comprised of both the idiosyncratic noise and the error term of the model, but it has a negligible L^2-norm. Note that $\boldsymbol{A}_j$ of the decomposition $\boldsymbol{E}_j = \boldsymbol{A}_j \boldsymbol{A}_j^T$ is far not unique, it can be post-multiplied with an $r \times r$ orthogonal matrix. With this,

$$\mathbf{X}_k \sim \sum_{j=1}^{k} \boldsymbol{B}_{kj} \boldsymbol{A}_j \boldsymbol{\xi}_j, \quad k = 1, 2, \dots, n \tag{5.19}$$

where $\boldsymbol{B}_{kk} = \boldsymbol{I}_k$. This approaches the following Wold decomposition of the d-dimensional process $\{\mathbf{X}_t\}$ with an r-dimensional ($r \le d$) innovation process $\{\boldsymbol{\xi}_t\}$:

$$\mathbf{X}_t = \sum_{j=0}^{\infty} \boldsymbol{B}_j \boldsymbol{\eta}_{t-j} = \sum_{j=0}^{\infty} \boldsymbol{B}_j \boldsymbol{A} \boldsymbol{\xi}_{t-j},$$

where $\lim_{k\to\infty} \boldsymbol{B}_{kj} = \boldsymbol{B}_j$ is $d \times d$ matrix; $\{\boldsymbol{\eta}_t\}$ is a d-dimensional white-noise sequence with covariance matrix $\boldsymbol{\Sigma}$ of rank r (actually, $\boldsymbol{\Sigma}$ it is the limit of the sequence $\boldsymbol{E}_n$), and $\{\boldsymbol{\xi}_t\}$ is an r-dimensional white-noise sequence with covariance matrix $\boldsymbol{I}_r$. Further, $\boldsymbol{\Sigma} = \boldsymbol{A}\boldsymbol{A}^T$ is the Gram-decomposition of the matrix $\boldsymbol{\Sigma}$ of rank r, where $\boldsymbol{A}$ is $d \times r$ (see Appendix B). Then the matrix sequence $\boldsymbol{B}_j \boldsymbol{A}$ plays the role of the $d \times r$ coefficient matrices in the multidimensional Wold decomposition of Section 4.4.

Note that here we use $nd \times nd$ block matrices, but the procedure, realized by Equations (5.17) and (5.18), iterates only with the $d \times d$ blocks of them, so the computational complexity of this algorithm is not significantly larger than that of the Kálmán's filtering of Section 5.5. However, in the next Section 5.4, we can decrease this computational complexity in the frequency domain.

5.4 Spectra of spectra

Let $\{\mathbf{X}_t\}$ be a d-dimensional, weakly stationary time series with real components and autocovariance matrices $\boldsymbol{C}(h)$, $h \in \mathbb{Z}$, $\boldsymbol{C}(-h) = \boldsymbol{C}^T(h)$. Consider the finite segment $\mathbf{X}_1, \dots, \mathbf{X}_n \in \mathbb{R}^d$ of it and the $nd \times nd$ covariance matrix $\mathfrak{C}_n$ of the compounded random vector $[\mathbf{X}_1^T, \dots, \mathbf{X}_n^T]^T \in \mathbb{R}^{nd}$, as introduced in Equation (1.3). This is a symmetric, positive semidefinite block Toeplitz matrix, the (i, j) block of which is $\boldsymbol{C}(j - i)$. The symmetry comes from the fact, that the (j, i) entry is $\boldsymbol{C}(i - j) = \boldsymbol{C}^T(j - i)$.

To characterize the eigenvalues of the block Toeplitz matrix $\mathfrak{C}_n$, we need the symmetric block circulant matrix $\mathfrak{C}_n^{(s)}$ that we consider for odd n, say $n = 2k + 1$ here (for even n, the calculations are similar); for the definition, see [53]. In fact, the rows of a circulant matrix are cyclic permutations of the preceding ones; whereas, in the block circulant case, when permuting, we take transposes of the blocks if those are not symmetric themselves, see the example of Equation (5.20). Spectra of block circulant matrices are well characterized, but $\mathfrak{C}_n$ is not block circulant, in general; this is why $\mathfrak{C}_n^{(s)}$ is constructed, by disregarding the autocovariances of order greater than $\frac{n}{2}$. This can be done only on the assumption that the sequences of autcovariances are (entrywise) absolutely summable.

The (i, j) block of $\mathfrak{C}_n^{(s)}$ for $1 \leq i \leq j \leq n$ is

$$\mathfrak{C}_n^{(s)}(\text{block}_i, \text{block}_j) = \begin{cases} \boldsymbol{C}(j-i), & j-i \leq k \\ \boldsymbol{C}(n-(j-i)), & j-i > k; \end{cases}$$

whereas, for $i > j$, it is

$$\mathfrak{C}_n^{(s)}(\text{block}_i, \text{block}_j) = \begin{cases} \boldsymbol{C}^T(i-j), & i-j \leq k \\ \boldsymbol{C}^T(n-(i-j)), & i-j > k. \end{cases}$$

In this way, $\mathfrak{C}_n^{(s)}$ is a symmetric block Toeplitz matrix, like $\mathfrak{C}_n$, and it is the same as $\mathfrak{C}_n$ within the blocks (i, j)s for which $|j - i| \leq k$ holds. However, $\mathfrak{C}_n^{(s)}$ is also a block circulant matrix that fits our purposes. For example, if $n = 7$ and $k = 3$, then we have

$$\mathfrak{C}_7^{(s)} := \begin{bmatrix} \boldsymbol{C}(0) & \boldsymbol{C}(1) & \boldsymbol{C}(2) & \boldsymbol{C}(3) & \boldsymbol{C}(3) & \boldsymbol{C}(2) & \boldsymbol{C}(1) \\ \boldsymbol{C}^T(1) & \boldsymbol{C}(0) & \boldsymbol{C}(1) & \boldsymbol{C}(2) & \boldsymbol{C}(3) & \boldsymbol{C}(3) & \boldsymbol{C}(2) \\ \boldsymbol{C}^T(2) & \boldsymbol{C}^T(1) & \boldsymbol{C}(0) & \boldsymbol{C}(1) & \boldsymbol{C}(2) & \boldsymbol{C}(3) & \boldsymbol{C}(3) \\ \boldsymbol{C}^T(3) & \boldsymbol{C}^T(2) & \boldsymbol{C}^T(1) & \boldsymbol{C}(0) & \boldsymbol{C}(1) & \boldsymbol{C}(2) & \boldsymbol{C}(3) \\ \boldsymbol{C}^T(3) & \boldsymbol{C}^T(3) & \boldsymbol{C}^T(2) & \boldsymbol{C}^T(1) & \boldsymbol{C}(0) & \boldsymbol{C}(1) & \boldsymbol{C}(2) \\ \boldsymbol{C}^T(2) & \boldsymbol{C}^T(3) & \boldsymbol{C}^T(3) & \boldsymbol{C}^T(2) & \boldsymbol{C}^T(1) & \boldsymbol{C}(0) & \boldsymbol{C}(1) \\ \boldsymbol{C}^T(1) & \boldsymbol{C}^T(2) & \boldsymbol{C}^T(3) & \boldsymbol{C}^T(3) & \boldsymbol{C}^T(2) & \boldsymbol{C}^T(1) & \boldsymbol{C}(0) \end{bmatrix}. \quad (5.20)$$

In the univariate ($d = 1$) case, when $n = 2k + 1$, by Kronecker products (with permutation matrices) it is well known, see e.g. [19, 11, 53], that the jth (real) eigenvalue of $\boldsymbol{C}_n^{(s)}$ is

$$\sum_{h=-k}^{k} c(h)\rho_j^h = c(0) + 2\sum_{h=1}^{k} c(h)\cos(h\omega_j),$$

where $\rho_j = e^{i\omega_j}$ is the jth primitive (complex) nth root of 1 and $\omega_j = \frac{2\pi j}{n}$ is the jth Fourier frequency ($j = 0, 1, \ldots, n-1$). Further, the eigenvector corresponding to the jth eigenvalue is $(1, \rho_j, \ldots, \rho_j^{n-1})^T$; it has norm $\sqrt{n}$. After normalizing with $\frac{1}{\sqrt{n}}$, we get a complete orthonormal set of eigenvectors (of complex coordinates).

When $\boldsymbol{C}(h)$s are $d \times d$ matrices, by inflation techniques and applying Kronecker products, we use blocks instead of entries and the eigenvectors also follow a block structure. In [19, 53], the eigenvalues and eigenvectors of a general symmetric block circulant matrix are characterized. We apply this result in our situation, when $n = 2k + 1$ is odd. In view of this, the spectrum of $\mathfrak{C}_n^{(s)}$ is the union of spectra of the matrices

$$\boldsymbol{M}_j = \boldsymbol{C}(0) + \sum_{h=1}^{k}[\boldsymbol{C}(h)\rho_j^h + \boldsymbol{C}^T(h)\rho_j^{-h}] = \boldsymbol{C}(0) + \sum_{h=1}^{k}[\boldsymbol{C}(h)e^{i\omega_j h} + \boldsymbol{C}^T(h)e^{-i\omega_j h}]$$

for $j = 0, 1, \ldots, n-1$; whereas, the eigenvectors are obtained by compounding the eigenvectors of these $d \times d$ matrices. So we need the spectral decomposition of the matrices $\boldsymbol{M}_0 = \boldsymbol{C}(0) + \sum_{h=1}^{k} [\boldsymbol{C}(h) + \boldsymbol{C}^T(h)]$ and

$$\boldsymbol{M}_j = \boldsymbol{C}(0) + \sum_{h=1}^{k} [(\boldsymbol{C}(h) + \boldsymbol{C}^T(h)) \cos(\omega_j h) + i(\boldsymbol{C}(h) - \boldsymbol{C}^T(h)) \sin(\omega_j h)]$$

for $j = 1, 2, \ldots, n-1$. Since $\boldsymbol{C}(h) + \boldsymbol{C}^T(h)$ is symmetric and $\boldsymbol{C}(h) - \boldsymbol{C}^T(h)$ is anti-symmetric with 0 diagonal, $\boldsymbol{M}_j$ is self-adjoint for each j and has real eigenvalues with corresponding orthonormal set of eigenvectors of possibly complex coordinates. Indeed, $\boldsymbol{M}_j$ may have complex entries if $j \neq 0$; actually, $\sum_{h=1}^{k} (\boldsymbol{C}(h) + \boldsymbol{C}^T(h)) \cos(\omega_j h)$ is the real and $\sum_{h=1}^{k} (\boldsymbol{C}(h) - \boldsymbol{C}^T(h)) \sin(\omega_j h)$ is the imaginary part of $\boldsymbol{M}_j$.

It is easy to see that $\boldsymbol{M}_{n-j} = \overline{\boldsymbol{M}_j}$ (entrywise conjugate), therefore, it has the same (real) eigenvalues as $\boldsymbol{M}_j$, but its (complex) eigenvectors are the (componentwise) complex conjugates of the eigenvectors of $\boldsymbol{M}_j$. We also need the following form of this matrix:

$$\begin{aligned}\boldsymbol{M}_{n-j} &= \boldsymbol{C}(0) + \sum_{h=1}^{k} [(\boldsymbol{C}(h) + \boldsymbol{C}^T(h)) \cos(\omega_j h) - i(\boldsymbol{C}(h) - \boldsymbol{C}^T(h)) \sin(\omega_j h)] \\ &= \boldsymbol{C}(0) + \sum_{h=1}^{k} [\boldsymbol{C}(h) e^{-i\omega_j h} + \boldsymbol{C}^T(h) e^{i\omega_j h}], \quad j = 1, \ldots, n-1.\end{aligned} \tag{5.21}$$

Summarizing, for odd $n = 2k+1$, the nd eigenvalues of $\mathfrak{C}_n^{(s)}$ are obtained as the union of the (real) eigenvalues of $\boldsymbol{M}_0$ and those of $\boldsymbol{M}_j$ ($j = 1, \ldots, k$) duplicated. Note that for even n, similar arguments hold with the difference that there the spectrum of $\mathfrak{C}_n^{(s)}$ is the union of the eigenvalues of $\boldsymbol{M}_0$ and $\boldsymbol{M}_{n-1}$, whereas the eigenvalues of $\boldsymbol{M}_1, \ldots, \boldsymbol{M}_{\frac{n}{2}-1}$ are duplicated.

The eigenvectors of $\mathfrak{C}_n^{(s)}$ are obtainable by compounding the d (usually complex) orthonormal eigenvectors of the $d \times d$ self-adjoint matrices $\boldsymbol{M}_0, \boldsymbol{M}_1, \ldots, \boldsymbol{M}_{n-1}$ as follows. For $j = 1, \ldots, k$: if $\mathbf{v}$ is a unit-length eigenvector of $\boldsymbol{M}_j$ with eigenvalue λ, then in [53] it is proved that the compound vector

$$(\mathbf{v}^T, \rho_j \mathbf{v}^T, \rho_j^2 \mathbf{v}^T, \ldots, \rho_j^{n-1} \mathbf{v}^T)^T \in \mathbb{C}^{nd}$$

is an eigenvector of $\mathfrak{C}_n^{(s)}$ with the same eigenvalue λ. It has squared norm

$$\mathbf{v}^* \mathbf{v} (1 + \rho_j \rho_j^{-1} + \rho_j^2 \rho_j^{-2} + \cdots + \rho_j^{n-1} \rho_j^{-(n-1)}) = n.$$

Therefore, the vector

$$\mathbf{w} = \frac{1}{\sqrt{n}} (\mathbf{v}^T, \rho_j \mathbf{v}^T, \rho_j^2 \mathbf{v}^T, \ldots, \rho_j^{n-1} \mathbf{v}^T)^T \in \mathbb{C}^{nd} \tag{5.22}$$

is a unit-norm eigenvector (of complex coordinates) of $\mathfrak{C}_n^{(s)}$.

Further, if

$$\mathbf{z} = \frac{1}{\sqrt{n}}(\mathbf{t}^T, \rho_\ell \mathbf{t}^T, \rho_\ell^2 \mathbf{t}^T, \dots, \rho_\ell^{n-1} \mathbf{t}^T)^T \in \mathbb{C}^{nd}$$

is another unit-norm eigenvector of $\mathfrak{C}_n^{(s)}$ compounded from a unit-norm eigenvector $\mathbf{t}$ of another $\boldsymbol{M}_\ell$ ($\ell \neq j$), then $\mathbf{w}$ and $\mathbf{z}$ are orthogonal, irrespective whether $\boldsymbol{M}_\ell$ has the same eigenvalue λ as $\boldsymbol{M}_j$ or not. Similar construction holds starting with the eigenvectors of $\boldsymbol{M}_0$.

Here for each $j = 0, 1, \dots, n-1$, there are d pairwise orthonormal eigenvectors (potential $\mathbf{v}$s) of $\boldsymbol{M}_j$, and the so obtained $\mathbf{w}$s are also pairwise orthonormal. Assume that the eigenvectors of $\boldsymbol{M}_j$ are enumerated in non-increasing order of its (real) eigenvalues, and the inflated $\mathbf{w}$s also follow this ordering, for $j = 0, 1, \dots, n-1$.

Choose a unit-norm eigenvector $\mathbf{v} \in \mathbb{C}^d$ of $\boldsymbol{M}_j$ with (real) eigenvalue λ. Then $\overline{\mathbf{v}} \in \mathbb{C}^d$ is the corresponding unit-norm eigenvector of $\boldsymbol{M}_{n-j}$ with the same eigenvalue λ. Consider the compounded $\mathbf{w} \in \mathbb{C}^{nd}$ and $\overline{\mathbf{w}} \in \mathbb{C}^{nd}$ obtained from them by Equation (5.22). We learned that they are orthonormal eigenvectors of $\mathfrak{C}_n^{(s)}$ corresponding to the eigenvalue λ with multiplicity (at least) two. From them, corresponding to this double eigenvalue λ, the new orthonormal pair of eigenvectors

$$\frac{\mathbf{w} + \overline{\mathbf{w}}}{\sqrt{2}} \quad \text{and} \quad -i\frac{\mathbf{w} - \overline{\mathbf{w}}}{\sqrt{2}} \tag{5.23}$$

is constructed, but they, in this order, occupy the original positions of $\mathbf{w}$ and $\overline{\mathbf{w}}$. They have real coordinates and unit norm; actually, their coordinates contain the $\sqrt{2}$ multiples the real and imaginary parts of the corresponding coordinates of $\mathbf{w}$. It is in accord with the fact that a real symmetric matrix, as $\mathfrak{C}_n^{(s)}$, must have an orthogonal system of eigenvectors with real coordinates too. We do not go in details, neither discuss defective cases.

Consider $\mathbf{u}_1, \dots, \mathbf{u}_{nd}$, the so obtained orthonormal set of eigenvectors (of real coordinates) of $\mathfrak{C}_n^{(s)}$ (in the above ordering), and denote by $\boldsymbol{U} = (\mathbf{u}_1, \dots, \mathbf{u}_{nd})$ the $nd \times nd$ (real) orthogonal matrix containing them columnwise. Let

$$\mathfrak{C}_n^{(s)} = \boldsymbol{U}\boldsymbol{\Lambda}^{(s)}\boldsymbol{U}^T \tag{5.24}$$

be the corresponding spectral decomposition. After this preparation, we are able to prove the following theorem.

Theorem 5.2. *Let $\{\mathbf{X}_t\}$ be d-dimensional weakly stationary time series of real components. Denoting by $\boldsymbol{C}(h) = [c_{ij}(h)]$ the $d\times d$ autocovariance matrices ($\boldsymbol{C}(-h) = \boldsymbol{C}^T(h)$, $h \in \mathbb{Z}$) in the time domain, assume that their entries are absolutely summable, i.e. $\sum_{h=0}^{\infty} |c_{pq}(h)| < \infty$ for $p, q = 1, \dots, d$. Then, the self-adjoint, positive semidefinite spectral density matrix $\boldsymbol{f}$ exists in the frequency domain, and it is defined by*

$$\boldsymbol{f}(\omega) = \frac{1}{2\pi}\sum_{h=-\infty}^{\infty} \boldsymbol{C}(h)e^{-ih\omega}, \quad \omega \in [0, 2\pi].$$

For odd $n = 2k+1$, consider $\mathbf{X}_1, \dots \mathbf{X}_n$ and the block Toeplitz matrix $\mathfrak{C}_n$ of (1.3); further, the Fourier frequencies $\omega_j = \frac{2\pi j}{n}$ for $j = 0, \dots, n-1$. Let $\boldsymbol{D}_n$ be the $dn \times dn$ diagonal matrix that contains the spectra of the matrices $\boldsymbol{f}(0), \boldsymbol{f}(\omega_1), \boldsymbol{f}(\omega_2), \dots, \boldsymbol{f}(\omega_k), \boldsymbol{f}(\omega_k), \dots, \boldsymbol{f}(\omega_2), \boldsymbol{f}(\omega_1)$ in its main diagonal, i.e.

$$\boldsymbol{D}_n = \mathrm{diag}(\mathrm{spec}\, \boldsymbol{f}(0), \mathrm{spec}\, \boldsymbol{f}(\omega_1), \dots, \mathrm{spec}\, \boldsymbol{f}(\omega_k), \mathrm{spec}\, \boldsymbol{f}(\omega_k), \dots, \mathrm{spec}\, \boldsymbol{f}(\omega_1)).$$

Here spec *denotes the eigenvalues of the affected matrix in non-increasing order if not otherwise stated. (The duplication is due to the fact that $\boldsymbol{f}(\omega_j) = \boldsymbol{f}(\omega_{n-j})$, $j = 1, \dots, k$, for real time series). Then, with the spectral decomposition* (5.24),

$$\boldsymbol{U}^T \mathfrak{C}_n \boldsymbol{U} - 2\pi \boldsymbol{D}_n \to \boldsymbol{O}, \quad n \to \infty,$$

i.e. the entries of the matrix $\boldsymbol{U}^T \mathfrak{C}_n \boldsymbol{U} - 2\pi \boldsymbol{D}_n$ tend to 0 uniformly as $n \to \infty$.

Proof. We saw that $\boldsymbol{U}^T \mathfrak{C}_n^{(s)} \boldsymbol{U} = \boldsymbol{\Lambda}^{(s)}$. Recall that the eigenvalues in the diagonal of $\boldsymbol{\Lambda}^{(s)}$ comprise the union of spectra of the matrices $\boldsymbol{M}_0$ and those of $\boldsymbol{M}_1, \dots, \boldsymbol{M}_{n-1}$, which are the same as the eigenvalues of $\boldsymbol{M}_0$ and those of $\boldsymbol{M}_{n-1}, \dots, \boldsymbol{M}_{n-k}$ of (5.21), duplicated. But these matrices are finite sub-sums (for $|h| \le k$) of the infinite summations

$$2\pi \boldsymbol{f}(\omega_j) = \sum_{h=-\infty}^{\infty} \boldsymbol{C}(h) e^{-ih\omega} = \boldsymbol{C}(0) + \sum_{h=1}^{\infty} [\boldsymbol{C}(h) e^{-i\omega_j h} + \boldsymbol{C}^T(h) e^{i\omega_j h}].$$

So, by the absolute summability of the autocovariances, and because the eigenvalues depend continuously on the underlying matrices, the pairwise distances between the eigenvalues of $\boldsymbol{M}_j$ and the corresponding eigenvalues of $2\pi \boldsymbol{f}(\omega_j)$ (both in non-increasing order) tend to 0 as $n \to \infty$, for $j = 0, 1, \dots, k$. Indeed, the absolute summability of the entries of $\boldsymbol{C}(h)$s implies that the diagonal entries of the diagonal matrix $\boldsymbol{\Lambda}^{(s)} - 2\pi \boldsymbol{D}_n$ are bounded in absolute value by

$$\max_{p,q \in \{1,\dots,d\}} \sum_{|h|>k} |c_{pq}(h)| \to 0, \quad n = 2k+1 \to \infty.$$

Therefore, the matrix $\boldsymbol{\Lambda}^{(s)} - 2\pi \boldsymbol{D}_n$ tends to the zero matrix entrywise uniformly as $n \to \infty$. It remains to show that the entries of $\mathbf{U}^T \mathfrak{C}_n \mathbf{U} - \mathbf{U}^T \mathfrak{C}_n^{(s)} \mathbf{U}$ tend to 0 uniformly as $n \to \infty$.

Before doing this, some facts should be clarified.

- The pth row sum of $\boldsymbol{M}_j$ is bounded by

$$\sum_{q=1}^{d} |c_{pq}(0)| + \sum_{q=1}^{d} \sum_{h=1}^{k} |c_{pq}(h)| + \sum_{q=1}^{d} \sum_{h=1}^{k} |c_{qp}(h)| \le d c_{pp}(0) + 2dL,$$

for $p \in \{1,\dots,d\}$ with $L = \max_{p,q\in\{1,\dots,d\}} \sum_{h=1}^{\infty} |c_{pq}(h)| > 0$, independently of n, because of the absolute summability of the entries of $\boldsymbol{C}(h)$. This is true for any $j \in \{0,1,\dots,n-1\}$. For simplicity, consider (any) one of the $\boldsymbol{M}_j$s, and denote it by $\boldsymbol{M} = [m_{pq}]_{p,q=1}^{d}$. Then

$$\|\boldsymbol{M}\|_\infty = \max_{p\in\{1,\dots,d\}} \sum_{q=1}^{d} |m_{pq}| \le d \max_{p\in\{1,\dots,d\}} c_{pp}(0) + 2dL = K.$$

As the spectral radius of $\boldsymbol{M}$ is at most $\|\boldsymbol{M}\|_\infty$, any eigenvalue λ of $\boldsymbol{M}$ is bounded in absolute value by K (independently of n).

- Recall that $\mathbf{u}$ is compounded via (5.22) and (5.23) from the primitive roots. Therefore, its coordinates are bounded by $\sqrt{\frac{2}{n}}$ in absolute value.

Now we are ready to show that

$$|\mathbf{u}_i^T \mathfrak{C}_n^{(s)} \mathbf{u}_j - \mathbf{u}_i^T \mathfrak{C}_n \mathbf{u}_j| \to 0, \quad n \to \infty$$

uniformly in $i,j \in \{1,\dots,nd\}$. Recall that in the $nd \times nd$ matrices $\mathfrak{C}_n^{(s)}$ and $\mathfrak{C}_n$ the (m,ℓ) blocks are the same if $|m-\ell| \le k$. Denote by $\mathbf{u}_{i,m}$ and $\mathbf{u}_{j,\ell}$ the mth and ℓth blocks of the unit-norm eigenvectors $\mathbf{u}_i$ and $\mathbf{u}_j$, respectively. Then

$$\begin{aligned}
&|\mathbf{u}_i^T(\mathfrak{C}_n^{(s)} - \mathfrak{C}_n)\mathbf{u}_j| \\
&= \Bigg| \sum_{m=1}^{k} \sum_{\ell=1}^{m} [\mathbf{u}_{i,\ell}^T (\boldsymbol{C}(m) - \boldsymbol{C}(n-m))\mathbf{u}_{j,n-m+\ell} \\
&+ \mathbf{u}_{i,n-m+\ell}^T (\boldsymbol{C}(n-m) - \boldsymbol{C}(m))\mathbf{u}_{j,\ell}] \Bigg| \\
&\le 2\sqrt{\frac{2}{n}}\sqrt{\frac{2}{n}} \sum_{m=1}^{k} m \left|\mathbf{1}_d^T (\boldsymbol{C}(m) - \boldsymbol{C}(n-m))\mathbf{1}_d\right| \\
&\le \frac{4}{n}\left(\sum_{m=1}^{k} m \sum_{p=1}^{d}\sum_{q=1}^{d} |c_{pq}(m)| + \sum_{m=1}^{k} m \sum_{p=1}^{d}\sum_{q=1}^{d} |c_{pq}(n-m)| \right) \\
&\le 4d^2 \left(\max_{p,q\in\{1,\dots,d\}} \sum_{m=1}^{k} \frac{m}{n}|c_{pq}(m)| + \max_{p,q\in\{1,\dots,d\}} \sum_{m=1}^{k} \frac{m}{n}|c_{pq}(n-m)| \right) \\
&\le 4d^2 \left(\max_{p,q\in\{1,\dots,d\}} \sum_{m=1}^{k} \frac{m}{n}|c_{pq}(m)| + \max_{p,q\in\{1,\dots,d\}} \sum_{m=n-k}^{n-1} \frac{k}{n}|c_{pq}(m)| \right),
\end{aligned}$$

where $\mathbf{1}_d \in \mathbb{R}^d$ is the vector of all 1 coordinates and so, the quadratic form $\mathbf{1}_d^T(\boldsymbol{C}(m) - \boldsymbol{C}(n-m))\mathbf{1}_d$ is the sum of the entries of $\boldsymbol{C}(m) - \boldsymbol{C}(n-m)$. In the last line, the second term converges to 0, since it is bounded by

$\sum_{m=k}^{\infty}|c_{pq}(m)|$ (indeed, $\sum_{m=n-k}^{n-1}\frac{k}{n}|c_{pq}(m)| \le \sum_{m=k}^{\infty}|c_{pq}(m)|$ as $k < n-k$), and together with n, k tends to ∞ too; further, it holds uniformly for all $p, q \in \{1, \dots, d\}$. The first term for every p, q pair also tends to 0 as $n \to \infty$ by the discrete version of the dominated convergence theorem (for series), see the forthcoming Lemma 5.1. Indeed, the summand is dominated by $|c_{pq}(m)|$ and $\sum_{m=1}^{\infty}|c_{pq}(m)| < \infty$; further, $\frac{m}{n}|c_{pq}(m)| \to 0$ as $n \to \infty$, for any fixed m. Consequently, $\sum_{m=1}^{\infty}\frac{m}{n}|c_{pq}(m)|$ tends to 0, and so does $\sum_{m=1}^{k}\frac{m}{n}|c_{pq}(m)|$ as $n \to \infty$. It holds uniformly for all p, q, and also for all i, j, so the proof is complete. □

Lemma 5.1. (Dominated convergence theorem for sums, discrete version). *Consider* $\sum_{m=1}^{\infty} f_n(m)$ *and assume that* $|f_n(m)| \le g(m)$ *with* $\sum_{m=1}^{\infty} g(m) < \infty$. *If* $\lim_{n\to\infty} f_n(m) = f(m)$ *exists* $\forall m \in \mathbb{N}$, *then*

$$\lim_{n\to\infty}\sum_{m=1}^{\infty} f_n(m) = \sum_{m=1}^{\infty} f(m).$$

Some important consequences of Theorem 5.2 follow.

5.4.1 Bounds for the eigenvalues of $\mathfrak{C}_n$

Proposition 5.3. *Analogously to the 1D statement (see [11], Proposition 4.5.3), the above theorem implies the following. Assume that for the spectra of the spectral densities* $\boldsymbol{f}$ *of the* d*-dimensional weakly stationary process* $\{\mathbf{X}_t\}$ *of real coordinates the following hold:*

$$m := \inf_{\omega\in[0,2\pi],q\in\{1,\dots,d\}} \lambda_q(\boldsymbol{f}(\omega)) \ge 0,$$
$$M := \sup_{\omega\in[0,2\pi],q\in\{1,\dots,d\}} \lambda_q(\boldsymbol{f}(\omega)) < \infty.$$

(Note that under the conditions of Theorem 5.2, $\boldsymbol{f}(\omega)$ *is continuous almost everywhere on* $[0, 2\pi]$, *so the above conditions are readily satisfied.)*

Then for the eigenvalues $\lambda_1 \le \lambda_2 \le \dots \le \lambda_{nd}$ *of the block Toeplitz matrix* $\mathfrak{C}_n$ *the following holds:*

$$2\pi m \le \lambda_1 \le \lambda_{nd} \le 2\pi M.$$

Proof. Let λ be an arbitrary eigenvalue of $\mathfrak{C}_n$ with a corresponding eigenvector $\mathbf{x} \in \mathbb{C}^{nd}$, $\mathbf{x}^* = [\mathbf{x}_1^*, \dots, \mathbf{x}_n^*]$, $\mathbf{x}_j \in \mathbb{C}^d$: $\mathfrak{C}_n\mathbf{x} = \lambda\mathbf{x}$. Take the spectral decomposition of the spectral density matrix $\boldsymbol{f}$:

$$\boldsymbol{f}(\omega) = \sum_{\ell=1}^{d} \lambda_\ell(\boldsymbol{f}(\omega)) \cdot \mathbf{u}_\ell(\boldsymbol{f}(\omega)) \cdot \mathbf{u}_\ell^*(\boldsymbol{f}(\omega)).$$

Then we can write that

$$\begin{aligned}
\lambda|\mathbf{x}|^2 &= \lambda\mathbf{x}^*\mathbf{x} = \mathbf{x}^*\mathfrak{C}_n\mathbf{x} \\
&= \mathbf{x}^* \cdot \int_{-\pi}^{\pi} \left[e^{-i(j-k)\omega} \boldsymbol{f}(\omega) \right]_{j,k=1}^{n} d\omega \cdot \mathbf{x} \\
&= \int_{-\pi}^{\pi} \sum_{j,k=1}^{n} e^{-i(j-k)\omega} \mathbf{x}_j^* \boldsymbol{f}(\omega) \mathbf{x}_k d\omega \\
&= \int_{-\pi}^{\pi} \sum_{j,k=1}^{n} e^{-i(j-k)\omega} \sum_{\ell=1}^{d} \lambda_\ell(\boldsymbol{f}(\omega)) \cdot \mathbf{x}_j^* \cdot \mathbf{u}_\ell(\boldsymbol{f}(\omega)) \cdot \mathbf{u}_\ell^*(\boldsymbol{f}(\omega)) \cdot \mathbf{x}_k \, d\omega \\
&= \int_{-\pi}^{\pi} \sum_{\ell=1}^{d} \lambda_\ell(\boldsymbol{f}(\omega)) \sum_{j,k=1}^{n} e^{-ij\omega} \mathbf{x}_j^* \cdot \mathbf{u}_\ell(\boldsymbol{f}(\omega)) \cdot \mathbf{u}_\ell^*(\boldsymbol{f}(\omega)) \cdot \mathbf{x}_k \cdot e^{ik\omega} \, d\omega \\
&= \int_{-\pi}^{\pi} \sum_{\ell=1}^{d} \lambda_\ell(\boldsymbol{f}(\omega)) \left| \sum_{j=1}^{n} e^{-ij\omega} \cdot \mathbf{x}_j^* \cdot \mathbf{u}_\ell(\boldsymbol{f}(\omega)) \right|^2 d\omega \\
&\leq M \sum_{j,k=1}^{n} \mathbf{x}_j^* \cdot \int_{-\pi}^{\pi} e^{-i(j-k)\omega} \sum_{\ell=1}^{d} \mathbf{u}_\ell(\boldsymbol{f}(\omega)) \cdot \mathbf{u}_\ell^*(\boldsymbol{f}(\omega)) \, d\omega \cdot \mathbf{x}_k \\
&= 2\pi M \sum_{j=1}^{n} \mathbf{x}_j^* \mathbf{x}_j = 2\pi M |\mathbf{x}|^2.
\end{aligned}$$

This proves that $\lambda \leq 2\pi M$ for any eigenvalue of $\mathfrak{C}_n$. The proof of the fact that $\lambda \geq 2\pi m$ is similar. □

5.4.2 Principal component transformation as discrete Fourier transformation

The complex principal component (PC) transform of the collection of random vectors $\mathbf{X} = (\mathbf{X}_1^T, \dots \mathbf{X}_n^T)^T$ of real coordinates is the random vector $\mathbf{Z} = (\mathbf{Z}_1^T, \dots, \mathbf{Z}_n^T)^T$ of complex coordinates obtained by

$$\mathbf{Z} = \boldsymbol{W}^* \mathbf{X}.$$

Here, analogously to (5.24), $\mathfrak{C}_n^{(s)}$ also has the spectral decomposition

$$\mathfrak{C}_n^{(s)} = \boldsymbol{W} \boldsymbol{\Lambda}^{(s)} \boldsymbol{W}^*,$$

where the unitary matrix $\boldsymbol{W} = (\mathbf{w}_1, \dots, \mathbf{w}_{nd})$, contains a complete orthonormal set of eigenvectors of $\mathfrak{C}_n^{(s)}$, columnwise. They usually have complex coordinates.

To relate the PC transformation to a discrete Fourier transformation, we also make PC transformations within the blocks. For this purpose we use the eigenvectors in the columns of $\boldsymbol{W}$ (of complex coordinates) in the ordering described in the preparation of Theorem 5.2. We utilize their block structure and also assume that they are already normalized to have a complete orthonormal system in $\mathbb{C}^{nd}$.

By Theorem 5.2, $\mathbb{E}\mathbf{Z}\mathbf{Z}^* \sim 2\pi \boldsymbol{D}_n$, so the coordinates of $\mathbf{Z}$ are asymptotically uncorrelated, for "large" n. Instead, we consider the blocks $\mathbf{Z}_j$s of it, and perform a "partial principal component transformation" (in d-dimension) of them. Let $\mathbf{w}_{1j}, \dots, \mathbf{w}_{dj}$ be the columns of $\mathbf{W}$ corresponding to the coordinates of $\mathbf{Z}_j$. In view of (5.22), $\mathbf{Z}_j$ can be written as

$$\mathbf{Z}_j = \frac{1}{\sqrt{n}}(\boldsymbol{V}_j^* \otimes \mathbf{r}^*)\mathbf{X},$$

where $\mathbf{r}^* = (1, \rho_j^{-1}, \rho_j^{-2}, \dots, \rho_j^{-(n-1)})$ and $\boldsymbol{V}_j$ is the $d \times d$ unitary matrix in the spectral decomposition $\boldsymbol{M}_j = \boldsymbol{V}_j \boldsymbol{\Lambda}_j \boldsymbol{V}_j^*$. Because of $\mathbb{E}\mathbf{Z}_j\mathbf{Z}_j^* = \boldsymbol{\Lambda}_j$ (apparently from the proof of Theorem 5.2), we have that

$$\mathbb{E}(\boldsymbol{V}_j\mathbf{Z}_j)(\boldsymbol{V}_j\mathbf{Z}_j)^* = \boldsymbol{V}_j \boldsymbol{\Lambda}_j \boldsymbol{V}_j^* = \boldsymbol{M}_j.$$

At the same time,

$$\boldsymbol{V}_j\mathbf{Z}_j = \frac{1}{\sqrt{n}}\boldsymbol{V}_j(\boldsymbol{V}_j^* \otimes \mathbf{r}^*)\mathbf{X} = \frac{1}{\sqrt{n}}(\boldsymbol{I}_d \otimes \mathbf{r}^*)\mathbf{X} = \frac{1}{\sqrt{n}}\sum_{t=1}^{n} \mathbf{X}_t e^{-it\omega_j}, \quad j = 1, \dots, n.$$

This is the discrete Fourier transform of $\mathbf{X}_1, \dots, \mathbf{X}_n$. It is in accord with the existence of the orthogonal increment process $\{\mathbf{Z}_\omega\}$ (see Chapter 1 and [11]) of which $\boldsymbol{V}_j\mathbf{Z}_j \sim \mathbf{Z}_{\omega_j}$ is the discrete analogue. Also, $\mathbf{Z}_1, \dots \mathbf{Z}_n$ are asymptotically pairwise orthogonal akin to $\boldsymbol{V}_1\mathbf{Z}_1, \dots, \boldsymbol{V}_n\mathbf{Z}_n$. Further,

$$\mathbb{E}(\boldsymbol{V}_j\mathbf{Z}_j)(\boldsymbol{V}_j\mathbf{Z}_j)^* \sim 2\pi \boldsymbol{f}(\omega_j),$$

and it is in accord with the fact that

$$\mathbb{E}\mathbf{Z}_j\mathbf{Z}_j^* \sim 2\pi \,\mathrm{diag\,spec}\, \boldsymbol{f}(\omega_j),$$

for $j = 0, 1, \dots, n-1$ when n is "large".

5.5 Kálmán's filtering

Given a linear dynamical system, with state equations and specified matrices, R.E. Kálmán gave a recursive algorithm, how to find prediction for the state variable $\mathbf{X}_t$ in the possession of newer and newer observations for the observable variable $\mathbf{Y}_t$. Starting at time 0, estimates $\widehat{\mathbf{X}}_{t+1|t}$ are found, while observing $\mathbf{Y}_t$, $t = 1, 2, \dots$. The point is that we only use the last observation

$\mathbf{Y}_t$ and the preceding estimate $\widehat{\mathbf{X}}_{t|t-1}$. During the recursion, we use the linearity of the state equations and the predictions, for which either normality is assumed, or we confine ourselves to the second moments of the underlying distributions, see Appendix C.

This so-called filtering technique is widely used in the engineering practice, when we can "get rid" of the noise, and also possess an algorithm to find the innovations of the observed $\{\mathbf{Y}_t\}$ process (we need not perform the block Cholesky decomposition of Section 5.3.2, but get the innovations recursively). The problem is that we merely have the output of a linear system that is burdened with noise, and usually not invertible; e.g. in case of telecommunication systems, when sensors can sense only noisy signals. It is not by accident that the research of R.E. Kálmán and R.S. Bucy followed the era of the information theoretical breakthroughs, e.g. the intensive use of the Shannon entropy.

Here we follow the discussion of R.E. Kálmán's original paper [32], where stationarity is not assumed, but the random vectors are Gaussian. (Sometimes we use simpler notation in accordance with the one used in the previous sections of this chapter.) Here the linear dynamical system is

$$\begin{aligned} \mathbf{X}_{t+1} &= \boldsymbol{A}_t \mathbf{X}_t + \mathbf{U}_t \\ \mathbf{Y}_t &= \boldsymbol{C}_t \mathbf{X}_t, \end{aligned} \tag{5.25}$$

where $\boldsymbol{A}_t$ and $\boldsymbol{C}_t$ are specified matrices; $\boldsymbol{A}_t$ is an $n \times n$ matrix, called *phase transition matrix*, and $\boldsymbol{C}_t$ is $p \times n$; further, $\mathbf{U}_t$ is an orthogonal noise process with $\mathbb{E}\mathbf{U}_t\mathbf{U}_s^T = \delta_{st}\boldsymbol{Q}_{\mathbf{U}}(t)$ and $\mathbb{E}\mathbf{X}_s^T\mathbf{U}_t = \mathbf{0}$ for $s \le t$. All the expectations are zeros, and all the random vectors have real components. Sometimes $\mathbf{U}_t$ is called *random excitation*, $\mathbf{X}_t$ is the n-dimensional hidden *state variable*, while $\mathbf{Y}_t$ is the p-dimensional *observable variable*. In the paper [32], $p \le n$ is assumed, but it is not a restriction. Even if $p = n$, the matrix $\boldsymbol{C}_t$ is not invertible, otherwise the process $\mathbf{X}_t$ is trivially observable, unless a noise term is added to $\boldsymbol{C}_t\mathbf{X}_t$ in the second equation (we will touch upon this possibility at the end of this section).

So the problem is the following: starting the observations at time 0, given $\mathbf{Y}_0, \dots, \mathbf{Y}_{t-1}$, we want to estimate $\mathbf{X}$ component-wise, with minimum mean square error. More precisely, if $\widehat{\mathbf{X}}$ denotes this estimate, then $\widehat{\mathbf{X}} = \mathrm{Proj}_{H_{t-1}(\mathbf{Y})}\mathbf{X}$, where $H_{t-1}(\mathbf{Y}) = \mathrm{Span}\,(\mathbf{Y}_0, \dots, \mathbf{Y}_{t-1})$ consists of the linear combinations of all the components of $\mathbf{Y}_0, \dots, \mathbf{Y}_{t-1}$ (with the notation of Appendix C, but here the indexing starts at 0) and the projection of $\mathbf{X}$ is meant component-wise as noted in Remark C.2. If we minimize the mean square error, the minimizer is the conditional expectation $\mathbb{E}(\mathbf{X} \,|\, \mathbf{Y}_0, \dots, \mathbf{Y}_{t-1})$, which is the linear function of the coordinates of the random vectors in the condition, whenever the underlying distribution is Gaussian.

If $\mathbf{X} = \mathbf{X}_t$, this is the prediction problem and we denote the optimal *one-step ahead prediction* of $\mathbf{X}_t$ by $\widehat{\mathbf{X}}_{t|t-1}$. In a similar vein, $\widehat{\mathbf{X}}_{t|t}$ solves the *filtration* problem, when we project onto $H_t(\mathbf{Y})$ for the prediction; finally, $\widehat{\mathbf{X}}_{t|t+h}$ solves the *smoothing* problem, when we project onto $H_{t+h}(\mathbf{Y})$ with

$h > 0$ integer. The first one-step ahead prediction problem can be generalized to the h-step ahead prediction of $\widehat{\mathbf{X}}_{t+h|t-1}$, $h > 0$ integer (not the same as the smoothing problem). The first problem is sometimes called extrapolation, whereas the second two interpolation, respectively, see also A.N. Kolmogorov [34] and N. Wiener [57]. Note that the problem itself is originated in the Wiener–Hopf problem.

As for the one-step ahead prediction problem, if $\mathbf{Y}_0, \dots, \mathbf{Y}_{t-1}$ are observed, i.e. $H_{t-1}(\mathbf{Y})$ is known, then the newly observed (measured) $\mathbf{Y}_t$ can be orthogonally decomposed as

$$\mathbf{Y}_t = \mathrm{Proj}_{H_{t-1}(\mathbf{Y})}\mathbf{Y}_t + \widetilde{\mathbf{Y}}_{t|t-1} = \overline{\mathbf{Y}}_{t|t-1} + \widetilde{\mathbf{Y}}_{t|t-1}, \tag{5.26}$$

where the orthogonal component $\widetilde{\mathbf{Y}}_{t|t-1} \in I_t(\mathbf{Y})$, and $I_t(\mathbf{Y})$ is the so-called *innovation subspace.* (Actually, the components of $\widetilde{\mathbf{Y}}_{t|t-1}$ span $I_t(\mathbf{Y})$). We shall make intensive use of this innovation. Assume that $I_t(\mathbf{Y})$ is not the sole $\mathbf{0}$ vector, otherwise observing $\mathbf{Y}_t$ does not give any additional information to $H_{t-1}(\mathbf{Y})$. If $\{\mathbf{Y}_t\}$ is weakly stationary, it means that the process is *regular.*

Equation (5.26) implies the decomposition of the corresponding subspaces like

$$H_t(\mathbf{Y}) = H_{t-1}(\mathbf{Y}) \oplus I_t(\mathbf{Y}) \tag{5.27}$$

that is the analogue of the multidimensional Wold decomposition in the case when the prediction is based on finite past measurements. The multidimensional Wold decomposition applies to the stationary and infinite past case. When $t \to \infty$, i.e. going to the future, we approach this situation.

Assume that $\widehat{\mathbf{X}}_{t|t-1}$ is already known. We shall give a recursion to find $\widehat{\mathbf{X}}_{t+1|t}$ by using the new value of $\mathbf{Y}_t$. In view of Equation (5.27), we proceed as follows:

$$\begin{aligned}\widehat{\mathbf{X}}_{t+1|t} &= \mathrm{Proj}_{H_t(\mathbf{Y})}\mathbf{X}_{t+1} = \mathrm{Proj}_{H_{t-1}(\mathbf{Y})}\mathbf{X}_{t+1} + \mathrm{Proj}_{I_t(\mathbf{Y})}\mathbf{X}_{t+1} \\ &= \boldsymbol{A}_t\mathrm{Proj}_{H_{t-1}(\mathbf{Y})}\mathbf{X}_t + \mathrm{Proj}_{H_{t-1}(\mathbf{Y})}\mathbf{U}_t + \boldsymbol{K}_t\widetilde{\mathbf{Y}}_{t|t-1} \\ &= \boldsymbol{A}_t\widehat{\mathbf{X}}_{t|t-1} + \boldsymbol{K}_t\widetilde{\mathbf{Y}}_{t|t-1},\end{aligned} \tag{5.28}$$

where we utilized Lemma C.3 of Appendix C and the fact that $\mathbf{U}_t \perp H_{t-1}(\mathbf{Y})$; we also used the first (state) equation of (5.25). Since $\mathrm{Proj}_{I_t(\mathbf{Y})}\mathbf{X}_{t+1}$ is a linear operation and results in a vector within $I_t(\mathbf{Y})$ (linear combination of the components of $\widetilde{\mathbf{Y}}_{t|t-1}$), its effect can be written as a matrix $\boldsymbol{K}_t$ times $\widetilde{\mathbf{Y}}_{t|t-1}$. This $n \times p$ matrix $\boldsymbol{K}_t$ is called *Kálmán gain* matrix.

(In fact, the notation $\boldsymbol{K}$ is first used in the paper [33] of Kálmán and Bucy.) In another context, when $\widehat{\mathbf{X}}_{t|t}$ is produced, then a strongly related matrix $\boldsymbol{L}_t$ emerges that in some places is also called gain matrix; however, $\boldsymbol{K}_t = \boldsymbol{A}_t\boldsymbol{L}_t$ as it will be shown later in this section. In the stationary case, the rank $r(\leq p)$ of the spectral density matrix of the process $\{\mathbf{Y}_t\}$ is equal to the dimension of the innovation subspace if we predict based on the infinite past (see [40, 47]). In this stationary, infinite past case, $\boldsymbol{K}_t$ has a limiting value at the passage to infinity and it is unique only if the spectral density matrix of the process $\{\mathbf{Y}_t\}$ is of full rank ($r = p$), or equivalently, if the $p \times p$ covariance matrix of the innovations is non-singular. If $t \to \infty$, we approach the infinite past based prediction, and so, if $\mathbb{E}[\widetilde{\mathbf{Y}}_{t|t-1}\widetilde{\mathbf{Y}}_{t|t-1}^T]$ has some near zero eigenvalues, this is an indication of a reduced rank spectral density matrix of $\{\mathbf{Y}_t\}$, see Section 5.3.2. In the nonstationary case too, even if there are innovations (the innovations are not zeros), the innovation subspace can be of reduced rank, in which case $\mathbb{E}[\widetilde{\mathbf{Y}}_{t|t-1}\widetilde{\mathbf{Y}}_{t|t-1}^T]$ is not invertible (we shall take its generalized inverse later if necessary).

To specify the Kálmán gain matrix $\boldsymbol{K}_t$, we have to write $\widetilde{\mathbf{Y}}_{t|t-1}$ in terms of $\widehat{\mathbf{X}}_{t|t-1}$ and $\mathbf{Y}_t$. For this purpose, let us project both sides of the second (observation) equation of (5.25), i.e. of $\mathbf{Y}_t = \boldsymbol{C}_t\mathbf{X}_t$, to $H_{t-1}(\mathbf{Y})$. We get that

$$\overline{\mathbf{Y}}_{t|t-1} = \boldsymbol{C}_t\widehat{\mathbf{X}}_{t|t-1}.$$

Taking the orthogonal decomposition (5.26) of $\mathbf{Y}_t$ into consideration yields that

$$\widetilde{\mathbf{Y}}_{t|t-1} = \mathbf{Y}_t - \overline{\mathbf{Y}}_{t|t-1} = \mathbf{Y}_t - \boldsymbol{C}_t\widehat{\mathbf{X}}_{t|t-1}. \tag{5.29}$$

We substitute this into the last line of Equation (5.28) and obtain that

$$\widehat{\mathbf{X}}_{t+1|t} = \boldsymbol{A}_t\widehat{\mathbf{X}}_{t|t-1} + \boldsymbol{K}_t\widetilde{\mathbf{Y}}_{t|t-1} = (\boldsymbol{A}_t - \boldsymbol{K}_t\boldsymbol{C}_t)\widehat{\mathbf{X}}_{t|t-1} + \boldsymbol{K}_t\mathbf{Y}_t.$$

With the notation

$$\boldsymbol{A}_t^* = \boldsymbol{A}_t - \boldsymbol{K}_t\boldsymbol{C}_t \tag{5.30}$$

for the updated transition matrix, we get the new linear dynamics:

$$\widehat{\mathbf{X}}_{t+1|t} = \boldsymbol{A}_t^*\widehat{\mathbf{X}}_{t|t-1} + \boldsymbol{K}_t\mathbf{Y}_t. \tag{5.31}$$

It is also important that Equations (5.28) and (5.31) give two equivalent formulas for the prediction of $\widehat{\mathbf{X}}_{t+1|t}$:

$$\widehat{\mathbf{X}}_{t+1|t} = \boldsymbol{A}_t\widehat{\mathbf{X}}_{t|t-1} + \boldsymbol{K}_t(\mathbf{Y}_t - \boldsymbol{C}_t\widehat{\mathbf{X}}_{t|t-1}) = \boldsymbol{A}_t^*\widehat{\mathbf{X}}_{t|t-1} + \boldsymbol{K}_t\mathbf{Y}_t. \tag{5.32}$$

We shall intensively use this equivalence.

The estimation error is also governed by the linear dynamical system. This error term is

$$\begin{aligned}\widetilde{\mathbf{X}}_{t+1|t} &= \mathbf{X}_{t+1} - \widehat{\mathbf{X}}_{t+1|t} = \boldsymbol{A}_t\mathbf{X}_t + \mathbf{U}_t - \boldsymbol{A}_t^*\widehat{\mathbf{X}}_{t|t-1} - \boldsymbol{K}_t\boldsymbol{C}_t\mathbf{X}_t \\ &= \boldsymbol{A}_t^*(\mathbf{X}_t - \widehat{\mathbf{X}}_{t|t-1}) + \mathbf{U}_t = \boldsymbol{A}_t^*\widetilde{\mathbf{X}}_{t|t-1} + \mathbf{U}_t,\end{aligned} \tag{5.33}$$

so $\boldsymbol{A}_t^*$ is not only the transition matrix in (5.31), but it is also the transition matrix of the linear dynamical system governing the error. By the equivalence, stated in Equation (5.32), we get another expression for the same error term:

$$\begin{aligned}\widetilde{\mathbf{X}}_{t+1|t} &= \mathbf{X}_{t+1} - \widehat{\mathbf{X}}_{t+1|t} \\ &= \boldsymbol{A}_t\mathbf{X}_t + \mathbf{U}_t - \boldsymbol{A}_t\widehat{\mathbf{X}}_{t|t-1} - \boldsymbol{K}_t(\mathbf{Y}_t - \boldsymbol{C}_t\widehat{\mathbf{X}}_{t|t-1}) \\ &= \boldsymbol{A}_t\widetilde{\mathbf{X}}_{t|t-1} + \mathbf{U}_t - \boldsymbol{K}_t(\mathbf{Y}_t - \boldsymbol{C}_t\widehat{\mathbf{X}}_{t|t-1}).\end{aligned} \tag{5.34}$$

In the heart of the algorithm there is a recursion for the propagation of the the covariance matrix of the above error term, which is defined as

$$\boldsymbol{P}(t) = \mathbb{E}[\widetilde{\mathbf{X}}_{t|t-1}\widetilde{\mathbf{X}}_{t|t-1}^T].$$

Then we shall write $\boldsymbol{P}(t+1)$ in terms of $\boldsymbol{P}(t)$ with the help of the two alternative Equations (5.33) and (5.34) for the same error term:

$$\begin{aligned}\boldsymbol{P}(t+1) &= \mathbb{E}[\widetilde{\mathbf{X}}_{t+1|t}\widetilde{\mathbf{X}}_{t+1|t}^T] \\ &= \mathbb{E}[(\boldsymbol{A}_t^*\widetilde{\mathbf{X}}_{t|t-1} + \mathbf{U}_t)(\boldsymbol{A}_t\widetilde{\mathbf{X}}_{t|t-1} + \mathbf{U}_t - \boldsymbol{K}_t(\mathbf{Y}_t - \boldsymbol{C}_t\widehat{\mathbf{X}}_{t|t-1}))^T] \\ &= \boldsymbol{A}_t^*\mathbb{E}[\widetilde{\mathbf{X}}_{t|t-1}\widetilde{\mathbf{X}}_{t|t-1}^T]\boldsymbol{A}_t^T + \boldsymbol{Q}_{\mathbf{U}}(t) = \boldsymbol{A}_t^*\boldsymbol{P}(t)\boldsymbol{A}_t^T + \boldsymbol{Q}_{\mathbf{U}}(t),\end{aligned} \tag{5.35}$$

where recall that $\boldsymbol{Q}_{\mathbf{U}}(t) = \mathbb{E}[\mathbf{U}_t\mathbf{U}_t^T]$ and we used that $\mathbf{U}_t$ is uncorrelated with $\mathbf{X}_t$ and, therefore, with $\widetilde{\mathbf{X}}_{t|t-1}$ too; further, that $\mathbf{Y}_t - \boldsymbol{C}_t\widehat{\mathbf{X}}_{t|t-1}$ is within the innovation subspace $I_t(\mathbf{Y})$.

It remains to find an explicit formula for $\boldsymbol{K}_t$, and thus, also for $\boldsymbol{A}_t^*$. Recall that $\boldsymbol{K}_t$ is the matrix of the linear operation $\mathrm{Proj}_{I_t(\mathbf{Y})}\mathbf{X}_{t+1}$, therefore by the geometry of projections:

$$\boldsymbol{K}_t = [\mathbb{E}\mathbf{X}_{t+1}\widetilde{\mathbf{Y}}_{t|t-1}^T][\mathbb{E}(\widetilde{\mathbf{Y}}_{t|t-1}\widetilde{\mathbf{Y}}_{t|t-1}^T]^+,$$

where $^+$ denotes the Moore–Penrose generalized inverse. See also the theory of simultaneous linear regressions in Appendix C, namely Equation (C.2); $\boldsymbol{A}^T$ there plays the role of our $\boldsymbol{K}_t$ here. Now we calculate the matrices in brackets. By the second equation of (5.25), that extends to $\widetilde{\mathbf{Y}}_t = \boldsymbol{C}_t\widetilde{\mathbf{X}}_t$, we get that

$$\mathbb{E}\widetilde{\mathbf{Y}}_{t|t-1}\widetilde{\mathbf{Y}}_{t|t-1}^T = \mathbb{E}(\boldsymbol{C}_t\widetilde{\mathbf{X}}_{t|t-1})(\boldsymbol{C}_t\widetilde{\mathbf{X}}_{t|t-1})^T = \boldsymbol{C}_t\boldsymbol{P}(t)\boldsymbol{C}_t^T.$$

By the first and second equation of (5.25) and the orthogonality of $\widehat{\mathbf{X}}_{t|t-1}$ and $\widetilde{\mathbf{X}}_{t|t-1}$:

$$\begin{aligned}\mathbb{E}\mathbf{X}_{t+1}\widetilde{\mathbf{Y}}_{t|t-1}^T &= \boldsymbol{A}_t\mathbb{E}\mathbf{X}_t\widetilde{\mathbf{Y}}_{t|t-1}^T = \boldsymbol{A}_t\mathbb{E}(\widehat{\mathbf{X}}_{t|t-1} + \widetilde{\mathbf{X}}_{t|t-1})(\boldsymbol{C}_t\widetilde{\mathbf{X}}_{t|t-1}^T) \\ &= \boldsymbol{A}_t\boldsymbol{P}(t)\boldsymbol{C}_t^T.\end{aligned} \tag{5.36}$$

Therefore,

$$\boldsymbol{K}_t = \boldsymbol{A}_t\boldsymbol{P}(t)\boldsymbol{C}_t^T[\boldsymbol{C}_t\boldsymbol{P}(t)\boldsymbol{C}_t^T]^+. \tag{5.37}$$

Instead of the Moore–Penrose generalized inverse, we use the regular inverse provided the matrix in brackets is invertible, i.e. the innovation subspace $I_t(\mathbf{Y})$ is of full dimension p, and $\boldsymbol{C}_t$ is of full rank p.

Then the recursion starts at $t = 1$, when the systems of p linear equations $\boldsymbol{C}_1\widehat{\mathbf{X}}_{1|0} = \overline{\mathbf{Y}}_{1|0}$ and $\boldsymbol{C}_1\mathbf{X}_1 = \mathbf{Y}_1$ should be solved for the coordinates of $\widehat{\mathbf{X}}_{1|0}$ and $\mathbf{X}_1$, respectively (the n coordinates are the unknowns). They obviously have a solution if $\boldsymbol{C}_1$ is of full rank. Here $\overline{\mathbf{Y}}_{1|0} = \mathbb{E}(\mathbf{Y}_1)$ if $\mathbf{Y}_0$ is a constant vector. Even if it is $\mathbf{0}$, the system has a nontrivial solution in the $p \leq n$ case.

Then $\widetilde{\mathbf{X}}_{1|0} = \mathbf{X}_1 - \widehat{\mathbf{X}}_{1|0}$. In the original paper of Kálmán [33], the following starting is suggested: $\widehat{\mathbf{X}}_{1|0} := \mathbf{0}$; $\widetilde{\mathbf{X}}_{1|0} := \mathbf{X}_1$; $\boldsymbol{P}(1) := \mathbb{E}[\mathbf{X}_1\mathbf{X}_1^T]$. This can be the product moment estimate from the training sample (possible past). See also Remark 5.7.

By using Remark C.1, we can summarize the above results and recursion as follows.

Proposition 5.4. *The optimal estimate* $\widehat{\mathbf{X}}_{t+1|t}$ *of* $\mathbf{X}_{t+1}$ *given* $\mathbf{Y}_0, \mathbf{Y}_1, \dots, \mathbf{Y}_t$ *is generated by the linear dynamical system*

$$\widehat{\mathbf{X}}_{t+1|t} = \boldsymbol{A}_t^*\widehat{\mathbf{X}}_{t|t-1} + \boldsymbol{K}_t\mathbf{Y}_t.$$

The estimation error term is given by

$$\widetilde{\mathbf{X}}_{t+1|t} = \boldsymbol{A}_t^*\widetilde{\mathbf{X}}_{t|t-1} + \mathbf{U}_t$$

and the propagated covariance matrix of the estimation error term is

$$\boldsymbol{P}(t) = \mathbb{E}[\widetilde{\mathbf{X}}_{t|t-1}\widetilde{\mathbf{X}}_{t|t-1}^T],$$

while the expected quadratic loss is $\mathrm{tr}\boldsymbol{P}(t)$. *The matrices involved are generated by the following recursion. Starting with*

$$\widehat{\mathbf{X}}_{1|0} = \mathrm{Proj}_{\mathbf{Y}_0}\mathbf{X}_1,\ \widetilde{\mathbf{X}}_1 = \mathbf{X}_1 - \widehat{\mathbf{X}}_{1|0},\ \boldsymbol{P}(1) = \mathbb{E}[\widetilde{\mathbf{X}}_{1|0}\widetilde{\mathbf{X}}_{1|0}^T] = \mathbb{E}[\mathbf{X}_1\mathbf{X}_1^T] - \mathbb{E}[\widehat{\mathbf{X}}_{1|0}\widehat{\mathbf{X}}_{1|0}^T],$$

for $t = 1, 2, \dots,$ *the steps of the following recursion are uniquely defined:*

- *Evaluate* $\boldsymbol{K}_t$ *by* (5.37): $\boldsymbol{K}_t = \boldsymbol{A}_t\boldsymbol{P}(t)\boldsymbol{C}_t^T[\boldsymbol{C}_t\boldsymbol{P}(t)\boldsymbol{C}_t^T]^+$.
- *Input* $\mathbf{Y}_t$. *Output*

$$\widehat{\mathbf{X}}_{t+1|t} = (\boldsymbol{A}_t - \boldsymbol{K}_t\boldsymbol{C}_t)\widehat{\mathbf{X}}_{t|t-1} + \boldsymbol{K}_t\mathbf{Y}_t.$$

- *Evaluate* $\boldsymbol{A}_t^*$ *by* (5.30): $\boldsymbol{A}_t^* = \boldsymbol{A}_t - \boldsymbol{K}_t\boldsymbol{C}_t$.
- *Eventually, calculate* $\boldsymbol{P}(t+1)$ *by* (5.35) *that completes the cycle:*

$$\boldsymbol{P}(t+1) = \boldsymbol{A}_t^*\boldsymbol{P}(t)\boldsymbol{A}_t^T + \boldsymbol{Q}_{\mathbf{U}}(t).$$

Note that $\boldsymbol{Q}_{\mathbf{U}}(t)$ *is known/given or estimated from a training sample. In the forthcoming Remark 5.11, a symmetric expression is also given for the matrix* $\boldsymbol{P}(t+1)$.

Some remarks are in order.

Remark 5.7. As for the starting, $\widehat{\mathbf{X}}_{1|0} = \mathrm{Proj}_{\mathbf{Y}_0}\mathbf{X}_1 = \hat{\boldsymbol{\Sigma}}_{\mathbf{XY}}\hat{\boldsymbol{\Sigma}}_{\mathbf{YY}}^+\mathbf{Y}_0$, by Lemma C.2, where the last training sample entry can be chosen for $\mathbf{Y}_0$. To initialize $\boldsymbol{P}(1)$, the whole training sample can be used. Another possibility is to start with $\widehat{\mathbf{X}}_{1|0} = \mathbf{0}$, see [29].

Remark 5.8. As a byproduct, the algorithm is able to get the innovations via Equation (5.29).

Remark 5.9. In some situations, the observation equation also contains a noise term, for example, in [11]; whereas, in [22] the stationary case is treated with this generalization that results only in minor changes in the algorithm.

In [33], Kálmán and Bucy consider the continuous-time case, but they write that even in this case, the assumption that every observed signal contains a white noise term, "is unnecessary when the random processes in question are sampled (discrete-time parameter), see [32]"; even in the continuous-time case, it "is no real restriction since it can be removed in various ways". However, the random excitation in the state (message) process "is quite basic; it is analogous to but somewhat less restrictive than the assumption of rational spectra in the conventional theory". Indeed, Kálmán uses only the regularity (causality) of the process if stationary, but not the rational spectral density. He mostly considers Gaussian processes that is not a restriction in the possession of second order processes, when we confine ourselves to the second moments.

In this case, the state equations have the form

$$\begin{aligned} \mathbf{X}_{t+1} &= \boldsymbol{A}_t \mathbf{X}_t + \mathbf{U}_t \\ \mathbf{Y}_t &= \boldsymbol{C}_t \mathbf{X}_t + \mathbf{W}_t, \end{aligned}$$

where $\mathbf{W}_t$ is independent of $\mathbf{X}_t$ and $\mathbf{U}_t$ (latter condition can be relaxed by introducing the covariance matrix between $\mathbf{U}_t$ and $\mathbf{W}_t$ as a given parameter, see [11]); further the covariance matrix of the zero expectation $\mathbf{W}_t$ is $\boldsymbol{Q}_{\mathbf{W}} = \mathbb{E}\mathbf{W}_t \mathbf{W}_t^T$ is also given.

The only difference in the calculations is that now

$$\begin{aligned} \mathbb{E}\widetilde{\mathbf{Y}}_{t|t-1} \widetilde{\mathbf{Y}}_{t|t-1}^T = \mathbb{E}(\boldsymbol{C}_t \widetilde{\mathbf{X}}_{t|t-1} + \mathbf{W}_t)(\boldsymbol{C}_t \widetilde{\mathbf{X}}_{t|t-1} + \mathbf{W}_t)^T = \boldsymbol{C}_t \boldsymbol{P}(t) \boldsymbol{C}_t^T \\ + \boldsymbol{Q}_{\mathbf{W}}(t), \end{aligned}$$

and so,

$$\boldsymbol{K}_t = \boldsymbol{A}_t \boldsymbol{P}(t) \boldsymbol{C}_t^T [\boldsymbol{C}_t \boldsymbol{P}(t) \boldsymbol{C}_t^T + \boldsymbol{Q}_{\mathbf{W}}(t)]^+.$$

Instead of $\mathbf{U}_t$ we may write $\boldsymbol{B}_t \mathbf{V}_t$ with some $n \times q$ matrix $\boldsymbol{B}_t$ with $q \leq n$ and q-dimensional orthogonal noise $\mathbf{V}_t$, i.e. $\mathbb{E}\mathbf{V}_t \mathbf{V}_t^T = \boldsymbol{Q}_{\mathbf{V}}(t)$ is a given diagonal matrix. Here instead of $\boldsymbol{Q}(t)$ the matrix $\boldsymbol{B}_t \boldsymbol{Q}_{\mathbf{V}}(t) \boldsymbol{B}_t^T$ enters into Equation (5.35). This approach mainly used in the stationary case, when a lower rank driving force (excitation) is assumed, but this is the topic of Dynamic Factor Analysis, see [15] and Section 5.6.

In the same vein, instead of $\mathbf{W}_t$ we may write $\boldsymbol{D}_t \mathbf{Z}_t$ with some $p \times s$ matrix $\boldsymbol{D}_t$ with $s \leq p$ and s-dimensional orthogonal noise $\mathbf{Z}_t$, i.e. $\mathbb{E}\mathbf{Z}_t \mathbf{Z}_t^T = \boldsymbol{Q}_{\mathbf{Z}}(t)$ is a given diagonal matrix. Here instead of $\boldsymbol{Q}_{\mathbf{W}}(t)$, the possibly reduced rank matrix $\boldsymbol{D}_t \boldsymbol{Q}_{\mathbf{Z}}(t) \boldsymbol{D}_t^T$ enters into the calculations.

Remark 5.10. Equation (5.31) gives rise to a predictive filtering, in the possession of the gain matrix $\boldsymbol{K}_t$. After this, the algorithm is also applicable to filtering. Indeed,

$$\widehat{\mathbf{X}}_{t+1|t} = \text{Proj}_{H_t(\mathbf{Y})}\mathbf{X}_{t+1} = \text{Proj}_{H_t(\mathbf{Y})}(\boldsymbol{A}_t\mathbf{X}_t + \mathbf{U}_t) = \boldsymbol{A}_t\widehat{\mathbf{X}}_{t|t}.$$

Now, provided $\boldsymbol{A}_t$ is invertible,

$$\widehat{\mathbf{X}}_{t|t} = \boldsymbol{A}_t^{-1}\widehat{\mathbf{X}}_{t+1|t}.$$

If $\boldsymbol{A}_t$ is not invertible, then we proceed as follows:

$$\widehat{\mathbf{X}}_{t|t} = \text{Proj}_{H_t(\mathbf{Y})}\mathbf{X}_t = \text{Proj}_{H_{t-1}(\mathbf{Y})}\mathbf{X}_t + \text{Proj}_{I_t(\mathbf{Y})}\mathbf{X}_t = \widehat{\mathbf{X}}_{t|t-1} + \boldsymbol{L}_t\widetilde{\mathbf{Y}}_{t|t-1}.$$

Now the gain matrix is $\boldsymbol{L}_t$, which is not the same as $\boldsymbol{K}_t$ (though, sometimes this is what called Kálmán gain matrix), can be determined with a similar calculation:

$$\boldsymbol{L}_t = [\mathbb{E}\mathbf{X}_t\widetilde{\mathbf{Y}}_{t|t-1}^T][\mathbb{E}(\widetilde{\mathbf{Y}}_{t|t-1}\widetilde{\mathbf{Y}}_{t|t-1}^T]^+.$$

The only difference between the formula for $\boldsymbol{K}_t$ and $\boldsymbol{L}_t$ that here we calculate the covariance between $\mathbf{X}_t$ and $\widetilde{\mathbf{Y}}_{t|t-1}^T$, but Equation (5.36) is at our disposal in this situation too. We get that

$$\mathbb{E}\mathbf{X}_t\widetilde{\mathbf{Y}}_{t|t-1}^T = \boldsymbol{P}(t)\boldsymbol{C}_t^T,$$

and so,

$$\boldsymbol{L}_t = \boldsymbol{P}(t)\boldsymbol{C}_t^T[\boldsymbol{C}_t\boldsymbol{P}(t)\boldsymbol{C}_t^T]^+.$$

Consequently,

$$\boldsymbol{K}_t = \boldsymbol{A}_t\boldsymbol{L}_t,$$

so we could first find

$$\boldsymbol{L}_t = \boldsymbol{P}(t)\boldsymbol{C}_t^T[\boldsymbol{C}_t\boldsymbol{P}(t)\boldsymbol{C}_t^T]^+$$

and then, $\boldsymbol{K}_t$. Therefore, in course of the iteration, the filtered process $\{\widehat{\mathbf{X}}_{t|t}\}$ can as well be obtained.

Remark 5.11. If we write the expression for $\boldsymbol{K}_t$, $\boldsymbol{L}_t$, and $\boldsymbol{A}_t^*$ into Equation (5.35), then we get

$$\begin{aligned}\boldsymbol{P}(t+1) &= \boldsymbol{A}_t^*\boldsymbol{P}(t){\boldsymbol{A}_t}^T + \boldsymbol{Q}_{\mathbf{U}}(t)\\ &= (\boldsymbol{A}_t - \boldsymbol{A}_t\boldsymbol{L}_t\boldsymbol{C}_t)\boldsymbol{P}(t){\boldsymbol{A}_t}^T + \boldsymbol{Q}(t)\\ &= (\boldsymbol{A}_t(\boldsymbol{I} - \boldsymbol{L}_t\boldsymbol{C}_t))\boldsymbol{P}(t){\boldsymbol{A}_t}^T + \boldsymbol{Q}_{\mathbf{U}}(t)\\ &= \boldsymbol{A}_t\boldsymbol{P}(t){\boldsymbol{A}_t}^T - \boldsymbol{A}_t\boldsymbol{L}_t\boldsymbol{C}_t\boldsymbol{P}(t){\boldsymbol{A}_t}^T + \boldsymbol{Q}_{\mathbf{U}}(t)\\ &= \boldsymbol{A}_t\boldsymbol{P}(t){\boldsymbol{A}_t}^T - \boldsymbol{A}_t\boldsymbol{P}(t)\boldsymbol{C}_t^T[\boldsymbol{C}_t\boldsymbol{P}(t)\boldsymbol{C}_t^T]^{-1}\boldsymbol{C}_t\boldsymbol{P}(t){\boldsymbol{A}_t}^T\\ &+ \boldsymbol{Q}_{\mathbf{U}}(t)\end{aligned} \tag{5.38}$$

which final formula shows that $\boldsymbol{P}(t+1)$ is indeed a symmetric matrix.

Remark 5.12. Assume that the underlying process is weakly stationary, and put $\boldsymbol{A}$ for $\boldsymbol{A}_t$, $\boldsymbol{C}$ for $\boldsymbol{C}_t$, and $\boldsymbol{Q}_{\mathrm{U}}$ for $\boldsymbol{Q}_{\mathrm{U}}(t)$. In this case, instead of the recursion, we get the fixed point iteration

$$\boldsymbol{P}_{t+1} = \boldsymbol{A}\boldsymbol{P}_t\boldsymbol{A}^T - \boldsymbol{A}\boldsymbol{P}_t\boldsymbol{C}^T[\boldsymbol{C}\boldsymbol{P}_t\boldsymbol{C}^T]^+\boldsymbol{C}\boldsymbol{P}_t\boldsymbol{A}^T + \boldsymbol{Q}_{\mathrm{U}},$$

where now $\boldsymbol{P}_t$ just denotes step t of the iteration. Note that [55] considers the question when the discrete matrix Riccati equation

$$\boldsymbol{P} = \boldsymbol{A}\boldsymbol{P}\boldsymbol{A}^T - \boldsymbol{A}\boldsymbol{P}\boldsymbol{C}^T[\boldsymbol{C}\boldsymbol{P}\boldsymbol{C}^T]^+\boldsymbol{C}\boldsymbol{P}\boldsymbol{A}^T + \boldsymbol{Q}_{\mathrm{U}}, \tag{5.39}$$

has a unique solution and so, the method of successive approximation, resembling the recursion in (5.38), is able to find it. (Actually, the Riccati operator is concave and has a unique fixed point under very general conditions.) With this $\boldsymbol{P}$, the limit of the sequence $\boldsymbol{K}_t$ is $\boldsymbol{K} = \boldsymbol{A}\boldsymbol{P}\boldsymbol{C}^T[\boldsymbol{C}\boldsymbol{P}\boldsymbol{C}^T]^+$ as $t \to \infty$, and $\boldsymbol{L}_t$ also has a limiting value $\boldsymbol{L} = \boldsymbol{P}\boldsymbol{C}^T[\boldsymbol{C}\boldsymbol{P}\boldsymbol{C}^T]^+$, when our sequence is weakly stationary.

The paper [33] gives guidance to the solution, mainly considers the continuous time case, and contains many applications in engineering and telecommunication. The authors also discuss the relation to differential equations and the Fisher information matrix.

We remark that in the possession of another error term $\mathbf{W}$ (but $\boldsymbol{Q}_{\mathrm{W}}$ does not depend on the time), Equation (5.39) has the slightly modified form

$$\boldsymbol{P} = \boldsymbol{A}\boldsymbol{P}\boldsymbol{A}^T - \boldsymbol{A}\boldsymbol{P}\boldsymbol{C}^T[\boldsymbol{C}\boldsymbol{P}\boldsymbol{C}^T + \boldsymbol{Q}_{\mathrm{W}}]^+\boldsymbol{C}\boldsymbol{P}\boldsymbol{A}^T + \boldsymbol{Q}_{\mathrm{U}},$$

though it does not change the type of the matrix Riccati equation.

In the stationary case, the stability of the matrix $\boldsymbol{A}$ should be assumed, as well as that of the new transition matrix, corresponding to $\boldsymbol{A}^*$, which is $\boldsymbol{A} - \boldsymbol{K}\boldsymbol{C}$. For further details, see [22, 51, 55].

5.6 Dynamic principal component and factor analysis

Here we confine ourselves to high dimensional weakly stationary processes that are usually of lower rank than their dimension or can be approximated with a lower rank process. In the time domain, we are looking for the convenient filters and for the matrices in the state equations too. In the frequency domain, we use the low rank approximation of the spectral density matrix at the Fourier frequencies. We summarize the findings based of the previous sections.

5.6.1 Time domain approach via innovations

First we use the method of innovations. If $\mathbf{X}_t$s have different dimensions, then denoting by d the minimal dimension, first we perform a static factor analysis on them, and start with the so obtained d-dimensional static factor process. We also deprive the process from trend and seasonality, and assume that it has a spectral density matrix of constant rank. If the process is also deprived of the singular part, then a regular process is at our disposal.

If $\mathbf{X}_t$ is regular, we learned that it can be expanded in terms of the d-dimensional innovations

$$\boldsymbol{\eta}_{t+1} := \mathbf{X}_{t+1} - \hat{\mathbf{X}}_{t+1},$$

where $\hat{\mathbf{X}}_{t+1}$ is the projection of $\mathbf{X}_{t+1}$ onto the subspace spanned by $\mathbf{X}_1, \dots, \mathbf{X}_t$, denoted by H_t. It can be done step by step as described in Section 5.3.2. If not regular, the prediction process gives the regular part of it.

We can as well reduce the dimension of the innovation process to $k < d$. This k-dimensional innovation process can be considered as a dynamic factor process, where $k \le r$, and r is the rank of the spectral density matrix of the process. As an alternative to the block Cholesky decomposition, the Kálmán filtering is also able to find the innovations, see Equation (5.29). In this way, instead of the decomposition of a huge block matrix, we operate with matrices of size comparable to the dimension of the process.

The above is also related to the minimal phase spectral factor. To find this and a reduced rank causal approximation of a process of rational spectral density, we refer to [40]. Another approach via singular autoregressions is discussed in [13].

We saw that a d-dimensional regular process $\{\mathbf{X}_t\}$, whose spectral density matrix $\boldsymbol{f}$ is of rank $r \le d$ has the variant of the multidimensional Wold decomposition:

$$\mathbf{X}_t = \sum_{j=0}^{\infty} \boldsymbol{B}_j \boldsymbol{\nu}_{t-j},$$

where $\boldsymbol{B}_j$s are $d \times r$ matrices (like dynamic factor loadings), and $\{\boldsymbol{\nu}_t\} \sim \mathrm{WN}(\boldsymbol{\Sigma})$ is r-dimensional white noise (like non-standardized minimal dynamic factors).

It is important that there is a one-to-one correspondence between $\boldsymbol{f}$ (frequency domain) and the $\boldsymbol{B}(z), \boldsymbol{\Sigma}$ pair (time domain):

$$\boldsymbol{B}(z) = \sum_{j=0}^{\infty} \boldsymbol{B}_j z^j, \quad |z| \le 1 \tag{5.40}$$

and $\boldsymbol{\Sigma}$ is the covariance matrix of $\boldsymbol{\nu}_t$. This correspondence is given by

$$\boldsymbol{f}(z) = \frac{1}{2\pi} \boldsymbol{B}(z) \boldsymbol{\Sigma} \boldsymbol{B}^*(z).$$

We can as well write $\boldsymbol{f}(z) = \frac{1}{2\pi} \boldsymbol{H}(z) \boldsymbol{H}^*(z)$, where $\boldsymbol{H}(z) = \boldsymbol{B}(z) \boldsymbol{\Sigma}^{1/2}$ is the transfer function and it is unique only up to unitary transformation. At the same time, the matrices $\boldsymbol{B}_j \boldsymbol{\Sigma}^{-1/2}$ are the impulse responses, also see (5.19). So by performing the expansion (5.40) at the Fourier frequencies, we can estimate the transfer function.

In Section 5.3.2, we gave an algorithm to this in the time domain, via block Cholesky decomposition, see (5.15). Then we can perform a static PCA on $\mathbf{\Sigma}$ with $k \le r$ principal components, that results in dynamic factors of dimension k. The choice of k is such that there are $n(r-k)$ negligible eigenvalues in the spectrum of $\mathfrak{C}_n$. By Theorem 5.2 , for "large" n, this is in accord with the existence of $r-k$ negligible eigenvalues of the spectral density matrix at all the n Fourier frequencies. Therefore, we proceed in the frequency domain.

5.6.2 Frequency domain approach

Let $\{\mathbf{X}_t\}$ be discrete time, d-dimensional, weakly stationary time series of zero expectation and spectral density matrix of constant rank. For given $0 < k \le d$ we are looking for the k-dimensional time series $\mathbf{Y}_t$ such that

$$\mathbf{Y}_t = \sum_j \boldsymbol{b}_{t-j}\mathbf{X}_j, \quad t \in \mathbf{Z},$$

where $\boldsymbol{b}_j$s are $k \times d$ matrices and $\boldsymbol{b}$ is the corresponding transfer function. (Here k is less than the rank of the process itself.)

Then approximate $\mathbf{X}_t$ with

$$\hat{\mathbf{X}}_t = \sum_j \boldsymbol{c}_{t-j}\mathbf{Y}_j, \quad t \in \mathbf{Z},$$

where the impulse responses $\boldsymbol{c}_j$s are $d \times k$ matrices, and $\boldsymbol{c}$ is the transfer function.

So $\hat{\mathbf{X}}$ is obtained from $\mathbf{X}$ with the time invariant filter

$$\boldsymbol{a}(\omega) = \boldsymbol{c}(\omega)\boldsymbol{b}(\omega).$$

The error of approximation is measured with

$$\mathbb{E}(\mathbf{X}_t - \hat{\mathbf{X}}_t)^*(\mathbf{X}_t - \hat{\mathbf{X}}_t).$$

Then Brillinger [8] in Theorem 9.3.1 states that the minimum is attained with the impulse responses

$$\boldsymbol{b}_j = \frac{1}{2\pi}\int_0^{2\pi} \boldsymbol{b}(\omega)e^{ij\omega}\,d\omega$$

and

$$\boldsymbol{c}_j = \frac{1}{2\pi}\int_0^{2\pi} \boldsymbol{c}(\omega)e^{ij\omega}\,d\omega,$$

where

$$\boldsymbol{c}(\omega) = (\mathbf{u}_1(\omega), \dots, \mathbf{u}_k(\omega))$$

contains columnwise the orthonormal eigenvectors corresponding to the k largest eigenvalues of the spectral density matrix $\boldsymbol{f}$ of $\{\mathbf{X}_t\}$. Further, $\boldsymbol{b}(\omega) = \boldsymbol{c}^*(\omega)$. (See Section 4.6 as well.) The approximation error is

$$\int_0^{2\pi} \sum_{j=k+1}^{d} \lambda_j(\omega)\, d\omega.$$

This is in Frobenius norm, in spectral norm it only depends on λ_{k+1}, but the best k-rank approximation is the same in any unitary invariant norm (that depends only on the eigenvalues). The larger the gap in the spectrum between the k largest and the other eigenvalues, the better the approximation is.

$\{\mathbf{Y}_t\}$ is called principal component process. Its spectral density matrix is diagonal with diagonal entries $\lambda_1(\omega), \dots \lambda_k(\omega)$. In Section 4.6 we proved that if the original process is regular, then its best k-rank approximation is regular too.

5.6.3 Best low-rank approximation in the frequency domain, and low-dimensional approximation in the time domain

Let $\{\mathbf{X}_t\}_{t=1}^n$ be the finite part of a d-dimensional process of real coordinates and constant rank $1 \le r \le d$. Its discrete Fourier transform, discussed in Section 5.4.2, is

$$\mathbf{T}_j = \mathbf{V}_j \mathbf{Z}_j = \frac{1}{\sqrt{n}} \sum_{t=1}^{n} \mathbf{X}_t e^{-it\omega_j}, \quad j = 0, \dots, n-1.$$

More precisely, $\mathbf{T}_0 = \frac{1}{\sqrt{n}} \sum_{t=1}^n \mathbf{X}_t$,

$$\mathbf{T}_j = \frac{1}{\sqrt{n}} \sum_{t=1}^{n} \mathbf{X}_t [\cos(t\omega_j) - i \sin(t\omega_j)],$$

and $\mathbf{T}_{n-j} = \overline{\mathbf{T}}_j$, for $j = 1, \dots, k$ $(n = 2k+1)$. Therefore,

$$\mathbf{Z}_j = \mathbf{V}_j^{-1} \mathbf{T}_j = \mathbf{V}_j^* \mathbf{T}_j, \quad j = 0, \dots, n-1.$$

It can easily be seen that $\mathbf{Z}_{n-j} = \overline{\mathbf{Z}}_j$.

To find the best m-rank approximation $(1 < m \le r)$ of the process in any unitary invariant norm (see Theorem B.8), we project the d-dimensional vector $\mathbf{T}_j$ onto the subspace spanned by the m leading eigenvectors of $\mathbf{V}_j$ (see e.g. [44] for the linear algebra justification for this). Important that the eigenvalues in $\mathbf{\Lambda}_j$ are in non-increasing order. Let us denote the eigenvectors corresponding to the m largest eigenvalues by $\mathbf{v}_{j1}, \dots, \mathbf{v}_{jm}$. Then

$$\widehat{\mathbf{T}}_j := \mathrm{Proj}_{\mathrm{Span}\{\mathbf{v}_{j1}, \dots, \mathbf{v}_{jm}\}} \mathbf{T}_j = \sum_{\ell=1}^{m} (\mathbf{v}_{j\ell}^* \mathbf{V}_j \mathbf{Z}_j) \mathbf{v}_{j\ell} = \sum_{\ell=1}^{m} Z_{j\ell} \mathbf{v}_{j\ell},$$

and $\widehat{\mathbf{T}}_{n-j} = \overline{\widehat{\mathbf{T}}}_j$, for for $j = 1, \dots, k$ (by the previous considerations), where $n = 2k + 1$. Further,

$$\widehat{\mathbf{T}}_0 := \sum_{\ell=1}^{m} Z_{0\ell} \mathbf{v}_{0\ell}.$$

So, for each j, the resulting vector is the linear combination of the vectors $\mathbf{v}_{j\ell}$s with the corresponding coordinates $Z_{j\ell}$s of $\mathbf{Z}_j$, $\ell = 1, \dots, m$.

Eventually, we find the m-rank approximation of $\mathbf{X}_t$ by inverse Fourier transformation:

$$\begin{aligned}
\widehat{\mathbf{X}}_t &:= \frac{1}{\sqrt{n}} \sum_{j=0}^{n-1} \widehat{\mathbf{T}}_j e^{it\omega_j} = \\
&= \frac{1}{\sqrt{n}} \left\{ \widehat{\mathbf{T}}_0 + \sum_{j=1}^{k} [(\widehat{\mathbf{T}}_j + \overline{\widehat{\mathbf{T}}}_j) \cos(t\omega_j) + i(\widehat{\mathbf{T}}_j - \overline{\widehat{\mathbf{T}}}_j) \sin(t\omega_j)] \right\} \\
&= \frac{1}{\sqrt{n}} \left\{ \widehat{\mathbf{T}}_0 + \sum_{j=1}^{k} [(2\mathrm{Re}(\widehat{\mathbf{T}}_j) \cos(t\omega_j) + i \cdot 2i \cdot \mathrm{Im}(\widehat{\mathbf{T}}_j) \sin(t\omega_j)] \right\} \\
&= \frac{1}{\sqrt{n}} \left\{ \widehat{\mathbf{T}}_0 + 2 \sum_{j=1}^{k} [\mathrm{Re}(\widehat{\mathbf{T}}_j) \cos(t\omega_j) - \mathrm{Im}(\widehat{\mathbf{T}}_j) \sin(t\omega_j)] \right\}.
\end{aligned}$$

Apparently, the vectors $\widehat{\mathbf{X}}_t$ ($t = 1, \dots, n$) all have real coordinates ($n = 2k+1$).

In this way, we have a lower rank process with spectral density of rank $m \le r$. Note that if the process is regular (e.g. it has a rational spectral density), then so is its low-rank approximation. The theory (e.g. [44]) guarantees that the "larger" the gap between the mth and $(m+1)$th eigenvalues (in non-increasing order) of the spectral density matrix, the "smaller" the approximation error is.

To back-transform the PC process into the time domain, note that

$$Z_{j\ell} = \mathbf{v}_{j\ell}^* \mathbf{T}_j, \quad \ell = 1, \dots m$$

defines the coordinates of an m-dimensional approximation of $\mathbf{T}_j$, $m \le r \le d$. This is the m-dimensional vector $\tilde{\mathbf{T}}_j = (Z_{j1}, \dots, Z_{jm})^T$. That is, we take the first m complex PCs in each blocks (it is important that the entries in the diagonal of each $\mathbf{\Lambda}_j$ are in non-increasing order). The other $d-m$ coordinates of $\mathbf{Z}_j$ are disregarded (they are taken zeros in the new coordinate system $\mathbf{v}_{j1}, \dots, \mathbf{v}_{jd}$). The proportion of the total variance explained by the first m principal components at the jth Fourier frequency is $\sum_{\ell=1}^{m} \lambda_{j\ell} / \sum_{\ell=1}^{d} \lambda_{j\ell}$.

Then the m-dimensional approximation of $\mathbf{X}_t$ by the PC process is as follows:

$$\tilde{\mathbf{X}}_t := \frac{1}{\sqrt{n}} \sum_{j=0}^{n-1} \tilde{\mathbf{T}}_j e^{it\omega_j} =$$

$$= \frac{1}{\sqrt{n}} \left\{ \tilde{\mathbf{T}}_0 + \sum_{j=1}^{k} [(\tilde{\mathbf{T}}_j + \overline{\tilde{\mathbf{T}}}_j) \cos(t\omega_j) + i(\tilde{\mathbf{T}}_j - \overline{\tilde{\mathbf{T}}}_j) \sin(t\omega_j)] \right\}$$

$$= \frac{1}{\sqrt{n}} \left\{ \tilde{\mathbf{T}}_0 + \sum_{j=1}^{k} [2\mathrm{Re}(\tilde{\mathbf{T}}_j) \cos(t\omega_j) + i \cdot 2i \cdot \mathrm{Im}(\tilde{\mathbf{T}}_j) \sin(t\omega_j)] \right\}$$

$$= \frac{1}{\sqrt{n}} \left\{ \tilde{\mathbf{T}}_0 + 2 \sum_{j=1}^{k} [\mathrm{Re}(\tilde{\mathbf{T}}_j) \cos(t\omega_j) - \mathrm{Im}(\tilde{\mathbf{T}}_j) \sin(t\omega_j)] \right\}$$

that again results in real coordinates. Equivalently, the m-dimensional PC process is:

$$\tilde{\mathbf{X}}_t = \frac{1}{\sqrt{n}} (\sum_{j=0}^{n-1} Z_{j1} e^{it\omega_j}, \dots, \sum_{j=0}^{n-1} Z_{jm} e^{it\omega_j})^T.$$

Note that to estimate the matrices $\boldsymbol{M}_j$s, only the estimates of the first $n/2$ autocovariances are needed. By the ergodicity considerations of Section 1.5.2, these can be estimated more accurately (using at least one half of the sample entries) as the remaining $n/2$ ones if n is "large".

Example 5.1. The previously detailed low-rank approximation is illustrated on a financial dataset [2] containing stock exchange log-returns: Istanbul stock exchange national 100 index, S&P 500 return index, stock market return index of Germany, UK, Japan, Brazil, the MSCI European index, and the MSCI emerging markets index; ranging from June 5, 2009 to February 22, 2011. It is a $d = 8$ dimensional time series dataset of length $n = 535$.

In Figure 5.1, the eigenvalue processes of the estimated M_j matrices are shown in the frequency domain. Based on this, the time series is of rank approximately 3, thus we can apply the outlined low-rank approximation with $m = 3$. In Figure 5.2, the individual variables of the original data and its rank 3 approximation are illustrated. There are calculated root mean square error (RMSE) values under each subplot. The 3 leading PC's, back-transformed to the time domain, are to be found in Figure 5.3.

5.6.4 Dynamic factor analysis

Standard factor analysis can be generalized to the case of a d-dimensional, real valued, vector stochastic process $\{\mathbf{X}_t\}$. Here $t \geq 0$ is the time, and our sample usually consists of observations at discrete time instances $t = 1, \dots, T$. In the

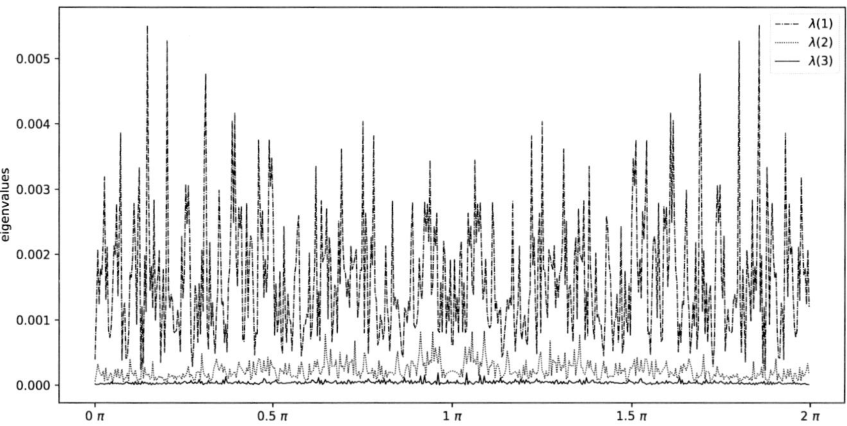

FIGURE 5.1
Eigenvalue processes of the estimated $\boldsymbol{M}_j$ $(j = 0, \dots, 534)$ matrices over $[0, 2\pi]$, ordered decreasingly in Example 5.1.

classical factor analysis approach, the data come from i.i.d. observations, and the dimension reduction happens in the so-called cross-sectional dimension, i.e. the number d of variables is decreased. In Dynamic Factor Analysis, the observations $\mathbf{X}_t$'s are not independent, and we want to compress the information, embodied by them, in the cross-sectional and the time dimension as well. Sometimes even the cross-sectional dimension d is large compared to the time span T.

Assume that $\{\mathbf{X}_t\}$ is *weakly stationary* with an absolutely continuous spectral distribution, i.e. it has the $d \times d$ spectral density matrix $\boldsymbol{f}_{\mathbf{X}}$. With the integer $1 \le k < d$, the *dynamic k-factor model* for $\mathbf{X}_t$ (see, e.g. [4]) is

$$\mathbf{X}_t = \boldsymbol{\mu} + \boldsymbol{B}(L)\mathbf{Z}_t + \mathbf{e}_t = \boldsymbol{\mu} + \boldsymbol{\chi}_t + \mathbf{e}_t \tag{5.41}$$

or with components,

$$X_t^i = \mu^i + b_{i1}(L)Z_t^1 + \cdots + b_{ik}(L)Z_t^k + e_t^i \tag{5.42}$$

where the k-dimensional stochastic process $\mathbf{Z}_t = (Z_t^1, \dots Z_t^k)^T$ is the *dynamic factor*, $\boldsymbol{\chi}_t$ is called *common component*, the d-dimensional stochastic process $\mathbf{e}_t = (e_t^1, \dots, e_t^d)^T$ is called *noise component*, and the $d \times k$ matrix $\boldsymbol{B}(L) = (b_{ij}(L))$, $i = 1, \dots, d$, $j = 1, \dots, k$, is the *transfer function*. Here L is the *lag operator* (backward shift) and $b_{ij}(L)$ is a square-summable one-sided filter, i.e. $b_{ij}(L) = b_{ij}(0) + b_{ij}(1)L + b_{ij}(2)L^2 + \dots$ with $\sum_{\ell=0}^{\infty} b_{ij}^2(\ell) < \infty$. Further, the components of (5.42) satisfy the following requirements:

$$\begin{aligned}
\mathbb{E}(\mathbf{Z}_t) = \mathbf{0}, \quad & \mathbb{E}(\mathbf{e}_t) = \mathbf{0}, \quad t \in \mathbb{Z} \\
\mathrm{Cov}(e_t^i, Z_s^j) = 0, \quad & i = 1, \dots, d, \quad j = 1, \dots, k, \quad t, s \in \mathbb{Z},\ s \le t. \\
\mathrm{Cov}(e_t^i, e_s^j) = 0, \quad & i, j = 1, \dots d, \quad i \ne j, \quad t, s \in \mathbb{Z},\ s < t.
\end{aligned}$$

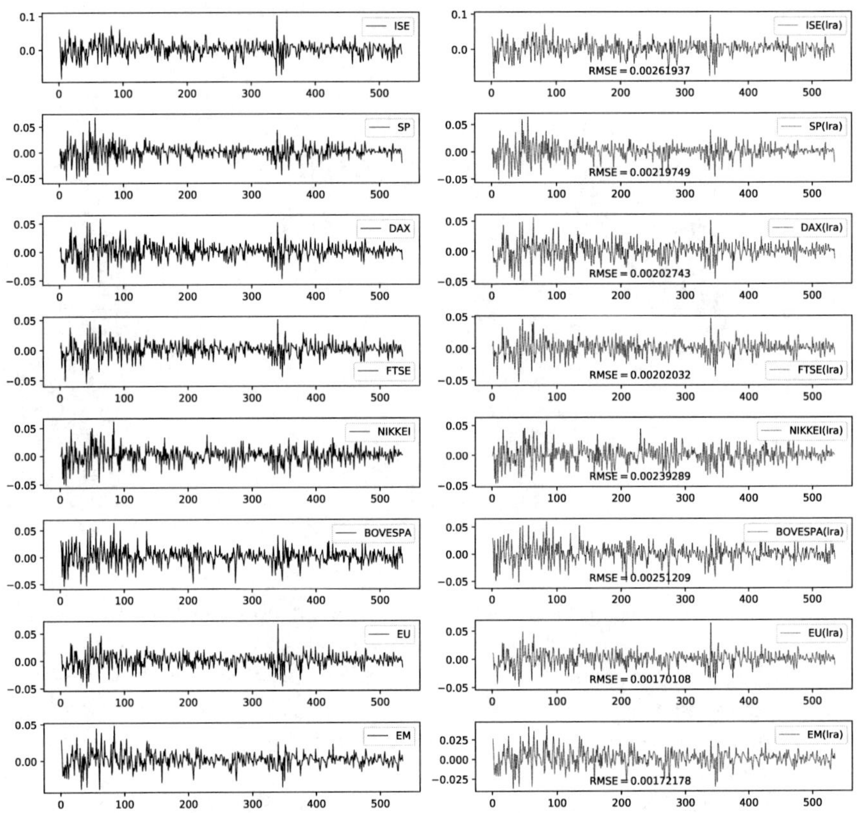

FIGURE 5.2
Approximation of the original time series by a rank 3 time series in Example 5.1.

If $\mathbf{Z}_t$ and $\mathbf{e}_t$ are also weakly stationary and they have rational spectral densities $\boldsymbol{f}_{\mathbf{Z}}$ and $\boldsymbol{f}_{\mathbf{e}}$, the model Equation (5.41) extends to the spectral density matrices:

$$\boldsymbol{f}_{\mathbf{X}}(\omega) = \boldsymbol{f}_{\boldsymbol{\chi}}(\omega) + \boldsymbol{f}_{\mathbf{e}}(\omega) = \boldsymbol{B}(e^{-i\omega})\boldsymbol{f}_{\mathbf{Z}}(\omega)\boldsymbol{B}(e^{-i\omega})^* + \boldsymbol{f}_{\mathbf{e}}(\omega), \quad \omega \in [-\pi, \pi]. \tag{5.43}$$

Very frequently, $\mathbf{Z}_t$ is assumed to be orthonormal WN($\boldsymbol{I}_k$) process. Then Equation (5.43) simplifies to

$$\boldsymbol{f}_{\mathbf{X}}(\omega) = \frac{1}{2\pi}\boldsymbol{B}(e^{-i\omega})\boldsymbol{B}(e^{-i\omega})^* + \boldsymbol{f}_{\mathbf{e}}(\omega).$$

The so-called *static* case occurs if, in addition, $\boldsymbol{B}$ is constant. Otherwise,

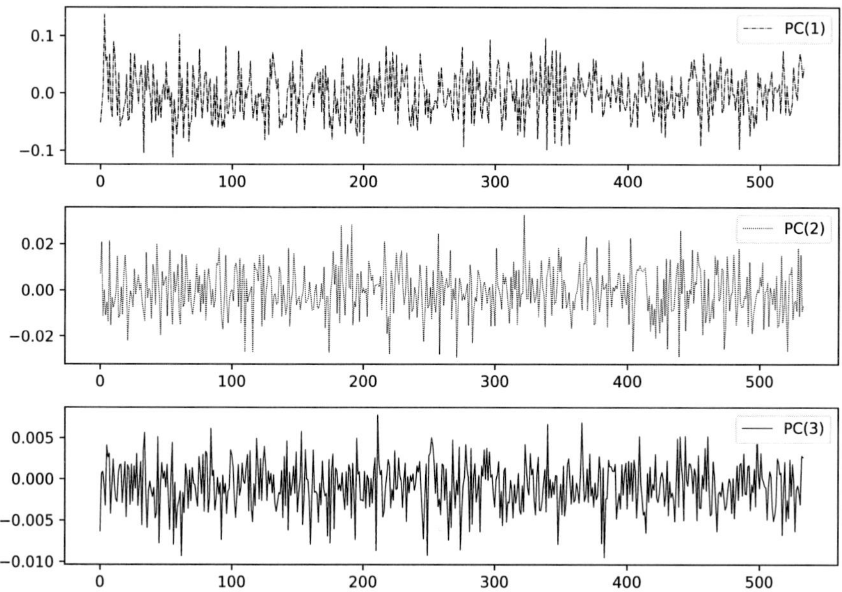

FIGURE 5.3
The 3 leading PC's of the stock exchange data in the time domain in Example 5.1.

Equation (5.41) is dynamic in that the latent variables Z_t^js can affect the observables X_t^is both contemporaneously and with lags.

Like in the standard factor model, neither $\boldsymbol{B}(L)$ nor $\mathbf{Z}_t$ are identified uniquely; and given the spectral density $\boldsymbol{f}_{\mathbf{X}}$, the spectra $\boldsymbol{f}_{\boldsymbol{\chi}}$ and $\boldsymbol{f}_{\mathbf{e}}$ are generically can be determined for $k \leq n - \sqrt{n}$ (reminiscent of the Lederman bound).

5.6.5 General Dynamic Factor Model

Let $\mathbf{X}_t$ be a *weakly stationary* time series ($t = 1, 2, \dots$) with an absolutely continuous spectral measure and the positive semidefinite *spectral density* matrix $\boldsymbol{f}_{\mathbf{X}}$.

Assume that $\boldsymbol{f}_{\mathbf{X}}(\omega)$ has *constant rank* r for a.e. $\omega \in [-\pi, \pi]$. If $\mathbf{X}_t$ is also regular (it always holds if $\boldsymbol{f}_{\mathbf{X}}$ is a rational spectral density matrix), then the multidimensional Wold decomposition is able to make it a one-sided $VMA(\infty)$ process. It is important that the dimension of the *innovation subspaces* is also r.

With the integer $1 \leq k \leq r$, the *k-factor GDFM*:

$$\mathbf{X}_t = \boldsymbol{\chi}_t + \mathbf{e}_t, \quad t = 1, 2, \dots$$

where now χ_t denotes the common component, $\mathbf{e}_t$ is the *idiosyncratic noise*, and all the expectations are zeros, for simplicity. Here χ_t is subordinated to $\mathbf{X}_t$, but has spectral density matrix of rank $k \le r$. For example, there are k uncorrelated signals (given by k distinct sources) detected by r sensors. Opposed to the static factors, this is not a low-rank approximation of the (zero-lag) auto-covariance matrix that provides the *static factors.*

Forni and Lippi [18] and Deistler et al. [13, 15] gave necessary and sufficient conditions for the existence of an underlying GDFM in terms of the expanding sequence of $n \times n$ spectral density matrices $\boldsymbol{f}^n_{\mathbf{X}}(\omega)$, $n \in \mathbb{N}$.

Theorem 5.3. *The nested sequence* $\{\mathbf{X}^n_t : n \in \mathbb{N}, t = 1, 2, \dots\}$ *can be represented by a sequence of k-factor GDFMs if and only if*

- *the k largest eigenvalues,* $\lambda^n_{\mathbf{X},1}(\omega) \ge \dots \ge \lambda^n_{\mathbf{X},k}(\omega)$ *of* $\boldsymbol{f}^n_{\mathbf{X}}(\omega)$ *diverge almost everywhere in* $[-\pi, \pi]$ *as* $n \to \infty$*;*
- *the* $(k+1)$*-th largest eigenvalue* $\lambda^n_{\mathbf{X},k+1}(\omega)$ *of* $\boldsymbol{f}^n_{\mathbf{X}}(\omega)$ *is uniformly bounded for* $\omega \in [-\pi, \pi]$ *(almost everywhere) and for all* $n \in \mathbb{N}$.

The theorem is rather theoretical; its message is that for large n and T (T is not necessarily larger than n) we can conclude for k from the spectral gap of the constant rank spectral density matrix. The estimate χ^n_t is consistent if $n, T \to \infty$. The idiosyncratic noise is less and less important when $n, T \to \infty$, and it may have slightly correlated components. Also, the largest eigenvalue of $\boldsymbol{f}^n_{\mathbf{e}}(\omega)$ is uniformly bounded for $\omega \in [-\pi, \pi]$ and for all $n \in \mathbb{N}$. As we learned in Chapter 4, a stationary process with a not full rank spectral density matrix may have some singular components. All these parts are included in the weakly dependent idiosyncratic noise.

Dynamic factor analysis is an unsupervised learning method, and with the lag-dependent factor loading matrices we are able to give meaning to the dynamic factors that embody the comovements between the components at different lags. For example, when we use a parametric method, see, e.g. [5], we are also able to give predictions for the dynamic factors (via autoregression) and, in turn, for the components of the time series too. There are also state space models that are able to estimate the parameter matrices via singular autoregression [13, 18]. The reduced rank approximation in Section 5.6.3 offers a first step, and the Yule–Walker equations can be solved for the reduced rank process.

5.7 Summary

First we have a 1D real valued time series $\{X_t\}$ which is not necessarily stationary, $\mathbb{E}(X_t) = 0$ ($t \in \mathbb{Z}$). Selecting a starting observation X_1 and with the notation $H_n = \text{Span}\{X_1, \dots, X_n\}$, X_{n+1} is predicted linearly based on random past values $X_1, \dots, X_n$ such that $\hat{X}_1 := 0$ and $e_n^2 = \mathbb{E}\eta_{n+1}^2 = \mathbb{E}(X_{n+1} - \hat{X}_{n+1})^2$ is minimized, $n = 1, 2, \dots$. By the general theory of Hilbert spaces, $\hat{X}_{n+1}$ is the projection of X_{n+1} onto the linear subspace H_n. The coefficients of the optimal linear predictor

$$\hat{X}_{n+1} = a_{n1}X_n + \cdots + a_{nn}X_1$$

can be obtained by solving the system of linear equations

$$\boldsymbol{C}_n \mathbf{a}_n = \mathbf{d}_n,$$

where $\mathbf{a}_n = (a_{n1}, \dots, a_{nn})^T$, $\boldsymbol{C}_n = [\mathrm{Cov}(X_i, X_j)]_{i,j=1}^n$ and $\mathbf{d}_n = (\mathrm{Cov}(X_{n+1}, X_n), \dots, \mathrm{Cov}(X_{n+1}, X_1))^T$. A solution (the projection) always exists, and it is unique if $\boldsymbol{C}_n$ is positive definite; then the unique solution is $\mathbf{a}_n = \boldsymbol{C}_n^{-1}\mathbf{d}_n$, otherwise, the generalized inverse of $\boldsymbol{C}_n$ comes into existence. However, in case of stationary processes, this is not an issue. The h-step ahead prediction is obtained from $\boldsymbol{C}_n \mathbf{a}_n = \mathbf{d}_n(h)$, where $\mathbf{d}_n(h) = (c(h), \dots, c(n + h - 1))^T$ in the stationary case.

As for the innovation $\boldsymbol{\eta}_n = (\eta_1, \dots, \eta_n)^T$, we have to find an $n \times n$ lower triangular matrix $\boldsymbol{L}_n$ such that $\mathbf{X}_n = \boldsymbol{L}_n \boldsymbol{\eta}_n$. Taking the covariance matrices on both sides, yields $\boldsymbol{C}_n = \boldsymbol{L}_n \boldsymbol{D}_n \boldsymbol{L}_n^T$. In this way, the LDL decomposition (a variant of the Cholesky decomposition) gives the prediction errors e_n^2s (diagonal entries of $\boldsymbol{D}_n$), and the entries of $\boldsymbol{L}_n$ below its main diagonal (the main diagonal is constantly 1). The situation further simplifies in the stationary case, when $\boldsymbol{C}_n$ is a Toeplitz matrix. However, $\boldsymbol{L}_n$ will not be Toeplitz, but asymptotically, it becomes more and more like a Toeplitz one, and the entries of $\boldsymbol{D}_n$ will be more and more similar to each other, i.e. to the limit $\sigma^2 = \lim_{n\to\infty} e_n^2$.

In particular, if $\{X_t\}$ is stationary, then $\boldsymbol{C}_n = [c(i - j)]_{i,j=1}^n$, so $\boldsymbol{C}_n$ is a Toeplitz matrix, and $d_n(j) = c(j)$, $j = 1, \dots, n$. Therefore, no double indexing is necessary, but $\mathbf{a}_n = (a_1, \dots, a_n)^T$. With it, the defining equation is exactly the same as the first n Yule–Walker equations for estimating the parameters of a stationary AR(n) process. The prediction error is $e_n^2 = \mathrm{Var}(\eta_{n+1})$. It can be written in many equivalent forms, e.g.

$$e_n^2 = c(0)(1 - r^2_{X_t,(X_{t-1},\dots,X_{t-n})}) = c(0) - \mathbf{d}_n^T \boldsymbol{C}_n^{-1} \mathbf{d}_n,$$

where $r^2_{X_t,(X_{t-1},\dots,X_{t-n})}$ is the squared multiple correlation coefficient between X_t and $(X_{t-1}, \dots, X_{t-n})$; it does not depend on t either, and obviously increases (does not decrease) with n, i.e. $e_1^2 \geq e_2^2 \geq \dots$. The mean square error can as well be written with the determinants of the consecutive Toeplitz matrices $\boldsymbol{C}_n$ and $\boldsymbol{C}_{n+1}$. If for some n, $|\boldsymbol{C}_n| \neq 0$, then

$$e_n^2 = c(0) - \mathbf{d}_n^T \boldsymbol{C}_n^{-1} \mathbf{d}_n = \frac{|\boldsymbol{C}_{n+1}|}{|\boldsymbol{C}_n|}.$$

If $|\boldsymbol{C}_n| = 0$ for some n, then $|\boldsymbol{C}_{n+1}| = |\boldsymbol{C}_{n+2}| = \cdots = 0$ too. The smallest index n for which this happens indicates that there is a linear relation between n consecutive X_js, but no linear relation between $n-1$ consecutive ones (by stationarity, this property is irrespective of the position of the consecutive random variables). This can happen only if some X_t linearly depends on $n-1$ preceding X_js. In this case $e_{n-1}^2 = 0$ and, of course $e_n^2 = e_{n+1}^2 = \dots = 0$ too. In any case, $e_1^2 \geq e_2^2 \geq \dots$ is a decreasing (non-increasing) nonnegative sequence, and in view of Equation (5.8),

$$|\boldsymbol{C}_1| = c(0), \quad |\boldsymbol{C}_n| = c(0)e_1^2 \dots e_{n-1}^2, \quad n = 2, 3, \dots,$$

so, provided $c(0) > 0$, $|\boldsymbol{C}_n| = 0$ holds if and only if $e_{n-1}^2 = 0$. Note that in this stationary case there is no sense of using generalized inverse if $|\boldsymbol{C}_n| = 0$, since then exact one-step ahead prediction with the $n-1$ long past can be done with zero error, and this property is manifested for longer past predictions too. Note that the previous LDL decomposition also implies that $|\boldsymbol{C}_n| = |\boldsymbol{D}_n| = c(0)e_1^2 \dots e_{n-1}^2$, $n = 2, 3, \dots$.

In case of a stationary, non-singular process, we can project X_{n+1} onto the infinite past $H_n^- = \overline{\text{span}}\,\{X_t : t \leq n\}$ and expand it in terms of an orthonormal system, see the Wold decomposition. This part will be the regular (causal) part of the process, whereas, the other, singular part, is orthogonal to it. Also, by stationarity, the one-step ahead prediction error σ^2 does not depend on n, and it is positive, since the process is non-singular.

Then we have a d-dimensional time series $\{\mathbf{X}_t\}$ with components $\mathbf{X}_t = (X_t^1, \dots, X_t^d)^T$, the state space is $\mathbb{R}^d$ and $\mathbb{E}(\mathbf{X}_t) = \mathbf{0}$. Select a starting observation $\mathbf{X}_1$ and

$$H_n := \text{span}\{X_t^j : t = 1, \dots, n;\ j = 1, \dots, d\}.$$

We want to linearly predict $\mathbf{X}_{n+1}$ based on random past values $\mathbf{X}_1, \dots, \mathbf{X}_n$. Analogously to the 1D situation, $\hat{\mathbf{X}}_1 := 0$ and $\hat{\mathbf{X}}_{n+1}$ is the best one-step ahead linear predictor that minimizes $\mathbb{E}(\mathbf{X}_{n+1} - \hat{\mathbf{X}}_{n+1})^2$. Now we solve the system of linear equations

$$\sum_{j=1}^{n} \boldsymbol{A}_{nj} \text{Cov}(\mathbf{X}_{n+1-j}, \mathbf{X}_{n+1-k}) = \text{Cov}(\mathbf{X}_{n+1}, \mathbf{X}_{n+1-k}). \quad k = 1, \dots, n.$$

When $\{\mathbf{X}_t\}$ is stationary, then it simplifies to

$$\sum_{j=1}^{n} \boldsymbol{A}_j \boldsymbol{C}(k-j) = \boldsymbol{C}(k), \quad k = 1, \dots, n.$$

This provides a system of $d^2 n$ linear equations with the same number of unknowns that always has a solution. Further, the solution does not depend on the selection of the time of the starting observation $\mathbf{X}_1$, and no double indexing of the coefficient matrices is necessary. If for some $n \geq 1$ the covariance matrix of $(\mathbf{X}_{n+1}^T, \dots, \mathbf{X}_1^T)^T$ is positive definite, then the matrix polynomial

(VAR polynomial) $\boldsymbol{\alpha}(z) = \boldsymbol{I} - \boldsymbol{A}_1 z - \cdots - \boldsymbol{A}_n z^n$ is causal in the sense that $|\boldsymbol{\alpha}(z)| \neq 0$ for $z \leq 1$; otherwise, block matrix techniques and reduction in the innovation subspaces is needed.

$\mathbf{X}_t$ can again be expanded in terms of the now d-dimensional innovations, i.e. the prediction error terms $\boldsymbol{\eta}_{n+1} = \mathbf{X}_{n+1} - \hat{\mathbf{X}}_{n+1}$. In this way, we get the innovations $\boldsymbol{\eta}_1, \ldots, \boldsymbol{\eta}_n$ that trivially have $\mathbf{0}$ expectation and form an orthogonal system in H_n. Actually, we have the recursive equations

$$\mathbf{X}_k = \sum_{j=1}^{k-1} \boldsymbol{B}_{kj} \boldsymbol{\eta}_j + \boldsymbol{\eta}_k, \quad k = 1, 2, \ldots, n.$$

Here the covariance matrix $\boldsymbol{E}_j = \mathbb{E} \boldsymbol{\eta}_j \boldsymbol{\eta}_j^T$ is a positive semidefinite matrix, but can be of reduced rank. At a passage to infinity, we obtain the multidimensional Wold decomposition. At the end, we have to perform the block Cholesky (LDL) decomposition:

$$\mathfrak{C}_n = \boldsymbol{L}_n \boldsymbol{D}_n \boldsymbol{L}_n^T,$$

where $\mathfrak{C}_n$ is $nd \times nd$ positive definite block Toeplitz matrix, $\boldsymbol{D}_n$ is $nm \times nm$ block diagonal and contains the positive semidefinite prediction error matrices $\boldsymbol{E}_1, \ldots, \boldsymbol{E}_n$ in its diagonal blocks, whereas $\boldsymbol{L}_n$ is $nd \times nd$ lower triangular with blocks $\boldsymbol{B}_{kj}$s below its diagonal blocks which are $d \times d$ identities. In view of this,

$$|\mathfrak{C}_n| = |\boldsymbol{D}_n| = \prod_{j=1}^{n} |\boldsymbol{E}_j|,$$

analogously to the 1D situation. We also prove that if the entries of the autocovariance matrices are absolutely summable, then the eigenvalues of $\mathfrak{C}_n$ asymptotically comprise the union of the spectra of the spectral density matrices at the n Fourier frequencies as $n \to \infty$.

When the $\boldsymbol{E}_j$s are of reduced rank, we can find a system $\boldsymbol{\xi}_1, \ldots, \boldsymbol{\xi}_n \in \mathbb{R}^r$ in the d-dimensional innovation subspaces that span the same subspace as $\boldsymbol{\eta}_1, \ldots, \boldsymbol{\eta}_n$. If the the spectral density matrix has $r < d$ structural eigenvalues in a General Dynamic Factor Model, then $\boldsymbol{\xi}_j \in \mathbb{R}^r$ is the principal component factor of $\boldsymbol{\eta}_j$ obtained from an r-factor model.

Note that here we use $d \times d$ block matrices in the calculations, so the computational complexity of the procedure is not significantly larger than that of the subsequent Kálmán's filtering for which we use the notation of R.E. Kálmán's original paper [32], where stationarity is not assumed, but the random vectors are Gaussian. The linear dynamical system is

$$\begin{aligned} \mathbf{X}_{t+1} &= \boldsymbol{A}_t \mathbf{X}_t + \mathbf{U}_t \\ \mathbf{Y}_t &= \boldsymbol{C}_t \mathbf{X}_t, \end{aligned}$$

where $\boldsymbol{A}_t$ and $\boldsymbol{C}_t$ are specified matrices; $\boldsymbol{A}_t$ is an $n \times n$ matrix, called phase transition matrix, and $\boldsymbol{C}_t$ is $p \times n$; further, $\mathbf{U}_t$ (random excitation) is an orthogonal noise process with $\mathbb{E}\mathbf{U}_t\mathbf{U}_s^T = \delta_{st}\boldsymbol{Q}_{\mathbf{U}}(t)$ and $\mathbb{E}\mathbf{X}_s^T\mathbf{U}_t = \mathbf{0}$ for $s \le t$. All the expectations are zeros, and all the random vectors have real components. $\mathbf{X}_t$ is the n-dimensional hidden state variable, while $\mathbf{Y}_t$ is the p-dimensional observable variable. In the paper [32], $p \le n$ is assumed, but it is not a restriction. Even if $p = n$, the matrix $\boldsymbol{C}_t$ is not invertible, otherwise the process $\mathbf{X}_t$ is trivially observable, unless a noise term is added to $\boldsymbol{C}_t\mathbf{X}_t$ in the second equation.

The problem is the following: starting the observations at time 0, given $\mathbf{Y}_0, \dots, \mathbf{Y}_{t-1}$, we want to estimate $\mathbf{X}$ component-wise, with minimum mean square error. If $\mathbf{X} = \mathbf{X}_t$, this is the prediction problem and we denote the optimal one-step ahead prediction of $\mathbf{X}_t$ by $\widehat{\mathbf{X}}_{t|t-1}$. If $\mathbf{Y}_0, \dots, \mathbf{Y}_{t-1}$ is observed and $\widehat{\mathbf{X}}_{t|t-1}$ is already known, then we give a recursion to find $\widehat{\mathbf{X}}_{t+1|t}$ by using the new value of $\mathbf{Y}_t$:

$$\widehat{\mathbf{X}}_{t+1|t} = \boldsymbol{A}_t\widehat{\mathbf{X}}_{t|t-1} + \boldsymbol{K}_t(\mathbf{Y}_t - \boldsymbol{C}_t\widehat{\mathbf{X}}_{t|t-1}),$$

where $\boldsymbol{K}_t$ is the Kálmán gain matrix:

$$\boldsymbol{K}_t = \boldsymbol{A}_t\boldsymbol{P}(t)\boldsymbol{C}_t^T[\boldsymbol{C}_t\boldsymbol{P}(t)\boldsymbol{C}_t^T]^-.$$

Here

$$\boldsymbol{P}(t) = \mathbb{E}\widetilde{\mathbf{X}}_{t|t-1}\widetilde{\mathbf{X}}_{t|t-1}^T$$

is the error covariance matrix that drives the process. For it, the recursion

$$\boldsymbol{P}(t+1) = \boldsymbol{A}_t\boldsymbol{P}(t)\boldsymbol{A}_t{}^T - \boldsymbol{A}_t\boldsymbol{P}(t)\boldsymbol{C}_t^T[\boldsymbol{C}_t\boldsymbol{P}(t)\boldsymbol{C}_t^T]^-\boldsymbol{C}_t\boldsymbol{P}(t)\boldsymbol{A}_t{}^T + \boldsymbol{Q}_{\mathbf{U}}(t)$$

holds, which makes rise to an iteration. The above equation results in a matrix Riccati equation for $\boldsymbol{P} = \boldsymbol{P}(t) = \boldsymbol{P}(t+1)$ if the process is stationary.

With the integer $1 \le k < r$, the dynamic k-factor mode (GDFM) is:

$$\mathbf{X}_t = \chi_t + \mathbf{e}_t, \quad t = 1, 2, \dots$$

where now χ_t denotes the common component, $\mathbf{e}_t$ is the n-dimensional idiosyncratic noise, and all the expectations are zeros, for simplicity. Here χ_t is subordinated to $\mathbf{X}_t$, but has spectral density matrix of rank $k < r$. For example, there are k uncorrelated signals (given by k distinct sources), detected by d sensors. Opposed to the static factors, this is not a low-rank approximation of the (zero-lag) auto-covariance matrix that provides the static factors.

Forni and Lippi [18] and Deistler et al. [15, 13] gave necessary and sufficient conditions for the existence of an underlying GDFM in terms of the observable $n \times n$ spectral densities $\boldsymbol{f}_{\mathbf{X}}^n(\omega)$, $n \in \mathbb{N}$. The nested sequence $\{\mathbf{X}_t^n : n \in \mathbb{N}, t = 1, 2, \dots\}$ can be represented by a sequence of GDFMs if and only if

- the first k eigenvalues, $\lambda_{\mathbf{X},1}^n(\omega) \ge \cdots \ge \lambda_{\mathbf{X},k}^n(\omega)$ (in non-increasing order), of $\boldsymbol{f}_{\mathbf{X}}^n(\omega)$ diverge almost everywhere in $[-\pi, \pi]$ as $n \to \infty$;

- the $(k+1)$-th eigenvalue $\lambda^n_{\mathbf{X},k+1}(\omega)$ of $\boldsymbol{f}^n_{\mathbf{X}}(\omega)$ is uniformly bounded for $\omega \in [-\pi, \pi]$ almost everywhere and for all $n \in \mathbb{N}$.

So we can conclude for k from the spectral gap. The estimate $\boldsymbol{\chi}^n_t$ is consistent if $n, T \to \infty$. The idiosyncratic noise is less and less important when n, T get larger and larger, and it may have slightly correlated components.

A

Tools from complex analysis

Details of the material and the proofs of the theorems discussed in this chapter can be found e.g. in [50]. The aim of this chapter to summarize the most important tools from complex analysis that we are using in this book.

A.1 Holomorphic (or analytic) functions

Let $\Omega \subset \mathbb{C}$ be an open set and $f : \Omega \to \mathbb{C}$ be a complex function. If $z_0 \in \Omega$ and if

$$\lim_{z \to z_0} \frac{f(z) - f(z_0)}{z - z_0}$$

exists then we denote this limit by $f'(z_0) \in \mathbb{C}$ and call it the *derivative* of f at z_0 and f is said to be *complex differentiable* at z_0. If f is complex differentiable at every $z_0 \in \Omega$, then we say that f is *holomorphic* (or *analytic*) in Ω.

Let $[\alpha, \beta] \subset \mathbb{R}$ be a closed interval. A *path* γ is a piecewise continuously differentiable function $\gamma : [\alpha, \beta] \to \mathbb{C}$. A *closed path* is a path such that $\gamma(\alpha) = \gamma(\beta)$. If $f : \Omega \to \mathbb{C}$ is a continuous complex function and the *range* γ^* of the path γ is in Ω, then the integral of f over γ is defined as

$$\int_\gamma f(z)dz := \int_\alpha^\beta f(\gamma(t))\gamma'(t)dt.$$

Let γ be a closed path and take $\Omega = \mathbb{C} \setminus \gamma^*$. Define

$$\mathrm{Ind}_\gamma(z) := \frac{1}{2\pi i} \int_\gamma \frac{d\zeta}{\zeta - z}, \qquad z \in \Omega.$$

Then Ind_γ is an integer-valued function in Ω which is constant in each component of Ω and which is 0 in the unbounded component of Ω. We call $\mathrm{Ind}_\gamma(z)$ the *index* or *winding number* of z with respect to γ.

The most important special class of closed paths is when there are exactly two components of Ω w.r.t. γ: one where the winding number is 1 and one where the winding number is 0. Such a closed path is called a *simple closed path*. If γ is a simple closed path in Ω, that is, such that

$$\Omega = \Omega_1 \cup \Omega_0 \cup \gamma^*, \quad \mathrm{Ind}_\gamma = 1 \quad \text{in} \quad \Omega_1, \quad \mathrm{Ind}_\gamma = 0 \quad \text{in} \quad \Omega_0,$$

then we may call Ω_1 the *interior* and Ω_0 the *exterior* of γ. This is the case e.g. when $\gamma(t) = e^{it}$, $0 \le t \le 2\pi$, so γ^* is the unit circle T. Then γ winds around the points of the open unit disc D exactly once counterclockwise. It is easy to see that in this example $\mathrm{Ind}_\gamma(z) = 1$ if $z \in D$ and $\mathrm{Ind}_\gamma(z) = 0$ if $z \in \{\zeta \in \mathbb{C} : |\zeta| > 1\}$.

The fundamental theorems of complex analysis are *Cauchy's theorem* and *Cauchy's formula.*

Theorem A.1. *Suppose that Ω is an open set and f is a holomorphic function in Ω. Then for any closed path γ in Ω such that* $\mathrm{Ind}_\gamma(z) = 0$ *for any $z \notin \Omega$, we have Cauchy's theorem*

$$\int_\gamma f(z)dz = 0.$$

Moreover, we also have Cauchy's formula

$$f(z) \cdot \mathrm{Ind}_\gamma(z) = \frac{1}{2\pi i}\int_\gamma \frac{f(\zeta)}{\zeta - z} d\zeta, \quad \forall z \in \Omega \setminus \gamma^*. \tag{A.1}$$

An important consequence of (A.1) is that if γ is a simple closed path, then the values of a holomorphic function f in the interior of γ are uniquely determined by the values of f on the boundary γ^*. The condition "$\mathrm{Ind}_\gamma(z) = 0$ for any $z \notin \Omega$" of the theorem prevents a situation that points where f is not holomorphic may influence the value of the integrals. (Imagine that Ω is an annulus and a closed path γ in Ω goes around the inner circle boundary of the annulus.)

Another important consequence is stated in the next theorem.

Theorem A.2. *Suppose that Ω is an open set. Then a function f is holomorphic in Ω if and only if it is* representable by power series in Ω *in the sense that for any $a \in \Omega$ there exists an $r > 0$ such that*

$$f(z) = \sum_{n=0}^{\infty} c_n (z-a)^n, \quad \forall z \in D(a,r) := \{\zeta : |\zeta - a| < r\}, \tag{A.2}$$

where each $c_n \in \mathbb{C}$.

It follows that if f is holomorphic in Ω, then f is arbitrary many times complex differentiable in Ω.

Recall that a function f which is representable by power series in an open set Ω is usually called an *analytic function.* We have just seen that the class of holomorphic and the class of analytic functions in an open set Ω coincide in complex analysis. Notice that the power series in (A.2) contains only non-negative powers of $(z - a)$. The domain of convergence of such a power series is always a disc $\{z : |z - a| < R\}$ and possibly some points of the boundary circle as well, where $R \in [0, \infty]$,

$$\frac{1}{R} = \limsup_{n\to\infty} |c_n|^{1/n}. \tag{A.3}$$

Here comes another important property of holomorphic functions.

Theorem A.3. *(The maximum modulus theorem)*

(a) Suppose that Ω is a connected open set, $D(a,r) \subset \Omega$, $r > 0$, and f is a holomorphic function in Ω. Then

$$|f(a)| \leq \max_t |f(a + re^{it})|.$$

Equality occurs here if and only if f is constant in Ω.

(b) Let Ω be a bounded connected open set and let K denote its closure. If the function f is continuous in K and holomorphic in Ω, then

$$|f(z)| \leq \sup_{\zeta \in \partial\Omega} |f(\zeta)|, \quad z \in \Omega,$$

where $\partial\Omega$ denotes the boundary of Ω. If equality holds at one point $z \in \Omega$, then f is constant.

Let f be a holomorphic function in the connected open set Ω, and define the *zero set* of f by

$$Z(f) := \{a \in \Omega : f(a) = 0\}.$$

If f is not identically 0 in Ω, then $Z(f)$ has no limit point in Ω and $Z(f)$ is at most countable. Moreover, then to each $a \in Z(f)$ there corresponds a unique positive integer m such that

$$f(z) = (z - a)^m g(z), \qquad z \in \mathbb{C},$$

where g is holomorphic in Ω and $g(a) \neq 0$. Then f is said to have a *zero of order* m at the point a.

Let Ω be an open set, $a \in \Omega$, and f be holomorphic in $\Omega \setminus \{a\}$. Then f is said to have an *isolated singularity* at a. If f can be defined at the point a so that the extended function is holomorphic in Ω, then the singularity is called *removable.* If there is a positive integer m and constants $c_{-1}, \ldots, c_{-m} \in \mathbb{C}$, $c_{-m} \neq 0$, such that

$$f(z) - Q(z) = f(z) - \sum_{k=1}^{m} c_{-k}(z - a)^{-k}$$

has a removable singularity at a, then f is said to have a *pole of order* m at a and $Q(z)$ is called the *principal part* of f at a. Then there exists an $r > 0$ such that f can be expressed by a two-sided power series around a as

$$f(z) = \sum_{k=-m}^{\infty} c_k (z - a)^k, \quad 0 < |z - a| < r.$$

Any other isolated singularity is called an *essential singularity.*

A function f is said to be *meromorphic* in an open set Ω if there is a subset $A \subset \Omega$ such that

1. A has no limit point in Ω,
2. f is holomorphic in $\Omega \setminus A$,
3. f has a pole at each point of A.

The set A can be at most countable. For each $a \in A$, the principal part of f at a has the form

$$Q_a(z) = \sum_{k=1}^{m(a)} c_{-k}^{(a)}(z-a)^{-k};$$

the coefficient $c_{-1}^{(a)}$ is called the residue of f at a:

$$\mathrm{Res}(f,a) := c_{-1}^{(a)}.$$

If γ is a closed path in $\Omega \setminus A$, then elementary integration shows that

$$\frac{1}{2\pi i}\int_\gamma Q_a(z)dz = \mathrm{Res}(f;a)\,\mathrm{Ind}_\gamma(a).$$

This simple fact can be used to show the following *Residue theorem.*

Theorem A.4. *Suppose that f is a meromorphic function in the open set Ω. Let A be the subset of points at which f has poles. If γ is a closed path in $\Omega \setminus A$ such that* $\mathrm{Ind}_\gamma(z) = 0$ *for all $z \notin \Omega$, then*

$$\frac{1}{2\pi i}\int_\gamma f(z)dz = \sum_{a\in A} \mathrm{Res}(f;a)\,\mathrm{Ind}_\gamma(a).$$

The next theorem is an application of the Residue theorem; useful to determine how many zeros a holomorphic function f has in the interior of a simple closed path.

Theorem A.5. *Assume that γ is a simple closed path in a connected open set Ω, such that* $\mathrm{Ind}_\gamma(z) = 0$ *for any $z \notin \Omega$.*

Let f be a holomorphic function in Ω and let $N_f(\gamma)$ denote the number of zeros of f in the interior Ω_1 of γ, counted according to their multiplicities. Assume that f has no zeros on γ^*. Then

$$N_f(\gamma) = \frac{1}{2\pi i}\int_\gamma \frac{f'(z)}{f(z)}dz = \mathrm{Ind}_{f\circ\gamma}(0).$$

This theorem is sometimes called the *"Argument principle."* This name can be explained by a heuristic argument. Since we assumed that f has no zeros on γ^*, along the closed path γ one can take

$$\log f(z) = \log\left(|f(z)|e^{i\arg f(z)}\right) = \log|f(z)| + i\arg f(z),$$

where $\arg f(z)$ denotes the multiple-valued argument (angle) of the complex number $f(z)$. By the chain rule, $(\log f(z))' = f'(z)/f(z)$, and so

$$\int_\gamma \frac{f'(z)}{f(z)} dz = \int_\gamma d\log|f(z)| + i\int_\gamma d\arg f(z)$$
$$= \log|f(\gamma(\beta))| - \log|f(\gamma(\alpha))| + i\Delta_\gamma \arg f(z) = i\Delta_\gamma \arg f(z),$$

since $\gamma(\beta) = \gamma(\alpha)$. Here $\Delta_\gamma \arg f(z)$ denotes the change of argument of f along the closed path γ, which divided by 2π gives the winding number $\mathrm{Ind}_{f\circ\gamma}(0)$ in the theorem.

A.2 Harmonic functions

If $f : \mathbb{C} \to \mathbb{C}$ is a complex function, we may write that $f(z) = u(x, y)+iv(x, y)$, where $u, v : \mathbb{R}^2 \to \mathbb{R}$ are real two-variable functions. If Ω is a plane open set, then f is holomorphic in Ω if and only if u and v are differentiable two-variable functions in Ω and the *Cauchy–Riemann equations* hold:

$$\partial_x u = \partial_y v, \quad \partial_y u = -\partial_x v, \quad (x, y) \in \Omega, \tag{A.4}$$

where ∂_x and ∂_y denote partial differentiation w.r.t. x and y, respectively. Then

$$f' = \frac{1}{2}(\partial_x u + \partial_y v) + \frac{i}{2}(\partial_x v - \partial_y u). \tag{A.5}$$

Another way of writing the above equalities can be obtained by introducing the differential operators

$$\partial := \frac{1}{2}(\partial_x - i\partial_y), \quad \bar\partial := \frac{1}{2}(\partial_x + i\partial_y).$$

Then (A.4) and (A.5) are equivalent to

$$\bar\partial f = 0, \qquad f' = \partial f. \tag{A.6}$$

(Applying ∂ and $\bar\partial$ goes like multiplication with complex numbers.)

Let Ω be a plane open set and let $u : \mathbb{R}^2 \to \mathbb{R}$ be a two-variable real function such that $\partial_{xx} u$ and $\partial_{yy} u$ exist at every point of Ω. Then the *Laplacian* of u is defined as

$$\Delta u := \partial_{xx} u + \partial_{yy} u.$$

The function u is called *harmonic* in Ω if it is continuous in Ω and $\Delta u = 0$ in Ω.

Similarly, the complex function $f : \mathbb{C} \to \mathbb{C}$, $f = u + iv$, is *harmonic* in Ω if it is continuous in Ω, $\partial_{xx} f$ and $\partial_{yy} f$ exist in Ω, and

$$\Delta f := \partial_{xx} f + \partial_{yy} f = \Delta u + i\Delta v = 0 \quad \text{in} \quad \Omega. \tag{A.7}$$

If f is a holomorphic function in Ω, then f has continuous derivatives of all orders, so $\partial_{xy}f = \partial_{yx}f$, moreover, $\Delta f = 4\partial\bar{\partial}f$. Since then $\bar{\partial}f = 0$ by (A.6), it follows that $\Delta f = 0$ in Ω. This shows that holomorphic functions are harmonic. Equation (A.7) shows that the real and imaginary parts of f are also harmonic, and by the Cauchy–Riemann equations (A.4) they are strongly related to each other; that is why u and v are called *harmonic conjugates*. For example, any harmonic function u in D is the real part of one and only one holomorphic function $f = u + iv$ such that $f(0) = u(0)$, and then $v(0) = 0$ and this v is also unique.

For $0 \le \rho < 1$; $t, \theta \in \mathbb{R}$ and $z = \rho e^{i\theta}$, the *Poisson kernel* is

$$P_\rho(\theta - t) := \operatorname{Re}\left[\frac{e^{it} + z}{e^{it} - z}\right] = \frac{1 - |z|^2}{|e^{it} - z|^2} = \frac{1 - \rho^2}{1 - 2\rho\cos(\theta - t) + \rho^2}.$$

Then

$$P_\rho(t) > 0, \quad \frac{1}{2\pi}\int_{-\pi}^{\pi} P_\rho(t)dt = 1 \quad (0 \le \rho < 1).$$

If $g \in L^1([-\pi, \pi])$, then

$$\tilde{G}(z) := \frac{1}{2\pi}\int_{-\pi}^{\pi} \frac{e^{it} + z}{e^{it} - z} g(t)dt$$

is a holomorphic function of $z = \rho e^{i\theta}$ in the open unit disc D. Hence the *Poisson integral*

$$G(\rho e^{i\theta}) := \frac{1}{2\pi}\int_{-\pi}^{\pi} P_\rho(\theta - t)g(t)dt \tag{A.8}$$

is the real part of a holomorphic function, so a harmonic real function in D, for any $g \in L^1([-\pi, \pi])$ real function. It implies that if $g \in L^1([-\pi, \pi])$ is complex valued, the Poisson integral $G(z)$ defined by (A.8) is a complex harmonic function. Moreover,

$$\lim_{\rho \to 1} G(\rho e^{i\theta}) = g(\theta) \quad \text{in} \quad L^1([-\pi, \pi]). \tag{A.9}$$

The *Dirichlet problem* is a famous and important problem of mathematics and physics. Here we discuss it in the unit disc. Assume that a continuous function g is given on T and it is required to find a harmonic function G in D, which is continuous on the closed unit disc $\bar{D}$ and has the boundary values g.

Theorem A.6. *We have a unique solution of the Dirichlet problem in the unit disc.*

(a) Assume that $g \in C(T)$. Let $G(e^{i\theta}) := g(e^{i\theta})$ on T and define $G(z)$ in D by the Poisson integral (A.8). *Then G is harmonic in D and $G \in C(\bar{D})$.*

(b) Conversely, suppose that $G \in C(\bar{D})$ and G is harmonic in D. Then G is the Poisson integral (A.8) *in D of its restriction to T.*

So far the Poisson integral was considered only in the unit disc. However, it is easy to extend it to an arbitrary disc. If g is a continuous complex or real function on the boundary of the open disc $D(a,R) := \{z : |z-a| < R\}$ and if g is defined by the Poisson integral

$$g(a+\rho e^{i\theta}) = \frac{1}{2\pi}\int_{-\pi}^{\pi} \frac{R^2-\rho^2}{R^2-2R\rho\cos(\theta-t)+\rho^2} g(a+Re^{it})dt \tag{A.10}$$

in $D(a,R)$, then g is continuous on the closed disc $\bar{D}(a,R)$ and harmonic in $D(a,R)$.

Conversely, if u is a harmonic real function in an open set Ω and if $\bar{D}(a,R) \subset \Omega$, then u satisfies (A.10) in $D(a,R)$ and there is a unique holomorphic function $f = u + iv$ in $D(a,R)$ such that $f(a) = u(a)$ and $v(a) = 0$. In sum, every real harmonic function is locally the real part of a holomorphic function. Consequently, every harmonic function has continuous partial derivatives of arbitrary order.

We say that a continuous complex or real function g has *the mean value property* in an open set Ω if we have

$$g(a) = \frac{1}{2\pi}\int_{-\pi}^{\pi} g(a+Re^{it})dt, \quad \forall \bar{D}(a,R) \subset \Omega.$$

The Poisson integral (A.10) shows that any harmonic complex or real function has the mean value property. In fact, much more is true, as shown by the next theorem.

Theorem A.7. *A continuous complex or real function has the mean value property in an open set Ω if and only if it is harmonic in Ω.*

A real-valued function u defined in a plane open set Ω is said to be *subharmonic* in Ω if it has the following four properties:

1. $-\infty \le u(z) < \infty$ for all $z \in \Omega$;
2. u is upper semicontinuous in Ω;
3. $u(a) \le \dfrac{1}{2\pi}\displaystyle\int_{-\pi}^{\pi} u(a+Re^{it})dt, \quad \forall \bar{D}(a,R) \subset \Omega$;
4. none of the integrals above is $-\infty$.

Clearly, every harmonic real function is also subharmonic.

Theorem A.8. *We have several useful criteria for subharmonic functions.*

(a) If u is subharmonic in the open set Ω and ϕ is a monotonically increasing convex function in $\mathbb{R}$, then $\phi \circ u$ is subharmonic in Ω.

(b) If Ω is a connected open set in the plane and f is a holomorphic function in Ω which is not identically 0, then $\log|f|$, $\log^+|f| := \max(0, \log|f|)$, and $|f|^p$ $(0 < p < \infty)$ are subharmonic in Ω.

The next theorem explains the term "subharmonic."

Theorem A.9. *Suppose that u is a continuous subharmonic function in a plane open set Ω, K is a compact subset of Ω, h is a continuous real function on K which is harmonic in the interior of K, and $u(z) \le h(z)$ at all boundary points of K. Then $u(z) \le h(z)$ for all $z \in K$.*

A.3 Hardy spaces

A.3.1 First approach

Details of the material and the proofs of the theorems discussed in this subsection can be found e.g. in Chapter 17 of [50].

Let D denote the open unit disc in $\mathbb{C}$ and T be its boundary circle. Let $f \in C(D)$ and define f_r on T by

$$f_r(e^{it}) := f(re^{it}) \quad (0 \le r < 1, \quad -\pi \le t \le \pi),$$

and

$$\|f_r\|_p := \left\{ \frac{1}{2\pi} \int_{-\pi}^{\pi} |f_r(e^{it})|^p dt \right\}^{1/p} \qquad (0 < p < \infty),$$

$$\|f_r\|_\infty := \sup_{-\pi \le t < \pi} |f_r(e^{it})|,$$

$$\|f_r\|_0 := \exp \int_{-\pi}^{\pi} \log^+ |f_r(e^{it})| \frac{dt}{2\pi},$$

where $\log^+(x) := \log x$ if $x \ge 1$, $\log^+(x) := 0$ if $x < 1$. If $H(D)$ denotes the set of holomorphic (analytic) functions in D, $f \in H(D)$, and $0 \le p \le \infty$, define

$$\|f\|_p := \sup_{0 \le r < 1} \|f_r\|_p.$$

For $0 < p \le \infty$, a *Hardy space* H^p is the class of all $f \in H(D)$ for which $\|f\|_p < \infty$. It is clear that $\|f_r\|$ is a non-decreasing function of r.

The *Nevanlinna class* N consists of all $f \in H(D)$ such that $\|f\|_0 < \infty$. An alternative definition, see e.g. [43, Definition 3.3.1], is that N consists of all $f \in H(D)$ such that

$$f = f_1/f_2, \qquad f_1 \in H^p, \qquad f_2 \in H^q, \qquad p, q > 0.$$

Clearly, $H^\infty \subset H^p \subset H^s \subset N$ if $0 < s < p < \infty$.

For $1 \le p \le \infty$, H^p is a Banach space. In particular, H^2 is a Hilbert-space, moreover, $f \in H^2$ if and only if

$$f \in H(D), \quad f(z) = \sum_{n=0}^{\infty} a_n z^n \quad (z \in D), \text{ where } \sum_{n=0}^{\infty} |a_n|^2 < \infty.$$

In fact, by Parseval's theorem applied to f_r with $r < 1$, we obtain that

$$\sum_{n=0}^{\infty} |a_n|^2 = \lim_{r\to 1} \sum_{n=0}^{\infty} |a_n|^2 r^{2n} = \lim_{r\to 1} \left\{ \frac{1}{2\pi} \int_{-\pi}^{\pi} \left| \sum_{n=0}^{\infty} a_n r^n e^{int} \right|^2 dt \right\}$$
$$= \lim_{r\to 1} \frac{1}{2\pi} \int_{-\pi}^{\pi} |f_r(e^{it})|^2 dt = \|f\|_2^2.$$

For $p < 1$, H^p is still a vector space, but the triangle inequality is no longer satisfied by $\|f\|_p$.

Theorem A.10. *We have several important relationships between a function in a Hardy class and its radial limit.*

(a) If $0 < p < \infty$ and $f \in H^p$, then the radial limits (boundary values) of $f_r(e^{it})$ as $r \to 1$ exists a.e. on T; it is denoted by $f^*(e^{it})$. It follows that

$$f^* \in L^p(T), \quad \lim_{r\to 1} \|f^* - f_r\|_{L^p(T)} = 0, \quad \|f^*\|_{L^p(T)} = \|f\|_{H^p}.$$

(b) If $f \in H^1$ then f is uniquely determined by its boundary value f^*, using either the Cauchy integral formula:

$$f(z) = \frac{1}{2\pi} \int_{-\pi}^{\pi} \frac{f^*(e^{it})}{1 - e^{-it}z} dt, \quad z \in D,$$

or the Poisson integral formula:

$$f(z) = \frac{1}{2\pi} \int_{-\pi}^{\pi} \frac{1 - |z|^2}{|e^{it} - z|^2} f^*(e^{it}) dt, \quad z \in D.$$

An *inner function* is a function $M \in H^\infty$ for which $|M^*| = 1$ a.e. on T. If ϕ is a positive measurable function on T such that $\log \phi \in L^1(T)$, and if

$$Q(z) = c \exp\left\{ \frac{1}{2\pi} \int_{-\pi}^{\pi} \frac{e^{it} + z}{e^{it} - z} \log \phi(e^{it}) dt \right\} \quad (z \in D)$$

then Q is called an *outer function*. Here c is a constant, $|c| = 1$.

The next theorem of this paragraph characterizes the inner functions; compare with the definition of outer functions.

Theorem A.11. *Suppose that c is a constant, $|c| = 1$. Also, suppose that B is a Blaschke product:*

$$B(z) = z^k \prod_{n=1}^{\infty} \frac{\alpha_n - z}{1 - \bar{\alpha}_n z} \frac{|\alpha_n|}{\alpha_n} \quad (z \in D),$$

where $k \geq 0$, $\{\alpha_n\}$ is a sequence in D such that

$$\sum_{n=1}^{\infty}(1 - |\alpha_n|) < \infty.$$

Further, let μ be a finite positive Borel measure on T which is singular w.r.t. Lebesgue measure. Then

$$M(z) = cB(z)\exp\left\{-\int_{-\pi}^{\pi}\frac{e^{it}+z}{e^{it}-z}d\mu(t)\right\} \quad (z \in D)$$

is an inner function, and every inner function is of this form.

Another theorem of this paragraph describes important properties of outer functions.

Theorem A.12. *Suppose Q is an outer function defined above. Then*

$$\lim_{r\to 1}|Q(re^{i\theta}| = \phi(e^{i\theta}) \quad a.e.\ on\ T$$

and $Q \in H^p$ if and only if $\phi \in L^p(T)$; in that case $\|Q\|_{H^p} = \|\phi\|_{L^p(T)}$.

The next theorem shows that every nonzero $f \in H^p$ function can be written as a product of an inner and an outer function.

Theorem A.13. *Suppose $0 < p \leq \infty$, $f \in H^p$, and f is not identically 0. Then $\log|f^*| \in L^1(T)$, the outer function*

$$Q_f(z) = \exp\left\{\frac{1}{2\pi}\int_{-\pi}^{\pi}\frac{e^{it}+z}{e^{it}-z}\log|f^*(e^{it})|dt\right\} \quad (z \in D)$$

is in H^p, and there is an inner function M_f such that $f = M_f Q_f$.

An important consequence is that $f^*(e^{i\theta}) \neq 0$ a.e. on T.

Consider now the Hilbert space $\ell^2(0,\infty)$ of all complex sequences

$$x = (x_0, x_1, x_2, x_3, \ldots),$$

with inner product $\langle x, y\rangle = \sum_{n=0}^{\infty} x_n \bar{y}_n$, and norm $\|x\| = \langle x, x\rangle^{1/2} < \infty$. The operator of right (forward) shift S on $\ell^2(0,\infty)$ is given by

$$Sx = (0, x_0, x_1, x_2, \ldots). \tag{A.11}$$

Clearly, S is a bounded linear operator, $\|S\| = 1$.

Associate with each $x \in \ell^2(0,\infty)$ the function

$$f(z) = \sum_{n=0}^{\infty} x_n z^n \quad (z \in D).$$

This defines a linear one-to-one mapping of $\ell^2(0,\infty)$ onto H^2. Define the inner product in H^2 by

$$\langle f, g\rangle := \frac{1}{2\pi}\int_{-\pi}^{\pi} f^*(e^{i\theta})\overline{g^*(e^{i\theta})}d\theta.$$

Parseval theorem shows that $\langle f, g\rangle = \langle x, y\rangle = \sum_{n=0}^{\infty} x_n \bar{y}_n$. Thus we have a Hilbert space isomorphism from $\ell^2(0,\infty)$ onto H^2 and the shift operator S turns into a multiplication by z:

$$(Sf)(z) = \sum_{n=0}^{\infty} x_n z^{n+1} = zf(z). \tag{A.12}$$

Observe the other important fact that in the case of $f \in H^2$, each Fourier coefficients of $f^*(e^{i\theta})$ are 0 when $n < 0$.

It is an important problem to describe the shift invariant subsets of $\ell^2(0,\infty)$. The above isomorphism is a very useful tool for that. Beurling's theorem below characterizes the shift invariant subspaces of H^2.

Theorem A.14. *For each inner function ϕ the space $\phi H^2 := \{\phi f : f \in H^2\}$ is a closed S-invariant subspace of H^2.*

If ϕ_1 and ϕ_2 are inner functions and if $\phi_1 H^2 = \phi_2 H^2$, then ϕ_1/ϕ_2 is constant.

Every closed S-invariant subspace Y of H^2, other than $\{0\}$, contains an inner function ϕ such that $Y = \phi H^2$.

A.3.2 Second approach

Details of the material discussed in this subsection can be found e.g. in Section II.12 of [20].

For $1 \le p \le \infty$ there is another approach to H^p spaces. Clearly, if $1 \le s \le p \le \infty$, then $L^1(T) \supset L^s(T) \supset L^p(T) \supset L^\infty(T)$. Each $f \in L^1(T)$ has well-defined Fourier coefficients given by

$$a_n := \frac{1}{2\pi}\int_{-\pi}^{\pi} f(e^{it})e^{-int}dt \quad (n \in \mathbb{Z}).$$

For $1 \le p \le \infty$ one can define a *Hardy space* $H^p(T)$ to be the closed subspace of $L^p(T)$ consisting of all functions for which $a_n = 0$ when $n < 0$. (The map $f \to a_n$ is continuous, this explains why the subspace is closed.)

Theorem A.15. *There exists an isometric isomorphism between H^p defined in Subsection A.3.1 and $H^p(T)$ defined above, when $1 \le p \le \infty$.*

Because of this theorem we do not distinguish $H^p(T)$ and H^p, when $p \ge 1$. H^2 is a Hilbert space; then we also define *co-Hardy spaces*

$$\bar{H}^2 := \{f \in L^2(T) : a_n = 0 \text{ if } n > 0\}, \quad \bar{H}_0^2 := \{f \in L^2(T) : a_n = 0 \text{ if } n \ge 0\}.$$

It follows that there is an orthogonal decomposition $L^2(T) = H^2 \oplus \bar{H}_0^2$. Also, $f \in H^2$ if and only if $\bar{f} \in \bar{H}^2$, where $\bar{f}$ is defined by $\bar{f}(e^{it}) := \overline{f(e^{it})}$.

Let us denote the space of square summable complex sequences on $\mathbb{Z}$ by $\ell^2(-\infty, \infty)$. One can define the Fourier transform $\mathcal{F}$ from $\ell^2(-\infty, \infty)$ onto $L^2(T)$ by

$$\mathcal{F}(\{a_n\}_{n\in\mathbb{Z}}) = \sum_{n\in\mathbb{Z}} a_n e^{int}.$$

Then $\mathcal{F}$ is a unitary map, that is, an isometric isomorphism from $\ell^2(-\infty, \infty)$ onto $L^2(T)$. Let S be the (bilateral) right shift in $\ell^2(-\infty, \infty)$. Then it defines a shift operator in $L^2(T)$ as well by the formula $(Sf)(e^{it}) = e^{it} f(e^{it})$, with which we have

$$\mathcal{F}S = S\mathcal{F}.$$

It is clear that $\mathcal{F}(\ell^2(0, \infty)) = H^2$ and the restricted operators $S|\ell^2(0, \infty)$ and $S|H^2$ are also unitarily equivalent and they agree with the operator S introduced in (A.11) and (A.12).

There is another useful characterization of outer functions introduced in Subsection A.3.1.

Theorem A.16. *A function $f \in H^2$ is outer if and only if the linear combinations of the functions $S^n f$ $(n \geq 0)$ are dense in H^2.*

B

Matrix decompositions and special matrices

We consider finite dimensional complex Euclidean spaces that are also Hilbert spaces. Linear operations between them can be described by matrices of complex entries. Vectors are treated as column-vectors and denoted by bold-face, lower-case letters. The inner product of the vectors $\mathbf{x}, \mathbf{y} \in \mathbb{C}^n$ is therefore written with matrix multiplication as $\mathbf{x}^*\mathbf{y}$, where * stands for the conjugate transpose of a complex vector, hence $\mathbf{x}^*$ is a row-vector. Matrices will be denoted by bold-face upper-case letters. An $m \times n$ matrix $\boldsymbol{A} = [a_{ij}]$ of complex entries a_{ij}'s corresponds to a $\mathbb{C}^n \to \mathbb{C}^m$ linear transformation (operator). Its adjoint, $\boldsymbol{A}^*$, is an $n \times m$ matrix, the entries of which are $a^*_{ij} = \overline{a_{ji}}$. An $n \times n$ matrix is called quadratic (or square) and it maps $\mathbb{C}^n$ into itself. The identity matrix is denoted by $\boldsymbol{I}$ or $\boldsymbol{I}_n$ if we want to refer to its size.

Definition B.1. The following types of matrices will be frequently used.

- The $n \times n$ complex matrix $\boldsymbol{A}$ is *self-adjoint (Hermitian)* if $\boldsymbol{A} = \boldsymbol{A}^*$. In particular, a real matrix with $\boldsymbol{A}^T = \boldsymbol{A}$ is called *symmetric.*
- The $n \times n$ real matrix $\boldsymbol{A}$ is *anti-symmetric* if $\boldsymbol{A}^T = -\boldsymbol{A}$.
- The $n \times n$ complex matrix $\boldsymbol{A}$ is *unitary* if $\boldsymbol{A}\boldsymbol{A}^* = \boldsymbol{A}^*\boldsymbol{A} = \boldsymbol{I}_n$. It means that both its rows and columns form a complete orthonormal system in $\mathbb{C}^n$.
- The $n \times r$ complex matrix ($r \le n$) is *sub-unitary* if its columns constitute a (usually not complete) orthonormal system in $\mathbb{C}^n$; consequently, $\boldsymbol{A}^*\boldsymbol{A} = \boldsymbol{I}_r$, whereas $\boldsymbol{A}\boldsymbol{A}^*$ is the matrix of the orthogonal projection onto the subspace spanned by the column vectors of $\boldsymbol{A}$.
- The $n \times n$ complex matrix $\boldsymbol{P}$ is *Hermitian projector* if it is self-adjoint and idempotent, i.e. $\boldsymbol{P}^2 = \boldsymbol{P}$.
- The complex matrix $\boldsymbol{A}$ is called *normal* if $\boldsymbol{A}\boldsymbol{A}^* = \boldsymbol{A}^*\boldsymbol{A}$.

The $n \times n$ matrix $\boldsymbol{A}$ has an inverse if and only if its determinant, $\det \boldsymbol{A} = |\boldsymbol{A}| \neq 0$, and its inverse is denoted by $\boldsymbol{A}^{-1}$. In this case, the linear transformation corresponding to $\boldsymbol{A}^{-1}$ undoes the effect of the $\mathbb{C}^n \to \mathbb{C}^n$ transformation corresponding to $\boldsymbol{A}$, i.e. $\boldsymbol{A}^{-1}\mathbf{y} = \mathbf{x}$ if and only if $\boldsymbol{A}\mathbf{x} = \mathbf{y}$ for any $\mathbf{y} \in \mathbb{C}^n$; equivalently, $\boldsymbol{A}\boldsymbol{A}^{-1} = \boldsymbol{A}^{-1}\boldsymbol{A} = \boldsymbol{I}_n$. A unitary matrix $\boldsymbol{A}$ is always invertible and $\boldsymbol{A}^{-1} = \boldsymbol{A}^*$.

It is important that in the case of an invertible (*regular*) matrix $\boldsymbol{A}$, the *range* (or image space) of $\boldsymbol{A}$ – denoted by $\mathcal{R}(\boldsymbol{A})$ – is the whole $\boldsymbol{C}^n$, and in exchange, the *kernel* of $\boldsymbol{A}$ (the subspace of vectors that are mapped into the zero vector by $\boldsymbol{A}$) consists of only the vector $\mathbf{0}$. Note that for an $m \times n$ matrix $\boldsymbol{A}$, its range is

$$\mathcal{R}(\boldsymbol{A}) = \text{Span}\{\mathbf{a}_1, \dots, \mathbf{a}_n\},$$

where $\mathbf{a}_1, \dots, \mathbf{a}_n$ are the column vectors of $\boldsymbol{A}$ for which fact the notation $\boldsymbol{A} = [\mathbf{a}_1, \dots, \mathbf{a}_n]$ will be used. The *rank* of $\boldsymbol{A}$ is the dimension of its range:

$$\text{rank}(\boldsymbol{A}) = \dim \mathcal{R}(\boldsymbol{A}),$$

and it also equals the maximum number of linearly independent rows, or equivalently, the maximum number of linearly independent columns of $\boldsymbol{A}$, or the maximal size of a nonzero minor (subdeterminant) of $\boldsymbol{A}$. Trivially, $\text{rank}(\boldsymbol{A}) \le \min\{m, n\}$; if equality is attained, we say that $\boldsymbol{A}$ has *full rank*. In the case of $m = n$, $\boldsymbol{A}$ is regular if and only if $\text{rank}(\boldsymbol{A}) = n$, and *singular* otherwise.

Eigenvalues and *eigenvectors* tell "everything" about a quadratic matrix. The complex number λ is an eigenvalue of the $n \times n$ complex matrix $\boldsymbol{A}$ with corresponding eigenvector $\mathbf{u} \neq \mathbf{0}$ if $\boldsymbol{A}\mathbf{u} = \lambda\mathbf{u}$. If $\mathbf{u}$ is an eigenvector of $\boldsymbol{A}$, it is easy to see that for $c \neq 0$, $c\mathbf{u}$ is also an eigenvector with the same eigenvalue. Therefore, it is better to speak about *eigen-directions* instead of eigenvectors; or else, we will consider specially normalized, e.g. unit-norm eigenvectors, when only the orientation is divalent. It is well known that an $n \times n$ matrix $\boldsymbol{A}$ has exactly n eigenvalues (with multiplicities) which are (complex) roots of the characteristic polynomial $|\boldsymbol{A} - \lambda\boldsymbol{I}|$. Knowing the eigenvalues, the corresponding eigenvectors are obtained by solving the system of linear equations $(\boldsymbol{A} - \lambda\boldsymbol{I})\mathbf{u} = \mathbf{0}$ which must have a non-trivial solution due to the choice of λ. In fact, there are infinitely many solutions (in case of a single eigenvalue, they are constant multiples of each other).

Normal matrices have the following important spectral property: to their eigenvalues there corresponds a complete orthonormal set of eigenvectors; choosing this as a new basis, the matrix becomes *diagonal* (all the off-diagonal entries are zeros). Here we only state the analogous version for Hermitian matrices.

Theorem B.1 (Hilbert–Schmidt theorem). *The $n \times n$ self-adjoint complex matrix $\boldsymbol{A}$ has real eigenvalues $\lambda_1 \ge \cdots \ge \lambda_n$ (with multiplicities), and the corresponding eigenvectors $\mathbf{u}_1, \dots, \mathbf{u}_n$ can be chosen so that they constitute a complete orthonormal set in $\mathbb{C}^n$.*

Theorem B.1 implies the following *Spectral Decomposition (SD)* of the $n \times n$ self-adjoint matrix $\boldsymbol{A}$:

$$\boldsymbol{A} = \sum_{i=1}^{n} \lambda_i \mathbf{u}_i \mathbf{u}_i^* = \boldsymbol{U}\boldsymbol{\Lambda}\boldsymbol{U}^*, \tag{B.1}$$

where $\mathbf{\Lambda} = \mathrm{diag}(\lambda_1, \dots, \lambda_n)$ is the diagonal matrix containing the eigenvalues — called *spectrum* — in its main diagonal, while $\boldsymbol{U} = [\mathbf{u}_1, \dots, \mathbf{u}_n]$ is the unitary matrix containing the corresponding unit-norm eigenvectors of $\boldsymbol{A}$ in its columns in the order of the eigenvalues. Of course, permuting the eigenvalues in the main diagonal of $\mathbf{\Lambda}$, and the columns of $\boldsymbol{U}$ accordingly, will lead to the same SD, however — if not otherwise stated — we will enumerate the real eigenvalues in non-increasing order. About the uniqueness of the above SD we can state the following: the unit-norm eigenvector corresponding to a single eigenvalue is unique (up to orientation), whereas to an eigenvalue with multiplicity m there corresponds a unique m-dimensional so-called *eigen-subspace* within which any orthonormal set can be chosen for the corresponding eigenvectors.

Definition B.2. The parsimonious SD of the $n \times n$ self-adjoint matrix $\boldsymbol{A}$ of rank r is

$$\boldsymbol{A} = \sum_{i=1}^{r} \lambda_i \mathbf{u}_i \mathbf{u}_i^* = \tilde{\boldsymbol{U}} \tilde{\mathbf{\Lambda}} \tilde{\boldsymbol{U}}^*, \tag{B.2}$$

where $\tilde{\boldsymbol{U}} = [\mathbf{u}_1, \dots, \mathbf{u}_r]$ is $n \times r$ sub-unitary matrix and $\tilde{\mathbf{\Lambda}} = \mathrm{diag}(\lambda_1, \dots, \lambda_r)$ is $r \times r$ diagonal matrix, where $\lambda_1 \geq \cdots \geq \lambda_r$ is the set of nonzero eigenvalues.

The quadratic form $\mathbf{x}^* \boldsymbol{A} \mathbf{x}$ with the SD of the self-adjoint $\boldsymbol{A}$ is

$$\mathbf{x}^* \boldsymbol{A} \mathbf{x} = \sum_{i=1}^{n} \lambda_i (\mathbf{x}^* \mathbf{u}_i) \overline{(\mathbf{x}^* \mathbf{u}_i)} = \sum_{i=1}^{n} \lambda_i |\mathbf{x}^* \mathbf{u}_i|^2$$

that is a real number. Some properties of the self-adjoint matrix $\boldsymbol{A}$ and of the quadratic forms generated by it follow:

- $\boldsymbol{A}$ is singular if and only if it has a 0 eigenvalue, and

  $$r = \mathrm{rank}(\boldsymbol{A}) = \mathrm{rank}(\mathbf{\Lambda}) = |\{i \,:\, \lambda_i \neq 0\}|;$$

  moreover, $\mathcal{R}(\boldsymbol{A}) = \mathrm{Span}\{\mathbf{u}_i \,:\, \lambda_i \neq 0\}$. Therefore, the SD of $\boldsymbol{A}$ simplifies to $\sum_{i=1}^{r} \lambda_i \mathbf{u}_i \mathbf{u}_i^*$.
- $\boldsymbol{A}$ is *positive (negative) definite* if $\mathbf{x}^* \boldsymbol{A} \mathbf{x} > 0$ ($\mathbf{x}^* \boldsymbol{A} \mathbf{x} < 0$), $\forall \mathbf{x} \neq \mathbf{0}$; equivalently, all the eigenvalues of $\boldsymbol{A}$ are positive (negative).
- $\boldsymbol{A}$ is *positive (negative) semidefinite* if $\mathbf{x}^* \boldsymbol{A} \mathbf{x} \geq 0$ ($\mathbf{x}^* \boldsymbol{A} \mathbf{x} \leq 0$), $\forall \mathbf{x} \in \mathbb{C}^n$; equivalently, all the eigenvalues of $\boldsymbol{A}$ are non-negative (non-positive).
- Note that the notion *non-negative (non-positive) definite* can be used instead of positive (negative) semidefinite. In some literature if $\boldsymbol{A}$ is called positive (negative) semidefinite, then it is understood that $\mathbf{x}^* \boldsymbol{A} \mathbf{x} = \mathbf{0}$ for at least one $\mathbf{x} \neq \mathbf{0}$; and so the spectrum of $\boldsymbol{A}$ contains the zero eigenvalue too.
- $\boldsymbol{A}$ is indefinite if $\mathbf{x}^* \boldsymbol{A} \mathbf{x}$ takes both positive and negative values (with different, non-zero $\mathbf{x}$'s); equivalently, the spectrum of $\boldsymbol{A}$ contains at least one positive and one negative eigenvalue.

- $|\boldsymbol{A}| = \det(\boldsymbol{A}) = \prod_{i=1}^{n} \lambda_i$ and $\operatorname{tr}(\boldsymbol{A}) = \sum_{i=1}^{n} \lambda_i$.

A canonical decomposition for a rectangular matrix is a useful tool.

Theorem B.2. *Let* $\boldsymbol{A}$ *be an* $m \times n$ *rectangular matrix of complex entries,* $\operatorname{rank}(\boldsymbol{A}) = r \leq \min\{m, n\}$*. Then there exist an orthonormal set* $(\mathbf{v}_1, \dots, \mathbf{v}_r) \subset \mathbb{C}^m$ *and* $(\mathbf{u}_1, \dots, \mathbf{u}_r) \subset \mathbb{C}^n$ *together with the positive real numbers* $s_1 \geq s_2 \geq \dots \geq s_r > 0$ *such that*

$$\boldsymbol{A}\mathbf{u}_j = s_j \mathbf{v}_j, \quad \boldsymbol{A}^* \mathbf{v}_j = s_j \mathbf{u}_j, \quad j = 1, 2, \dots, r. \tag{B.3}$$

The elements $\mathbf{v}_j \in \mathbb{C}^m$ and $\mathbf{u}_j \in \mathbb{C}^n$ $(j = 1, \dots, r)$ in (B.3) are called *relevant singular vector pairs* (or *left and right singular vectors*) corresponding to the *singular value* s_j $(j = 1, 2, \dots, r)$. The transformations in (B.3) give a one-to-one mapping between $\mathcal{R}(\boldsymbol{A})$ and $\mathcal{R}(\boldsymbol{A}^*)$, all the other vectors of $\mathbb{C}^n$ and $\mathbb{C}^m$ are mapped into the zero vector of $\mathbb{C}^m$ and $\mathbb{C}^n$, respectively. However, the left and right singular vectors can appropriately be completed into a complete orthonormal set $\{\mathbf{v}_1, \dots, \mathbf{v}_m\} \subset \mathbb{C}^m$ and $\{\mathbf{u}_1, \dots \mathbf{u}_n\} \subset \mathbb{C}^n$, respectively, such that, the so introduced extra vectors in the kernel subspaces in $\mathbb{C}^m$ and $\mathbb{C}^n$ are mapped into the zero vector of $\mathbb{C}^n$ and $\mathbb{C}^m$, respectively. With the unitary matrices $\boldsymbol{V} = (\mathbf{v}_1, \dots, \mathbf{v}_m)$ and $\boldsymbol{U} = (\mathbf{u}_1, \dots \mathbf{u}_n)$, the following *singular value decomposition (SVD)* of $\boldsymbol{A}$ and $\boldsymbol{A}^*$ holds:

$$\boldsymbol{A} = \boldsymbol{V}\boldsymbol{S}\boldsymbol{U}^* = \sum_{i=1}^{r} s_i \mathbf{v}_i \mathbf{u}_i^* \quad \text{and} \quad \boldsymbol{A}^* = \boldsymbol{U}\boldsymbol{S}^*\boldsymbol{V}^* = \sum_{i=1}^{r} s_i \mathbf{u}_i \mathbf{v}_i^*, \tag{B.4}$$

where $\boldsymbol{S}$ is an $m \times n$ so-called *generalized diagonal matrix* which contains the singular values $s_1, \dots, s_r$ in the first r positions of its main diagonal (starting from the upper left corner) and zeros otherwise. We remark that there are other equivalent forms of the above SVD depending on, whether $m < n$ or $m \geq n$. For example, in the $m < n$ case, $\boldsymbol{V}$ can be an $m \times m$ unitary, $\boldsymbol{S}$ an $m \times m$ diagonal, and $\boldsymbol{U}$ an $n \times m$ sub-unitary matrix with the same relevant entries. About the uniqueness of the SVD the following can be stated: to a single positive singular value there corresponds a unique singular vector pair (of course, the orientation of the left and right singular vectors can be changed at the same time). To a positive singular value of multiplicity $k > 1$ a k-dimensional left and right so-called *isotropic subspace* corresponds, within which, any k-element orthonormal sets can embody the left and right singular vectors with orientation such that the requirements in (B.3) are met.

We also remark that the singular values of a self-adjoint matrix are the absolute values of its real eigenvalues. In case of a positive eigenvalue, the left and right singular vectors are the same (they coincide with the corresponding eigenvector with any, but the same orientation). In case of a negative eigenvalue, the left and right side singular vectors are opposite (any of them is the corresponding eigenvector which have a divalent orientation). In case of a zero singular value the orientation is immaterial, as it does not contribute to the

SVD of the underlying matrix. Numerical algorithms for SD and SVD of real matrices are presented in [26, 61].

Assume that the $m \times n$ complex matrix $\boldsymbol{A}$ of rank r has SVD (B.4). It is easy to see that the matrices $\boldsymbol{A}\boldsymbol{A}^*$ and $\boldsymbol{A}^*\boldsymbol{A}$ are self-adjoint, positive semidefinite matrices of rank r, and their SD is

$$\boldsymbol{A}\boldsymbol{A}^* = \boldsymbol{V}(\boldsymbol{S}\boldsymbol{S}^*)\boldsymbol{V}^* = \sum_{j=1}^{r} s_j^2 \mathbf{v}_j \mathbf{v}_j^* \quad \text{and} \quad \boldsymbol{A}^*\boldsymbol{A} = \boldsymbol{U}(\boldsymbol{S}^*\boldsymbol{S})\boldsymbol{U}^* = \sum_{j=1}^{r} s_j^2 \mathbf{u}_j \mathbf{u}_j^*,$$

where the diagonal matrices $\boldsymbol{S}\boldsymbol{S}^*$ and $\boldsymbol{S}^*\boldsymbol{S}$ both contain the numbers $s_1^2, \dots, s_r^2$ in the leading positions of their main diagonals as non-zero eigenvalues.

These facts together also imply that the only positive singular value of a sub-unitary matrix is the 1 with multiplicity of its rank.

By means of SD and SVD we are able to define so-called *generalized inverses* of singular quadratic or rectangular matrices.

Definition B.3. The $m \times n$ complex matrix $\boldsymbol{X}$ is a generalized inverse of the $n \times m$ complex matrix $\boldsymbol{A}$ if $\boldsymbol{A}\boldsymbol{X}\boldsymbol{A} = \boldsymbol{A}$.

A generalized inverse $\boldsymbol{X}$ satisfying $\boldsymbol{A}\boldsymbol{X}\boldsymbol{A} = \boldsymbol{A}$ is denoted by $\boldsymbol{A}^-$. In fact, any matrix that undoes the effect of the underlying linear transformation between the ranges of $\boldsymbol{A}^*$ and $\boldsymbol{A}$ will do. A generalized inverse is far not unique as any transformation operating on the kernels can be added. However, the following *pseudoinverse (Moore–Penrose inverse)* is unique and, in case of a quadratic matrix, it coincides with the usual inverse if exists.

Definition B.4. The $m \times n$ complex matrix $\boldsymbol{X}$ is the pseudoinverse (in other words, the Moore–Penrose inverse) of the $n \times m$ complex matrix $\boldsymbol{A}$ if it satisfies all of the following conditions:

$$\begin{aligned} \boldsymbol{A}\boldsymbol{X}\boldsymbol{A} &= \boldsymbol{A}, \\ \mathbf{X}\boldsymbol{A}\mathbf{X} &= \boldsymbol{X}, \\ (\boldsymbol{A}\boldsymbol{X})^* &= \boldsymbol{A}\boldsymbol{X}, \\ (\boldsymbol{X}\boldsymbol{A})^* &= \boldsymbol{X}\boldsymbol{A}. \end{aligned}$$

It can be proven that there uniquely exists a pseudoinverse satisfying the conditions in the above definition, and it is denoted by $\boldsymbol{A}^+$. Actually, it can be obtained from the SVD (B.4) of $\boldsymbol{A}$ as follows:

$$\boldsymbol{A}^+ = \boldsymbol{U}\boldsymbol{S}^+\boldsymbol{V}^* = \sum_{j=1}^{r} \frac{1}{s_j} \mathbf{u}_j \mathbf{v}_j^*,$$

where $\boldsymbol{S}^+$ is the $m \times n$ generalized diagonal matrix containing the reciprocals of the non-zero singular values, otherwise zeros, in its main diagonal.

In particular, the Moore–Penrose inverse of the $n \times n$ self-adjoint matrix with SD (B.1) is

$$\boldsymbol{A}^+ = \sum_{j=1}^{r} \frac{1}{\lambda_j} \mathbf{u}_j \mathbf{u}_j^* = \boldsymbol{U}\boldsymbol{\Lambda}^+ \boldsymbol{U}^*,$$

where $\boldsymbol{\Lambda}^+ = \operatorname{diag}(\frac{1}{\lambda_1}, \dots, \frac{1}{\lambda_r}, 0, \dots, 0)$ is the diagonal matrix containing the reciprocals of the non-zero eigenvalues, otherwise zeros, in its main diagonal.

Note that any analytic function f of the self-adjoint matrix $\boldsymbol{A}$ can be defined by its SD, $\boldsymbol{A} = \boldsymbol{U}\boldsymbol{\Lambda}\boldsymbol{U}^*$, in the following way:

$$f(\boldsymbol{A}) := \boldsymbol{U} f(\boldsymbol{\Lambda}) \boldsymbol{U}^*$$

where $f(\boldsymbol{\Lambda}) = \operatorname{diag}(f(\lambda_1), \dots, f(\lambda_n))$, of course, only if every eigenvalue is in the domain of f. In this way, for a positive semidefinite $\boldsymbol{A}$, its square root is

$$\boldsymbol{A}^{1/2} := \boldsymbol{U}\boldsymbol{\Lambda}^{1/2}\boldsymbol{U}^*,$$

and for a regular $\boldsymbol{A}$ its inverse is obtained by applying the $f(x) = x^{-1}$ function to it:

$$\boldsymbol{A}^{-1} = \boldsymbol{U}\boldsymbol{\Lambda}^{-1}\boldsymbol{U}^*.$$

For a singular $\boldsymbol{A}$, the Moore–Penrose inverse is obtained by using $\boldsymbol{\Lambda}^+$ instead of $\boldsymbol{\Lambda}^{-1}$. Accordingly, for a positive definite matrix, its $-1/2$ power is defined as the square root of $\boldsymbol{A}^{-1}$.

Now a special type of a matrix is introduced.

Definition B.5. We say that the $n \times n$ self-adjoint complex matrix $\boldsymbol{G} = (g_{ij})$ is a *Gram matrix (Gramian)* if its entries are inner products; i.e., there is a dimension $d > 0$ and vectors $\mathbf{x}_1, \dots, \mathbf{x}_n \in \mathbb{C}^d$ such that

$$g_{jk} = \mathbf{x}_j^* \mathbf{x}_k, \quad j, k = 1, \dots n.$$

Proposition B.1. *The self-adjoint matrix* $\boldsymbol{G}$ *is a Gramian if and only if it is positive semidefinite.*

Proof. If $\boldsymbol{G}$ is a Gram-matrix, then it can be decomposed as $\boldsymbol{G} = \boldsymbol{A}\boldsymbol{A}^*$, where $\boldsymbol{A}^* = [\mathbf{x}_1, \dots, \mathbf{x}_n]$ with its generating vectors $\mathbf{x}_1, \dots, \mathbf{x}_n \in \mathbb{C}^d$. Therefore,

$$\mathbf{x}^* \boldsymbol{G} \mathbf{x} = \mathbf{x}^* \boldsymbol{A}\boldsymbol{A}^* \mathbf{x} = (\boldsymbol{A}^* \mathbf{x})^* (\boldsymbol{A}^* \mathbf{x}) = \|\boldsymbol{A}^* \mathbf{x}\|^2 \ge 0, \quad \forall \mathbf{x} \in \mathbb{C}^n.$$

Conversely, if $\boldsymbol{G}$ is positive semidefinite with rank $r \le n$, then its SD — using (B.2) — can be written as $\boldsymbol{G} = \sum_{j=1}^{r} \lambda_j \mathbf{u}_j \mathbf{u}_j^*$ with its positive real eigenvalues $\lambda_1 \ge \dots \ge \lambda_r > 0$. Let the $n \times r$ matrix $\boldsymbol{A}$ be defined as follows:

$$\boldsymbol{A} = \tilde{\boldsymbol{U}} \tilde{\boldsymbol{\Lambda}}^{1/2} = [\sqrt{\lambda_1}\mathbf{u}_1, \dots, \sqrt{\lambda_r}\mathbf{u}_r], \tag{B.5}$$

where $\tilde{\boldsymbol{U}}$ and $\tilde{\boldsymbol{\Lambda}}$ are as defined in Definition B.2.

Then the row vectors of the matrix $\boldsymbol{A}$ will be the vectors $\mathbf{x}_j \in \mathbb{C}^r$ ($j = 1, \dots n$) reproducing $\boldsymbol{G}$. Of course, the decomposition $\boldsymbol{G} = \boldsymbol{A}\boldsymbol{A}^*$ is far not unique: first of all, instead of $\boldsymbol{A}$ the matrix $\boldsymbol{A}\boldsymbol{Q}$ will also do, where $\boldsymbol{Q}$ is an arbitrary $r \times r$ unitary matrix (obviously, $\mathbf{x}_j$'s can be rotated); and $\mathbf{x}_j$'s can also be put in a higher ($d > r$) dimension with attaching any (but the same) number of zero coordinates to them. □

Now matrix norms are summarized.

Definition B.6. The *spectral norm* (or *operator norm*) of an $m \times n$ complex matrix $\boldsymbol{A}$ of rank r, with singular values $s_1 \geq \cdots \geq s_r > 0$, is

$$\|\boldsymbol{A}\| := \max_{|\mathbf{x}|=1} |\boldsymbol{A}\mathbf{x}| = s_1,$$

where $|\cdot|$ is the Euclidean (L^2) norm. Then for square matrices $\boldsymbol{A}$ and $\boldsymbol{B}$ we have $\|\boldsymbol{A}\boldsymbol{B}\| \leq \|\boldsymbol{A}\| \, \|\boldsymbol{B}\|$.

The *Frobenius norm*, denoted by $\|.\|_F$, is

$$\|\boldsymbol{A}\|_F := \left(\sum_{i=1}^{m} \sum_{j=1}^{n} |a_{ij}|^2 \right)^{1/2} = \sqrt{\operatorname{tr}(\boldsymbol{A}\boldsymbol{A}^*)} = \sqrt{\operatorname{tr}(\boldsymbol{A}^*\boldsymbol{A})} = \left(\sum_{i=1}^{r} s_i^2 \right)^{1/2}.$$

For a self-adjoint matrix $\boldsymbol{A}$,

$$\|\boldsymbol{A}\| = \max_{|\mathbf{x}|=1} |\boldsymbol{A}\mathbf{x}| = \max_i |\lambda_i| \quad \text{and} \quad \|\boldsymbol{A}\|_F = \left(\sum_{i=1}^{r} \lambda_i^2 \right)^{1/2}.$$

Obviously, for a matrix $\boldsymbol{A}$ of rank r,

$$\|\boldsymbol{A}\| \leq \|\boldsymbol{A}\|_F \leq \sqrt{r} \|\boldsymbol{A}\|. \tag{B.6}$$

More generally, a matrix norm is called *unitary invariant* if

$$\|\boldsymbol{A}\|_{\text{un}} = \|\boldsymbol{Q}\boldsymbol{A}\boldsymbol{R}\|_{\text{un}}$$

with any $m \times m$ and $n \times n$ unitary matrices $\boldsymbol{Q}$ and $\boldsymbol{R}$, respectively. It is easy to see that a unitary invariant norm of a matrix merely depends on its singular values (or eigenvalues if it is self-adjoint). For example, the spectral and Frobenius norms are such.

Next, the spectral radius of a quadratic matrix is defined and related to its so-called natural norms.

Definition B.7. The *spectrum* $\sigma(\boldsymbol{A})$ of the matrix $\boldsymbol{A} \in \mathbb{C}^{n \times n}$ is the set of all its eigenvalues λ_j, $j = 1, \dots, n$. The *spectral radius* of the matrix $\boldsymbol{A} \in \mathbb{C}^{n \times n}$ is

$$\rho(\boldsymbol{A}) = \max\{|\lambda| : \lambda \in \sigma(\boldsymbol{A})\}.$$

Note that for a self-adjoint matrix $\rho(\boldsymbol{A}) = \|\boldsymbol{A}\|$, where $\|\boldsymbol{A}\|$ is the spectral norm of $\boldsymbol{A}$.

Definition B.8. A *natural matrix norm* (or *matrix norm induced by a vector norm*) of a matrix $\boldsymbol{A} \in \mathbb{C}^{n\times n}$ is defined as

$$\|\boldsymbol{A}\|_p = \max_{\|\mathbf{x}\|_p=1} \|\boldsymbol{A}\mathbf{x}\|_p,$$

where $\|\mathbf{x}\|_p$ can be any L^p vector norm in $\mathbb{C}^n$, $1 \le p \le \infty$.

Note that $\|\boldsymbol{A}\|_2 = \|\boldsymbol{A}\|$, the previous spectral norm.

Lemma B.1. (Theorem 11.1.3 of [48]). *Between the spectral radius and any natural norm of the quadratic, complex matrix* $\boldsymbol{A}$, *the relation*

$$\rho(\boldsymbol{A}) \le \|\boldsymbol{A}\|_p$$

holds.

Proof. Let $\lambda^* := \rho(\boldsymbol{A})$, and let $\mathbf{x}^*$ denote a unit-norm eigenvector corresponding to λ^*. Then

$$\|\boldsymbol{A}\|_p = \max_{\|\mathbf{x}\|_p=1} \|\boldsymbol{A}\mathbf{x}\|_p \ge \|\boldsymbol{A}\mathbf{x}^*\|_p = \|\lambda^*\mathbf{x}^*\|_p = \rho(\boldsymbol{A})\,\|\mathbf{x}^*\|_p = \rho(\boldsymbol{A}),$$

and that proves the lemma. □

Here we quote some important facts about spectral radius. The first statement can be proved using the Jordan normal form of $\boldsymbol{A}$, the other statements easily follow from the first.

Lemma B.2. *For any matrix* $\boldsymbol{A} \in \mathbb{C}^{n\times n}$ *and its spectral norm* $\|\boldsymbol{A}\| = \|\boldsymbol{A}\|_2$ *we have*

(1) $\rho(\boldsymbol{A}) = \lim_{k\to\infty} \|\boldsymbol{A}^k\|^{1/k} = \inf_{k\ge 1} \|\boldsymbol{A}^k\|^{1/k}$;

(2) if $\rho(\boldsymbol{A}) < 1$, *then for any* c, $\rho(\boldsymbol{A}) < c < 1$, *there exists a constant* K *such that*

$$\|\boldsymbol{A}^j\| \le Kc^j \quad (j \ge 0);$$

(3) $\rho(\boldsymbol{A}) < 1 \quad \Leftrightarrow \quad \lim_{k\to\infty} \boldsymbol{A}^k = 0$;

(4) $\rho(\boldsymbol{A}) > 1 \quad \Leftrightarrow \quad \lim_{k\to\infty} \|\boldsymbol{A}^k\| = \infty$;

(5) $\rho(\boldsymbol{A}) = 1 \quad \Rightarrow \quad \|\boldsymbol{A}^k\| \ge 1$ *for any* $k \ge 1$.

It is obvious that all the complex eigenvalues of an $n \times n$ complex matrix $\boldsymbol{A}$ are within the closed circle of radius $\rho(\boldsymbol{A})$ around the origin of the complex plane. However, the subsequent Gersgorin disc theorem gives a finer allocation of them.

Theorem B.3 (Gersgorin disc theorem). *Let $\boldsymbol{A}$ be an $n \times n$ matrix of entries $a_{ij} \in \mathbb{C}$. The Gersgorin disks of $\boldsymbol{A}$ are the following regions of the complex plane:*

$$D_i = \{z \in \mathbb{C} : |z - a_{ii}| \le \sum_{j \ne i} |a_{ij}|\}, \quad i = 1, \dots n.$$

Let $\lambda_1, \dots, \lambda_n$ denote the eigenvalues of $\boldsymbol{A}$. Then

$$\{\lambda_1, \dots, \lambda_n\} \subset \cup_{i=1}^n D_i.$$

Furthermore, any connected component of the set $\cup_{i=1}^n D_i$ contains as many eigenvalues of $\boldsymbol{A}$ as the number of discs that form this component.

Theorem B.4. (Cayley–Hamilton theorem). *For any $n \times n$ complex matrix $\boldsymbol{A}$, $p_n(\boldsymbol{A}) = \boldsymbol{O}$ (the $n \times n$ zero matrix), where p_n is the characteristic polynomial of $\boldsymbol{A}$, i.e., $p_n(z) = |\boldsymbol{A} - z\boldsymbol{I}|$ (nth degree polynomial of z).*

Now some perturbation results follow for self-adjoint matrices.

Theorem B.5. (Weyl perturbation theorem). *Let $\boldsymbol{A}$ and $\boldsymbol{B}$ be $n \times n$ self-adjoint matrices. Then*

$$|\lambda_j(\boldsymbol{A}) - \lambda_j(\boldsymbol{B})| \le \|\boldsymbol{A} - \boldsymbol{B}\|, \quad j = 1, \dots, n$$

in spectral norm, where the eigenvalues of $\boldsymbol{A}$ and $\boldsymbol{B}$ are enumerated in non-increasing order.

A theorem for the perturbation of spectral subspaces (sometimes called Davis–Kahan theorem) is stated here for self-adjoint matrices.

Theorem B.6. *Let $\boldsymbol{A}$ and $\boldsymbol{B}$ be self-adjoint matrices; S_1 and S_2 are subsets of $\mathbb{R}$ such that $dist(S_1, S_2) = \delta > 0$. Let $\boldsymbol{P_A}(S_1)$ and $\boldsymbol{P_B}(S_2)$ be orthogonal projections onto the subspace spanned by the eigenvectors of the matrix in the lower index, corresponding to the eigenvalues within the subset in the argument. Then with any unitary invariant norm:*

$$\|\boldsymbol{P_A}(S_1)\boldsymbol{P_B}(S_2)\| \le \frac{c}{\delta}\|\boldsymbol{A} - \boldsymbol{B}\|$$

where c is a constant.

The statement is true for any unitary invariant norm. In case of the Frobenius norm, $c = 1$ will always do.

We also need the following simple lemma.

Lemma B.3. *If $\boldsymbol{A}$ and $\boldsymbol{B}$ are self-adjoint, positive semidefinite quadratic matrices of the same size, then $\boldsymbol{AB}$ has real nonnegative eigenvalues.*

Proof. Though $\boldsymbol{AB}$ is usually not self-adjoint, it is still diagonalizable as follows. The eigenvalue–eigenvector equation for the matrix $\boldsymbol{AB}$ is:

$$\boldsymbol{AB}\mathbf{x} = \lambda \mathbf{x}$$

that is equivalent to

$$(\boldsymbol{A}^{1/2}\boldsymbol{B}\boldsymbol{A}^{1/2})(\boldsymbol{A}^{-1/2}\mathbf{x}) = \lambda(\boldsymbol{A}^{-1/2}\mathbf{x}),$$

where $\boldsymbol{A}^{1/2}\boldsymbol{B}\boldsymbol{A}^{1/2} = (\boldsymbol{A}^{1/2}\boldsymbol{B}^{1/2})(\boldsymbol{A}^{1/2}\boldsymbol{B}^{1/2})^*$ is a Gram matrix (Definition B.5), so positive semidefinite. Each of its nonnegative real eigenvalue λ is also an eigenvalue of $\boldsymbol{AB}$ with eigenvector that is obtained with premultiplying its eigenvector with $\boldsymbol{A}^{1/2}$. □

Definition B.9. The quadratic matrix $\boldsymbol{A} = [a_{jk}]$ is of Toeplitz type if it has the same entries along its main diagonal and along all lines parallel to the main diagonal. In other words, the value of the entry a_{jk} depends only on $|j-k|$, $\forall j, k$.

Definition B.10. The quadratic matrix $\boldsymbol{A} = [a_{jk}]$ is of Hankel type if it has the same entries along its anti-diagonal and along all lines parallel to its anti-diagonal. In other words, the value of the entry a_{jk} depends only on $j+k$, $\forall j, k$.

Block Toeplitz and block Hankel matrices are defined analogously: they are of Toeplitz and of Hankel type in terms of their blocks considered as entries, respectively.

Without proofs, we enlist some notable matrix decompositions, see [26].

- The *Gram decomposition* of the self-adjoint, positive semidefinite, $n \times n$ matrix $\boldsymbol{G}$ is the decomposition $\boldsymbol{G} = \boldsymbol{A}\boldsymbol{A}^*$ in the proof of Proposition B.1. As we saw, it is not unique, and the minimal size of $\boldsymbol{A}$ is $n \times r$, where $r = \text{rank}(\boldsymbol{A})$.

  The Gram decomposition $\boldsymbol{G} = \boldsymbol{A}\boldsymbol{A}^*$ with the $\boldsymbol{A}$ of equation (B.5) is called *parsimonious Gram decomposition.*

- The *QR-decomposition* of a complex $m \times n$ matrix $\boldsymbol{A}$ is

  $$\boldsymbol{A} = \boldsymbol{Q}\boldsymbol{R},$$

  where the matrix $\boldsymbol{Q}$ is $m \times m$ unitary, whereas $\boldsymbol{R}$ is $m \times n$ generalized upper triangular matrix (there are 0 entries below its main diagonal, starting at its upper left corner). This (not necessarily unique) decomposition always exists, and can be derived by applying the Gram–Schmidt orthogonalization procedure to the column vectors of $\boldsymbol{A}$.

  The related QR-transformation (Francis, 1961) uses the QR-decomposition for an iteration converging to $\boldsymbol{\Lambda}$ of the SD $\boldsymbol{A} = \boldsymbol{U}\boldsymbol{\Lambda}\boldsymbol{U}^*$ of the self-adjoint matrix $\boldsymbol{A}$. The iteration is as follows:

  $$\boldsymbol{A}_0 = \boldsymbol{A}, \quad \boldsymbol{Q}_0 = \boldsymbol{Q}, \quad \boldsymbol{R}_0 = \boldsymbol{R},$$

  and for $t = 1, 2, \dots$, if $\boldsymbol{A}_{t-1} = \boldsymbol{Q}_{t-1}\boldsymbol{R}_{t-1}$, then $\boldsymbol{A}_t := \boldsymbol{R}_{t-1}\boldsymbol{Q}_{t-1}$. Then $\lim_{t\to\infty} \boldsymbol{A}_t = \boldsymbol{\Lambda}$ in L^2-norm.

- The *LDL-decomposition* of the complex $n \times n$ self-adjoint matrix $\boldsymbol{A}$ is

$$\boldsymbol{A} = \boldsymbol{L}\boldsymbol{D}\boldsymbol{L}^*,$$

where $\boldsymbol{L}$ is $n \times n$ lower triangular with 1s along its main diagonal, and $\boldsymbol{D}$ is $n \times n$ diagonal matrix (with nonnegative diagonal entries).

Moreover, the LDL-decomposition is nested in the following sense: if $\boldsymbol{A}_k$, $\boldsymbol{L}_k$, and $\boldsymbol{D}_k$ denote the $k \times k$ submatrices of the underlying matrices formed by their first k rows and columns, then

$$\boldsymbol{A}_k = \boldsymbol{L}_k\boldsymbol{D}_k\boldsymbol{L}_k^*$$

is also LDL-decomposition for $k = 1, 2, \ldots, n$.

- The *LU-decomposition* of the complex $n \times n$ matrix $\boldsymbol{A}$ is

$$\boldsymbol{A} = \boldsymbol{L}\boldsymbol{U},$$

where $\boldsymbol{L}$ is $n \times n$ lower, and $\boldsymbol{U}$ is $n \times n$ upper triangular matrix. It can be arranged that each diagonal entry of $\boldsymbol{L}$ is 1. This decomposition is sometimes called *Cholesky decomposition.*

If $\boldsymbol{A}$ is self-adjoint, then the LU-decomposition can be obtained from the LDL-decomposition via manipulations with the nonnegative entries of $\boldsymbol{D}$.

The related LR-transformation (Rutishauser, 1958, see [61]) uses the LU-decomposition for an iteration converging to $\boldsymbol{\Lambda}$ of the SD $\boldsymbol{A} = \boldsymbol{U}\boldsymbol{\Lambda}\boldsymbol{U}^*$ of the self-adjoint matrix $\boldsymbol{A}$. Here LR denotes left-right, which is the same as LU (lower-upper). The iteration is as follows:

$$\boldsymbol{A}_0 = \boldsymbol{A} = \boldsymbol{L}\boldsymbol{R}, \quad \boldsymbol{L}_0 = \boldsymbol{L}, \quad \boldsymbol{R}_0 = \boldsymbol{R},$$

and for $t = 1, 2, \ldots$, if $\boldsymbol{A}_{t-1} = \boldsymbol{L}_{t-1}\boldsymbol{R}_{t-1}$, then $\boldsymbol{A}_t := \boldsymbol{R}_{t-1}\boldsymbol{L}_{t-1}$. Eventually, $\lim_{t\to\infty} \boldsymbol{R}_t = \lim_{t\to\infty} \boldsymbol{A}_t = \boldsymbol{\Lambda}$ and $\lim_{t\to\infty} \boldsymbol{L}_t = \boldsymbol{I}_n$ in L^2-norm.

Note that the above matrix decompositions work for block-matrices similarly. The computational complexity is increased with the understanding that here matrix multiplications are substituted for entry-wise multiplications.

Block-matrices sometimes arise as Kronecker-products.

Definition B.11. Let $\boldsymbol{A}$ be $p \times n$ and $\boldsymbol{B}$ be $q \times m$ complex matrix. Their Kronecker product, denoted by $\boldsymbol{A} \otimes \boldsymbol{B}$, is the following $pq \times nm$ block-matrix: it has p block rows and n block columns; each block is a $q \times m$ matrix such that the block indexed by (j, k) is the matrix $a_{jk}\boldsymbol{B}$ $(j = 1, \ldots, p;\ k = 1, \ldots, n)$.

This product is associative, for the addition distributive, but usually not commutative. If $\boldsymbol{A}$ is $n \times n$ and $\boldsymbol{B}$ is $m \times m$ quadratic matrix, then

$$\det(\boldsymbol{A} \otimes \boldsymbol{B}) = (\det \boldsymbol{A})^m \cdot (\det \boldsymbol{B})^n;$$

further, if both are regular, then so is their Kronecker-product. Namely,

$$(\boldsymbol{A} \otimes \boldsymbol{B})^{-1} = \boldsymbol{A}^{-1} \otimes \boldsymbol{B}^{-1}.$$

It is also useful to know that — provided $\boldsymbol{A}$ and $\boldsymbol{B}$ are self-adjoint — the spectrum of $\boldsymbol{A} \otimes \boldsymbol{B}$ consists of the real numbers

$$\alpha_j \beta_k \quad (j = 1, \ldots, n;\ k = 1, \ldots, m),$$

where α_j's and β_k's are the eigenvalues of $\boldsymbol{A}$ and $\boldsymbol{B}$, respectively. Definition B.11 naturally extends to vectors: the Kronecker-product of vectors $\mathbf{a} \in \mathbb{C}^n$ and $\mathbf{b} \in \mathbb{C}^m$ is a vector $\mathbf{a} \otimes \mathbf{b} \in \mathbb{C}^{nm}$.

The eigenvalues of other types of block-matrices are characterized in the following theorem.

Theorem B.7. (Theorem 5.3.1 of [48]). *Let $\boldsymbol{A}$ be a $d \times d$ self-adjoint matrix with spectral decomposition*

$$\boldsymbol{A} = \sum_{k=1}^{d} a_k \mathbf{u}_k \mathbf{u}_k^*;$$

the analytic functions $g_{ij}(z)$ for $i, j = 1, \ldots, n$ satisfy

$$g_{ij}(z) = \overline{g_{ji}(z)},$$

and the eigenvalues $a_1, \ldots, a_k$ are within the convergence region of every $g_{ij}(z)$.

Denoting the spectral decomposition of the self-adjoint matrix $[g_{ij}(a_k)]_{i,j=1}^n$ with

$$[g_{ij}(a_k)]_{i,j=1}^n = \sum_{\ell=1}^{n} \lambda_\ell^{(k)} \mathbf{v}_\ell^{(k)} \mathbf{v}_\ell^{(k)*}, \quad k = 1, \ldots, d,$$

the spectral decomposition of the $nd \times nd$ block matrix $[\boldsymbol{A}_{ij}]_{i,j=1}^n = [g_{ij}(\boldsymbol{A})]_{i,j=1}^n$ is

$$[\boldsymbol{A}_{ij}]_{i,j=1}^n = \sum_{\ell=1}^{n} \sum_{k=1}^{d} \lambda_\ell^{(k)} (\mathbf{u}_k \otimes \mathbf{v}_\ell^{(k)})(\mathbf{u}_k \otimes \mathbf{v}_\ell^{(k)})^*.$$

In [48] it is noted that if $\boldsymbol{A}$ is normal and the matrices $[g_{ij}(a_k)]_{i,j=1}^n$ are as well all normal, the statement also holds irrespective whether g_{ij}s are analytic.

Next we discuss low rank approximations of a matrix.

Theorem B.8. *Let $\boldsymbol{A} \in \mathbb{C}^{m \times n}$ with SVD $\boldsymbol{A} = \sum_{i=1}^{r} s_i \mathbf{u}_i \mathbf{v}_i^*$, where r is the rank of $\boldsymbol{A}$ and $s_1 \geq \cdots \geq s_r > 0$. Then for any $1 \leq k < r$ such that $s_k > s_{k+1}$ we have*

$$\min \|\boldsymbol{A} - \boldsymbol{B}\| = s_{k+1} \quad \text{and} \quad \min \|\boldsymbol{A} - \boldsymbol{B}\|_F = \left(\sum_{i=k+1}^{r} s_i^2 \right)^{1/2},$$

where the minima are taken for all matrices $\boldsymbol{B} \in \mathbb{C}^{m \times n}$ of rank k. Both minima are attained with the matrix $\boldsymbol{B} = \boldsymbol{A}_k := \sum_{i=1}^{k} s_i \mathbf{u}_i \mathbf{v}_i^*$.

Note that $\boldsymbol{A}_k$ is called the *best rank k approximation* of $\boldsymbol{A}$, and the aforementioned theorem guarantees that it is the best approximation both in spectral and Frobenius norm. In fact, it is true for any unitary invariant norm:

$$\min_{\substack{\boldsymbol{B} \text{ is } m\times n \\ \text{rank}(\boldsymbol{B})=k}} \|\boldsymbol{A}-\boldsymbol{B}\|_{\text{un}} = \|\boldsymbol{A}-\boldsymbol{A}_k\|_{\text{un}}.$$

Corollary B.1. *Let $\boldsymbol{A} \in \mathbb{C}^{n\times n}$ be a self-adjoint, positive semidefinite matrix with SD $\boldsymbol{A} = \sum_{j=1}^{r} \lambda_j \mathbf{u}_j \mathbf{u}_j^*$, where r is the rank of $\boldsymbol{A}$ and the eigenvalues are $\lambda_1 \geq \cdots \geq \lambda_r > 0$. Then for any $1 \leq k < r$ such that $\lambda_k > \lambda_{k+1}$ we have*

$$\min \|\boldsymbol{A}-\boldsymbol{B}\| = \lambda_{k+1} \quad \text{and} \quad \min \|\boldsymbol{A}-\boldsymbol{B}\|_F = \left(\sum_{i=k+1}^{r} \lambda_i^2\right)^{1/2},$$

where the minima are taken for all self-adjoint, positive semidefinite matrices $\boldsymbol{B} \in \mathbb{C}^{n\times n}$ of rank k. Both minima are attained with the best rank k approximation

$$\boldsymbol{A}_k = \sum_{j=1}^{k} \lambda_j \mathbf{u}_j \mathbf{u}_j^* = \tilde{\boldsymbol{U}}_k \tilde{\boldsymbol{\Lambda}}_k \tilde{\boldsymbol{U}}_k^*$$

of $\boldsymbol{A}$, where $\tilde{\boldsymbol{U}}_k = [\mathbf{u}_1, \ldots, \mathbf{u}_k]$ is $n \times k$ sub-unitary and $\tilde{\boldsymbol{\Lambda}}_k = \text{diag}(\lambda_1, \ldots, \lambda_k)$ is $k \times k$ diagonal matrix.

In particular,

$$\boldsymbol{A} = \boldsymbol{A}_r = \tilde{\boldsymbol{U}}_r \tilde{\boldsymbol{\Lambda}}_r \tilde{\boldsymbol{U}}_r^*,$$

of which we can take the square-root:

$$\boldsymbol{A}^{1/2} = \tilde{\boldsymbol{U}}_r \tilde{\boldsymbol{\Lambda}}_r^{1/2} \tilde{\boldsymbol{U}}_r^*,$$

where $\tilde{\boldsymbol{\Lambda}}_r^{1/2} = \text{diag}(\sqrt{\lambda_1}, \ldots, \sqrt{\lambda_r})$. However, with any matrix $\boldsymbol{M} = \tilde{\boldsymbol{U}}_r \tilde{\boldsymbol{\Lambda}}_r^{1/2} \boldsymbol{Q}$, where $\boldsymbol{Q}$ is $r \times r$ unitary, the decomposition $\boldsymbol{A} = \boldsymbol{M}\boldsymbol{M}^*$ holds. In particular it holds with $\boldsymbol{M} = \tilde{\boldsymbol{U}}_r \tilde{\boldsymbol{\Lambda}}_r^{1/2} = [\sqrt{\lambda_1}\mathbf{u}_1, \ldots, \sqrt{\lambda_r}\mathbf{u}_r]$, see the parsimonious Gram-decomposition (B.5).

Theorem B.8 is proved for self-adjoint matrices in [44] with the Frobenius norm, but it can be easily extended to any unitary invariant norm, see also [45].

C

Best prediction in Hilbert spaces

Let $L^2(\Omega, \mathcal{A}, \mathbb{P})$ be the Hilbert space of real valued random variables with zero expectation and finite variance; the inner product is the covariance, and $\mathbb{P}$ denotes the joint distribution of all of them. We use subspaces of this, related to multivariate, weakly stationary processes.

Let $\mathbf{X} = \{\mathbf{X}_t\}_{t\in\mathbb{Z}}$ be a weakly stationary, d-dimensional time series, with real-valued coordinates, $\mathbb{E}\mathbf{X}_t = \mathbf{0}$. By weak stationarity, $\mathbf{X}_t$s all have the same covariance matrix $\boldsymbol{C}(0)$. Note that to any weakly stationary process there corresponds a Gaussian one with the same second moments, so it is not a restriction to confine ourselves to Gaussian processes. Sometimes we speak in terms of so-called *second order processes* that are determined by their first and second moments. When the expectations are $\mathbf{0}$s, the pairwise covariances characterize the process, and predictions can be discussed in terms of projections in Hilbert spaces.

So in the Gaussian case, $\mathbf{X}_t$s all have the same d-variate Gaussian distribution, but they are defined on different d-dimensional marginals of $\mathbb{P}$. In particular, their first, second, etc. autocovariances, $\boldsymbol{C}(1), \boldsymbol{C}(2), \dots$ characterize their joint distribution. Therefore, this can be regarded as a special random field that extends in space and time (cross-sectionally and longitudinally), i.e. the parameters of the random process contain both (discrete) time and (d-dimensional) space locations, see [23, 25, 24].

Corresponding to the above weakly stationary process, throughout the book we consider the following subspaces of $L^2(\Omega, \mathcal{A}, \mathbb{P})$:

$$\begin{aligned}
H(\mathbf{X}) &= \overline{\text{span}}\{X_k^i \mid k \in \mathbb{Z},\ i = 1, \dots, d\} \\
H_t^-(\mathbf{X}) &= \overline{\text{span}}\{X_k^i \mid k \le t,\ i = 1, \dots, d\} \\
H_t(\mathbf{X}) &= \overline{\text{span}}\{X_k^i \mid 1 \le k \le t,\ i = 1, \dots, d\} \\
&= \text{Span}\,\{X_k^i\ :\ 1 \le k \le t,\ i = 1, \dots, d\},
\end{aligned}$$

as the last one is finite dimensional. Obviously, $H_t(\mathbf{X}) \subseteq H_t^-(\mathbf{X}) \subseteq H(\mathbf{X})$.

In any Hilbert space, the following *Projection Theorem* holds true.

Theorem C.1. *Let $\mathcal{M}$ be a closed subspace of the Hilbert space $\mathcal{H}$. Then for any $Y \in \mathcal{H}$, there is a unique element $\hat{Y} \in \mathcal{M}$ such that*

$$\|Y - \hat{Y}\| \le \|Y - Z\|, \quad \forall Z \in \mathcal{M}$$

and

$$Y - \hat{Y} \perp Z, \quad \forall Z \in \mathcal{M}.$$

This unique $\hat{Y}$ is called the projection of Y onto $\mathcal{M}$ and denoted by $\text{Proj}_{\mathcal{M}} Y$. By the Pythagorean Theorem,

$$\|Y\|^2 = \|\hat{Y}\|^2 + \|Y - \hat{Y}\|^2.$$

The Projection Theorem is widely used in statistics, in the context of multivariate parallel, in other words, simultaneous or joint response regressions. We distinguish between nonparametric and parametric regression, where the former applies to arbitrary multivariate distributions, whereas, the latter to Gaussian ones. More generally, we speak in terms of random vectors which are not necessarily instances of a multidimensional time series. Let $\mathbf{X}$ be q- and $\mathbf{Y}$ be a p-dimensional random vector. In the Gaussian, zero mean case, their joint distribution is characterized by their $q \times p$ cross-covariance matrix $\mathbb{E}\mathbf{X}\mathbf{Y}^T$, and the orthogonality (independence) of them means that each component of $\mathbf{X}$ is uncorrelated with each component of $\mathbf{Y}$, i.e. $\mathbb{E}\mathbf{X}\mathbf{Y}^T = \boldsymbol{O}$. Note that $\mathbb{E}\mathbf{Y}\mathbf{X}^T = [\mathbb{E}\mathbf{X}\mathbf{Y}^T]^T$, whereas $\mathbb{E}\mathbf{X}\mathbf{X}^T$ is the usual (positive semidefinite) covariance matrix of $\mathbf{X}$.

Consider the multiple response regression problem, when each component of $\mathbf{Y}$ is projected onto the subspace generated by the coordinates of $\mathbf{X}$. First we deal with the general (nonparametric) situation.

Lemma C.1. *Let* $\mathbf{Y}$ *be an* $\mathbb{R}^p$*-valued and* $\mathbf{X}$ *be an* $\mathbb{R}^q$*-valued random vector with components of zero expectations. Then for every* $f : \mathbb{R}^q \to \mathbb{R}^p$ *measurable function for which the expectation below exists and is finite, the conditional expectation of* $\mathbf{Y}$ *conditioned on* $\mathbf{X}$ *minimizes the error covariance matrix*

$$\mathbb{E}(\mathbf{Y} - f(\mathbf{X}))(\mathbf{Y} - f(\mathbf{X}))^T \geq E(\mathbf{Y} - \mathbb{E}(\mathbf{Y}|\mathbf{X}))(\mathbf{Y} - \mathbb{E}(\mathbf{Y}|\mathbf{X}))^T,$$

in the sense that the difference of the left and right hand side matrices is positive semidefinite.

Proof. With the notation $f^*(\mathbf{X}) = \mathbb{E}(\mathbf{Y}|\mathbf{X})$ we have that

$$\begin{aligned}
&\mathbb{E}(\mathbf{Y} - f(\mathbf{X}))(\mathbf{Y} - f(\mathbf{X}))^T \\
&= \mathbb{E}[(\mathbf{Y} - f^*(\mathbf{X})) + (f^*(\mathbf{X}) - f(\mathbf{X}))][(\mathbf{Y} - f^*(\mathbf{X})) + (f^*(\mathbf{X}) - f(\mathbf{X}))]^T \\
&= \mathbb{E}(\mathbf{Y} - f^*(\mathbf{X}))(\mathbf{Y} - f^*(\mathbf{X}))^T + \mathbb{E}(f^*(\mathbf{X}) - f(\mathbf{X}))(f^*(\mathbf{X}) - f(\mathbf{X}))^T \\
&+ \mathbb{E}(\mathbf{Y} - f^*(\mathbf{X}))(f^*(\mathbf{X}) - f(\mathbf{X}))^T + \mathbb{E}(f^*(\mathbf{X}) - f(\mathbf{X}))(\mathbf{Y} - f^*(\mathbf{X}))^T \\
&= \mathbb{E}(\mathbf{Y} - f^*(\mathbf{X}))(\mathbf{Y} - f^*(\mathbf{X}))^T + \mathbb{E}(f^*(\mathbf{X}) - f(\mathbf{X}))(f^*(\mathbf{X}) - f(\mathbf{X}))^T.
\end{aligned}$$

In the last step we used that $\mathbb{E}(\mathbf{Y} - f^*(\mathbf{X}))(f^*(\mathbf{X}) - f(\mathbf{X}))^T = \boldsymbol{O}$ is the zero matrix, akin to its transpose $\mathbb{E}(f^*(\mathbf{X}) - f(\mathbf{X}))(\mathbf{Y} - f^*(\mathbf{X}))^T$. So it suffices to prove only for the first one:

$$\begin{aligned}
\mathbb{E}(\mathbf{Y} - f^*(\mathbf{X}))(f^*(\mathbf{X}) - f(\mathbf{X}))^T &= \mathbb{E}[\mathbb{E}(\mathbf{Y} - f^*(\mathbf{X}))(f^*(\mathbf{X}) - f(\mathbf{X}))^T \,|\, \mathbf{X}] \\
&= \mathbb{E}[\mathbb{E}(\mathbf{Y} - f^*(\mathbf{X}) \,|\, \mathbf{X})(f^*(\mathbf{X}) - f(\mathbf{X}))^T] \\
&= \mathbb{E}[(\mathbb{E}(\mathbf{Y}|\mathbf{X}) - f^*(\mathbf{X}))(f^*(\mathbf{X}) - f(\mathbf{X}))^T] \\
&= \boldsymbol{O}
\end{aligned}$$

since $\mathbb{E}(\mathbf{Y}|\mathbf{X}) - f^*(\mathbf{X}) = \mathbf{0}$. As $\mathbb{E}(f^*(\mathbf{X}) - f(\mathbf{X}))(f^*(\mathbf{X}) - f(\mathbf{X}))^T$ is positive semidefinite (being a covariance matrix), it follows that

$$\mathbb{E}(\mathbf{Y} - f(\mathbf{X}))(\mathbf{Y} - f(\mathbf{X}))^T - \mathbb{E}(\mathbf{Y} - f^*(\mathbf{X}))(\mathbf{Y} - f^*(\mathbf{X}))^T$$

is positive semidefinite that was to be proved. □

Remark C.1. The conditional expectation $f^*(\mathbf{X}) = \mathbb{E}(\mathbf{Y}|\mathbf{X})$ also minimizes

$$\mathbb{E}\|\mathbf{Y} - f(\mathbf{X})\|^2 = \sum_{i=1}^{p} \mathbb{E}[Y^i - f_i(\mathbf{X})]^2,$$

where f_is are the coordinate functions of f. Indeed, we can minimize the p terms separately. Applying Lemma C.1 to the univariate case, we get that the minimizer of the ith term is $f_i^*(\mathbf{X}) = \mathbb{E}(Y^i|\mathbf{X})$. As this is the ith coordinate of $f^*(\mathbf{X}) = \mathbb{E}(\mathbf{Y}|\mathbf{X})$, the minimum of $\mathbb{E}\|\mathbf{Y} - f(\mathbf{X})\|^2$ is attained at the same $f^*(\mathbf{X})$ that minimizes the covariance matrix $\mathbb{E}(\mathbf{Y} - f(\mathbf{X}))(\mathbf{Y} - f(\mathbf{X}))^T$ in the sense of Lemma C.1. Note that $\mathbb{E}\|\mathbf{Y} - f(\mathbf{X})\|^2$ is the trace of the error covariance matrix.

Therefore, it suffices to investigate univariate non-parametric regression estimations. If they are mean square consistent, can be constructed with a sequence $f_i^{(n)}$ (for example, with local averaging) such that the mean square error

$$\mathbb{E}[f_i^{(n)}(\mathbf{X}) - f_i^*(\mathbf{X})]^2 \to 0, \quad n \to \infty,$$

for $i = 1, \dots, p$. This implies that in the p-variate case, for the sequence $f^{(n)}(\mathbf{X}) = (f_1^{(n)}(\mathbf{X}), \dots, f_p^{(n)}(\mathbf{X}))^T$

$$\mathbb{E}\|f^{(n)}(\mathbf{X}) - f^*(\mathbf{X})\|^2 \to 0, \quad n \to \infty \tag{C.1}$$

holds, exhibiting a kind of mean square consistency in the multiple target situation. Since $\mathbb{E}\|f^{(n)}(\mathbf{X}) - f^*(\mathbf{X})\|^2$ is the trace of the $p \times p$ symmetric, positive semidefinite error covariance matrix $\boldsymbol{E}_n = \mathbb{E}(f^{(n)}(\mathbf{X}) - f^*(\mathbf{X}))(f^{(n)}(\mathbf{X}) - f^*(\mathbf{X}))^T$, Equation (C.1) is equivalent to $\|\boldsymbol{E}_n\|_2 \to 0$ as $n \to \infty$ (the spectral norm $\|\boldsymbol{E}_n\|_2$ is the largest eigenvalue of $\boldsymbol{E}_n$). Conversely, if $\|\boldsymbol{E}_n\|_2 \to 0$, then $\mathrm{tr}\boldsymbol{E}_n \to 0$, and by the Cauchy–Schwarz inequality, $\boldsymbol{E}_n \to \boldsymbol{O}$ too. See [7, 46], also [27].

Now we concentrate on linear estimates that are the best in the above sense too if the underlying distribution is multivariate Gaussian. Therefore, the forthcoming estimation can be called parametric simultaneous (joint response) regression, and can be described by matrices. Here we use the second order property of $\mathbb{P}$: the pairwise inner products of the random variables in $L^2(\Omega, \mathcal{A}, \mathbb{P})$ are determined by their covariances. If we consider subspaces, then the relation of a p- and a q-dimensional subspace can be described by all possible pq pairs of the pairwise covariances, i.e. by the cross-covariance matrices.

Lemma C.2. *Let* $\mathbf{Y} \in \mathbb{R}^p$ *and* $\mathbf{X} \in \mathbb{R}^q$ *be random vectors on a joint probability space with existing second moments and zero expectation. Then the* $q \times p$ *matrix* $\boldsymbol{A}$ *minimizing* $\mathbb{E}\|\mathbf{Y} - \boldsymbol{A}^T\mathbf{X}\|^2$ *is*

$$\boldsymbol{A} = [\mathbb{E}\mathbf{X}\mathbf{X}^T]^{-}[\mathbb{E}\mathbf{X}\mathbf{Y}^T], \tag{C.2}$$

where we use generalized inverse if the covariance matrix $\mathbb{E}\mathbf{X}\mathbf{X}^T$ *of* $\mathbf{X}$ *is singular (see Appendix B). If it is positive definite, then we get a unique minimizer* $\boldsymbol{A}$ *with the unique inverse matrix* $[\mathbb{E}\mathbf{X}\mathbf{X}^T]^{-1}$.

Proof. Observe that minimizing

$$\mathbb{E}\|\mathbf{Y} - \boldsymbol{A}^T\mathbf{X}\|^2 = \sum_{i=1}^{p} \mathbb{E}(Y^i - \mathbf{a}_i^T\mathbf{X})^2$$

with respect to $\boldsymbol{A} = [\mathbf{a}_1, \dots, \mathbf{a}_p]$ falls apart into the following p minimization tasks, with respect to the q-dimensional column vectors of $\boldsymbol{A}$:

$$\min_{\mathbf{a}_i} \mathbb{E}(Y^i - \mathbf{a}_i^T\mathbf{X})^2, \quad i = 1, \dots, p.$$

The ith task, with the coordinates of $\mathbf{X} = (X^1, \dots, X^q)^T$ and $\mathbf{a}_i^T = (a_{1i}, \dots, a_{qi})$, is equivalent to

$$\mathbb{E}(Y^i - \sum_{k=1}^{q} a_{ki}X^k)^2 \to \min.$$

Take the derivative with respect to a_{ji} and make it equal to 0. (We assume regularity, i.e. that the differentiation and taking the expectation can be interchanged, which is true if the underlying distribution is Gaussian.)

$$2\mathbb{E}[(-X_j)(Y^i - \sum_{k=1}^{q} a_{ki}X^k)] = 0, \quad j = 1, \dots, q.$$

After rearranging, we have the system of equations

$$\sum_{k=1}^{q} a_{ki}\mathbb{E}(X^jX^k) = \mathbb{E}(X^jY^i), \quad j = 1, \dots, q.$$

This can be condensed into the well-known system of *Gauss normal equations* from the classical theory of multivariate regression:

$$[\mathbb{E}\mathbf{X}\mathbf{X}^T]\mathbf{a}_i = [\mathbb{E}\mathbf{X}Y^i], \quad i = 1, \dots, p.$$

(Actually, the original equations of Gauss apply to the sample version, and do not contain expectations.) Since this system of linear equations is consistent (the vector $\mathbb{E}\mathbf{X}Y^i$ is in the column space of $\mathbb{E}\mathbf{X}\mathbf{X}^T$), it always has a solution in the general form:

$$\mathbf{a}_i = [\mathbb{E}\mathbf{X}\mathbf{X}^T]^-[\mathbb{E}\mathbf{X}Y^i], \quad i = 1, \dots, p.$$

Here $[\mathbb{E}\mathbf{X}\mathbf{X}^T]^-$ is the generalized inverse of the matrix in brackets.

Therefore the matrix $\boldsymbol{A}$ giving the optimum is

$$\boldsymbol{A} = [\mathbb{E}\mathbf{X}\mathbf{X}^T]^-[\mathbb{E}\mathbf{X}\mathbf{Y}^T],$$

that is unique only if $\mathbb{E}\mathbf{X}\mathbf{X}^T$ is invertible (positive definite), otherwise (if $\mathbb{E}\mathbf{X}\mathbf{X}^T$ is singular, positive semidefinite) infinitely many versions of the generalized inverse give infinitely many convenient $\boldsymbol{A}$s (see Appendix B). Albeit, with different linear combinations of the coordinates of $\mathbf{X}_i$s, these always provide the same optimal linear prediction for $\mathbf{Y}$ as follows:

$$\widehat{\mathbf{Y}} = \begin{bmatrix} \widehat{Y}^1 \\ \widehat{Y}^2 \\ \vdots \\ \widehat{Y}^p \end{bmatrix} = \begin{bmatrix} \mathbf{a}_1^T\mathbf{X} \\ \mathbf{a}_2^T\mathbf{X} \\ \vdots \\ \mathbf{a}_p^T\mathbf{X} \end{bmatrix}.$$

□

Remark C.2. It can easily be checked that $\mathbb{E}(\mathbf{Y} - \widehat{\mathbf{Y}})\widehat{\mathbf{Y}}^T = \boldsymbol{O}$. By the proof of Lemma C.1, it follows that $\mathbf{Y} - \widehat{\mathbf{Y}}$ is orthogonal to any element of $\mathcal{M}$ spanned by p-tuples of the coordinates of the vector $\mathbf{X}$ (in the sense that their cross-covariance is the zero matrix). Therefore, it bears the properties of a projection in a wider sense.

Hence, the statement of Lemma C.2 also follows by applying the Projection Theorem C.1 simultaneously. We know that the $q \times p$ matrix $\boldsymbol{A}$, giving the minimum of $\mathbb{E}\|\mathbf{Y} - \boldsymbol{A}^T\mathbf{X}\|^2$, is such that $\boldsymbol{A}^T\mathbf{X} = \mathrm{Proj}_{\mathcal{M}}\mathbf{Y} = (\mathrm{Proj}_{\mathcal{M}}Y^1, \dots, \mathrm{Proj}_{\mathcal{M}}Y^p)^T$ denotes coordinate-wise projection. But $\mathbf{Y} - \mathrm{Proj}_{\mathcal{M}}\mathbf{Y}$ is orthogonal to any vector in $\mathcal{M}$, which has the form $\boldsymbol{B}^T\mathbf{X}$ with a $q \times p$ matrix $\boldsymbol{B}$. Therefore,

$$\mathbb{E}[(\mathbf{Y} - \boldsymbol{A}^T\mathbf{X})(\boldsymbol{B}^T\mathbf{X})^T] = \boldsymbol{O}, \quad \forall \boldsymbol{B}_{q\times p}.$$

Equivalently,

$$[\mathbb{E}(\mathbf{Y}\mathbf{X}^T) - \boldsymbol{A}^T\mathbb{E}(\mathbf{X}\mathbf{X}^T)]\boldsymbol{B} = \boldsymbol{O}, \quad \forall \boldsymbol{B}_{q\times p}.$$

This implies that the matrix in brackets is the zero matrix, which fact after transposing and using that $\mathbb{E}(\mathbf{X}\mathbf{X}^T)$ is symmetric (it is the usual covariance matrix of $\mathbf{X}$) gives again the system of Gauss normal equations in concise form:

$$[\mathbb{E}(\mathbf{X}\mathbf{X}^T)]\boldsymbol{A} = \mathbb{E}(\mathbf{X}\mathbf{Y}^T).$$

Remark C.3. We can also estimate the attainable minimum error. When $p = 1$, then with the notations

$$\boldsymbol{C} := \mathbb{E}\mathbf{X}\mathbf{X}^T \quad \text{and} \quad \mathbf{d} := \mathbb{E}\mathbf{X}Y \quad (i = 1, \dots, p)$$

by the theory of multivariate regression [44] we have that

$$\mathbb{E}(Y-\widehat{Y})^2 = \mathrm{Var}(Y)(1-r_{Y\mathbf{X}}^2) = \mathrm{Var}(Y) - \mathbf{d}^T \boldsymbol{C}^{-1}\mathbf{d}, \tag{C.3}$$

where we assumed that $\boldsymbol{C}$ is positive definite and $r_{Y\mathbf{X}}$ denotes the multiple correlation between Y and the components of $\mathbf{X}$.

Adapting this for a p-dimensional $\mathbf{Y}$ we get that

$$\mathbb{E}\|\mathbf{Y}-\widehat{\mathbf{Y}}\|^2 = \sum_{i=1}^{p} \mathbb{E}(Y^i-\widehat{Y}^i)^2 = \|\mathbf{Y}\|^2 - \sum_{i=1}^{p} \mathbf{d}_i^T \boldsymbol{C}^{-1}\mathbf{d}_i,$$

where $\mathbf{d}_i = \mathbb{E}\mathbf{X}Y^i$, $i=1,\dots,p$.

Lemma C.3. *Let $\mathbf{Y}\in\mathbb{R}^p$ and $\mathbf{X}\in\mathbb{R}^q$ be random vectors on a joint probability space with existing second moments and zero expectation, and let* $\mathrm{Proj}_{\mathcal{M}}\mathbf{Y}$ *denote the best linear prediction of $\mathbf{Y}$ based on p-tuples of linear combinations of the coordinates of $\mathbf{X}$, denoted by $\mathcal{M}$, as in Lemma C.2. Then with any $p\times p$ matrix $\boldsymbol{\Phi}$,*

$$\mathrm{Proj}_{\mathcal{M}}(\boldsymbol{\Phi}\mathbf{Y}) = \boldsymbol{\Phi}\mathrm{Proj}_{\mathcal{M}}\mathbf{Y}.$$

Proof. We saw that $\mathrm{Proj}_{\mathcal{M}}\mathbf{Y} = \boldsymbol{A}^T\mathbf{X}$, where by (C.2) $\boldsymbol{A} = [\mathbb{E}\mathbf{X}\mathbf{X}^T]^-[\mathbb{E}\mathbf{X}\mathbf{Y}^T]$, and we use generalized inverse — if the covariance matrix $\mathbb{E}\mathbf{X}\mathbf{X}^T$ of $\mathbf{X}$ is singular. Then

$$\begin{aligned}\mathrm{Proj}_{\mathcal{M}}(\boldsymbol{\Phi}\mathbf{Y}) &= \{[\mathbb{E}\mathbf{X}\mathbf{X}^T]^-[\mathbb{E}\mathbf{X}(\boldsymbol{\Phi}\mathbf{Y})^T]\}^T\mathbf{X} = [\mathbb{E}(\boldsymbol{\Phi}\mathbf{Y}\mathbf{X}^T)][\mathbb{E}\mathbf{X}\mathbf{X}^T]^-\mathbf{X}\\ &= \boldsymbol{\Phi}[\mathbb{E}(\mathbf{Y}\mathbf{X}^T)][\mathbb{E}\mathbf{X}\mathbf{X}^T]^-\mathbf{X} = \boldsymbol{\Phi}\mathrm{Proj}_{\mathcal{M}}\mathbf{Y}.\end{aligned}$$

□

The above lemma shows that this projection is linear in $\mathbf{Y}$ and it commutes with $\boldsymbol{\Phi}$. In the Gaussian case, obviously, we have that

$$\mathrm{Proj}_{\mathcal{M}}(\boldsymbol{\Phi}\mathbf{Y}) = \mathbb{E}(\boldsymbol{\Phi}\mathbf{Y}\,|\,\mathbf{X}) = \boldsymbol{\Phi}\mathbb{E}(\mathbf{Y}\,|\,\mathbf{X}) = \boldsymbol{\Phi}\mathrm{Proj}_{\mathcal{M}}(\mathbf{Y})$$

by the properties of the conditional expectation.

Now go back to time series. In particular, if we look for the best linear prediction of the p-dimensional random vector $\mathbf{Y}$ based on the segment $\mathbf{X}_1,\dots,\mathbf{X}_t$ of a d-dimensional time series, then with the $q=dt$ dimensional vector $\mathbf{X}=[\mathbf{X}_1^T,\dots,\mathbf{X}_t^T]^T$, the above formula adapts as

$$\begin{aligned}\widehat{\mathbf{Y}} &= \mathrm{Proj}_{H_t(\mathbf{X})}\mathbf{Y} = \boldsymbol{A}^T\mathbf{X}\\ &= \begin{bmatrix}\mathbf{a}_{11}^T & \dots & \mathbf{a}_{1t}^T\\ \mathbf{a}_{21}^T & \dots & \mathbf{a}_{2t}^T\\ \vdots & \vdots & \vdots\\ \mathbf{a}_{p1}^T & \dots & \mathbf{a}_{pt}^T\end{bmatrix}\begin{bmatrix}\mathbf{X}_1\\ \mathbf{X}_2\\ \vdots\\ \mathbf{X}_t\end{bmatrix} = \begin{bmatrix}\mathbf{a}_{11}^T\mathbf{X}_1+\dots+\mathbf{a}_{1t}^T\mathbf{X}_t\\ \mathbf{a}_{21}^T\mathbf{X}_1+\dots+\mathbf{a}_{2t}^T\mathbf{X}_t\\ \vdots\\ \mathbf{a}_{p1}^T\mathbf{X}_1+\dots+\mathbf{a}_{pt}^T\mathbf{X}_t\end{bmatrix},\end{aligned}$$

where the columns of the $q\times p$ matrix $\boldsymbol{A}$ are partitioned into t segments of length d; or equivalently, the columns of the $p \times q$ matrix $\boldsymbol{A}^T$ are partitioned into $p \times d$ matrices $\boldsymbol{A}_1^T, \dots, \boldsymbol{A}_t^T$ like

$$\boldsymbol{A} = \begin{bmatrix} \mathbf{a}_{11} & \dots & \mathbf{a}_{p1} \\ \mathbf{a}_{12} & \dots & \mathbf{a}_{p2} \\ \vdots & \vdots & \vdots \\ \mathbf{a}_{1t} & \dots & \mathbf{a}_{pt} \end{bmatrix}, \quad \boldsymbol{A}^T = \begin{bmatrix} \boldsymbol{A}_1^T & \dots & \boldsymbol{A}_t^T \end{bmatrix}, \quad \boldsymbol{A}_j^T = \begin{bmatrix} \mathbf{a}_{1j}^T \\ \mathbf{a}_{2j}^T \\ \vdots \\ \mathbf{a}_{pj}^T \end{bmatrix},$$

$j = 1, \dots, t$. With this, $\widehat{\mathbf{Y}}$ is the linear combination of $\mathbf{X}_1, \dots, \mathbf{X}_t$ with matrices $\boldsymbol{A}_1^T, \dots, \boldsymbol{A}_t^T$, i.e.

$$\widehat{\mathbf{Y}} = \boldsymbol{A}_1^T \mathbf{X}_1 + \dots + \boldsymbol{A}_t^T \mathbf{X}_t.$$

Remark C.4. Observe that the pdt-dimensional linear space generated by the linear combinations $\boldsymbol{A}_1^T \mathbf{X}_1 + \dots + \boldsymbol{A}_t^T \mathbf{X}_t$ of $\mathbf{X}_1, \dots, \mathbf{X}_t$ with $p \times d$ matrices $\boldsymbol{A}_1^T, \dots, \boldsymbol{A}_t^T$ is the p-tuple Cartesian product of $H_t(\mathbf{X})$, which is dt-dimensional and contains scalar linear combinations of all the d coordinates of $\mathbf{X}_1, \dots, \mathbf{X}_t$. So, in case of p simulteneous regressions, $\mathbf{X}_1, \dots, \mathbf{X}_t$ are linearly combined with $p \times d$ matrices, the rows of which give scalar linear combinations that define the individual regressions. Just the solution is organized in matrix form, which is more suitable for our purposes.

So far, the time series was not necessarily stationary. When it is so, then the covariance matrix of the compounded vector $\mathbf{X} = [\mathbf{X}_1^T, \dots, \mathbf{X}_t^T]^T$ is

$$\mathbb{E}\mathbf{X}\mathbf{X}^T = \begin{bmatrix} \boldsymbol{C}(0) & \boldsymbol{C}(1) & \dots & \boldsymbol{C}_(t-1) \\ \boldsymbol{C}(-1) & \boldsymbol{C}(0) & \dots & \boldsymbol{C}(t-2) \\ \vdots & \vdots & \vdots & \vdots \\ \boldsymbol{C}(1-t) & \boldsymbol{C}(2-t) & \dots & \boldsymbol{C}(0) \end{bmatrix}$$

which is the symmetric (due to $\boldsymbol{C}(-k) = \boldsymbol{C}^T(k)$), positive semidefinite block Toeplitz matrix discussed in the context of VARMA processes. It is also positive definite whenever the process $\{\mathbf{X}_t\}$ is regular (see the Wold decomposition).

Going further, with $\mathbf{Y} = \mathbf{X}_{t+1}$, we get the one-step ahead prediction $\widehat{\mathbf{X}}_{t+1} = \boldsymbol{A}^T \mathbf{X}$, where the optimal $dt \times d$ matrix $\boldsymbol{A}$ is

$$\boldsymbol{A} = \begin{bmatrix} \boldsymbol{C}(0) & \boldsymbol{C}(1) & \dots & \boldsymbol{C}_(t-1) \\ \boldsymbol{C}(-1) & \boldsymbol{C}(0) & \dots & \boldsymbol{C}(t-2) \\ \vdots & \vdots & \vdots & \vdots \\ \boldsymbol{C}(1-t) & \boldsymbol{C}(2-t) & \dots & \boldsymbol{C}(0) \end{bmatrix}^{-} \begin{bmatrix} \boldsymbol{C}(1) \\ \boldsymbol{C}(2) \\ \vdots \\ \boldsymbol{C}(t) \end{bmatrix}.$$

When the above block Toeplitz matrix is not singular, $\boldsymbol{A}$ is the unique solution of the system of equations

$$\begin{bmatrix} \boldsymbol{C}(0) & \boldsymbol{C}(1) & \dots & \boldsymbol{C}_(t-1) \\ \boldsymbol{C}(-1) & \boldsymbol{C}(0) & \dots & \boldsymbol{C}(t-2) \\ \vdots & \vdots & \vdots & \vdots \\ \boldsymbol{C}(1-t) & \boldsymbol{C}(2-t) & \dots & \boldsymbol{C}(0) \end{bmatrix} \boldsymbol{A} = \begin{bmatrix} \boldsymbol{C}(1) \\ \boldsymbol{C}(2) \\ \vdots \\ \boldsymbol{C}(t) \end{bmatrix}$$

that are exactly the first t Yule–Walker equations introduced in the context of VARMA processes.

Note that the theory naturally extends to complex valued random variables with inner product used in Chapter 1.

D

Tools from algebra

Algebraic tools are important for proving many statements in linear system theory. Here we recall some tools from algebra that we are using. Readers who are not interested in these algebraic tools and their applications in the proofs may skip this. Otherwise, it is supposed that the interested reader is familiar with some basic concepts of algebra, like semigroups, groups, and fields.

A *ring* R is a set with two operations called addition and multiplication. An important example is $\mathbb{C}[z]$, the ring of complex polynomials. By definition, with respect to addition R is a commutative (Abelian) group, with respect to multiplication R is a semi-group with a multiplicative unit 1, and the two distributive laws hold. R is called commutative if $xy = yx$ for all $x, y \in R$. For example, $\mathbb{C}[z]$ is a commutative ring.

Given two rings R_1 and R_2, a *ring homomorphism* is a map $\phi : R_1 \to R_2$ such that

$$\phi(x + y) = \phi(x) + \phi(y), \quad \phi(xy) = \phi(x)\phi(y), \quad \phi(1) = 1;$$

$\phi(0) = 0$ follows automatically.

A subset J of a ring R is a *left ideal* if it is an additive subgroup of R and $RJ = J$. A *right ideal* is similar with $JR = J$. An *ideal* is simultaneously a left and right ideal. For example, fixing a non-zero polynomial $p(z) \in \mathbb{C}[z]$, $J = p(z)\mathbb{C}[z]$ is an ideal. If we have two nonzero elements $x, y \in R$ such that $xy = 0$, then we say that x and y are zero divisors. A *principal ideal domain* R is a commutative ring with no zero divisors and in which every ideal is *principal*, that is, of the form aR, where $a \in R$. For example, $\mathbb{C}[z]$ is a principal ideal domain.

A generalization of the notion of vector spaces is the one of *modules*. The reason for introducing this notion here is that we want to multiply our 'vectors' not only by complex scalars, but by complex polynomials and $\mathbb{C}[z]$ is only a ring, not a field. Let R be a ring. A *left R-module M* is a set in which an addition is defined, plus a multiplication from the left by elements of R. By definition, with respect to addition M is a commutative group, and with respect to multiplication it satisfies

$$r(x + y) = rx + ry, \quad (r + s)x = rx + sx, \quad r(sx) = (rs)x, \quad 1x = x$$

for all $r, s \in R$ and $x, y \in M$. *Right R-modules* are defined similarly, just we multiply by the elements of R from the right. In our main examples the elements of a module are commutative with respect to multiplication by polynomials from the ring $\mathbb{C}[z]$, so they can be simply called $\mathbb{C}[z]$-*modules.*

If M_1 and M_2 are two left R-modules, a map $\phi : M_1 \to M_2$ is an R-*module homomorphism*, if

$$\phi(x+y) = \phi(x) + \phi(y), \quad \phi(rx) = r\phi(x),$$

for any $x, y \in M$ and $r \in R$.

Our main example for a $\mathbb{C}[z]$-module is the set of *polynomial matrices* $P(z) \in \mathbb{C}[z]^{j\times k}$ whose entries are complex polynomials in the indeterminate z, where j and k are arbitrary fixed positive integers. It is often advantageous if we treat z as a complex number, though in several cases it can signify a left shift of a sequence over the integers $\mathbb{Z}$ as well.

We can also write $P(z)$ as a *matrix polynomial* with complex matrix coefficients

$$P(z) = \sum_{r=0}^{m} P_r z^r \in \mathbb{C}^{j\times k}[z], \quad P_r \in \mathbb{C}^{j\times k},$$

where m is the highest degree of the polynomial entries of $P(z)$. Thus $\mathbb{C}^{j\times k}[z]$ and $\mathbb{C}[z]^{j\times k}$ are isomorphic. In general, an isomorphism, denoted $\cong$, is a one-to-one and onto homomorphism.

Important special cases are the $\mathbb{C}[z]$-module of column matrices with polynomial entries denoted by $\mathbb{C}[z]^j \cong \mathbb{C}^j[z] = \mathbb{C}^{j\times 1}[z] \cong \mathbb{C}[z]^{j\times 1}$ and of the row matrices with polynomial entries denoted by $\mathbb{C}[z]^{1\times k} \cong \mathbb{C}^{1\times k}[z]$.

Given a $j\times \ell$ polynomial matrix $P(z)$ and a $j\times k$ polynomial matrix $R(z)$, $P(z)$ is *left divisible* by $R(z)$ if there exists a $k \times \ell$ polynomial matrix $Q(z)$ such that

$$P(z) = R(z)\, Q(z).$$

The $j \times k$ polynomial matrix $R(z)$ is a *common left divisor* of the $j \times \ell_r$ polynomial matrices $P_r(z)$ $(r = 1, \dots, s)$ if each $P_r(z)$ is left divisible by $R(z)$. A *greatest common left divisor (gcld)* $R(z)$ is a common left divisor such that any other common left divisor is a left divisor of $R(z)$ as well.

A $j \times j$ polynomial matrix $U(z)$ is *unimodular* if there exists $U^{-1}(z) \in \mathbb{C}^{j\times j}[z]$ such that

$$U(z)\, U^{-1}(z) = I_j = U^{-1}(z)\, U(z),$$

where I_j is the $j \times j$ identity matrix. Equivalently, $U(z)$ is unimodular if and only if $\det U(z) = \alpha \in \mathbb{C}$, $\alpha \neq 0$, since

$$U^{-1}(z) := \operatorname{adj}(U(z))/\det U(z),$$

where $\operatorname{adj}(U(z))$, the adjugate matrix (transpose of the cofactor matrix) of $U(z)$, is a polynomial matrix.

$P_1(z)$ and $P_2(z)$ are *left coprime* if their greatest common left divisor is a unimodular matrix. Observe that any $j \times j$ unimodular matrix is a left divisor of each $j \times \ell$ polynomial matrix $P(z)$:

$$P(z) = U(z)\,(U^{-1}(z)P(z)).$$

One can similarly define the corresponding notions of *right divisibility.*

We say that a polynomial matrix $P(z) = [p_{jk}(z)] \in \mathbb{C}^{p\times q}[z]$ and an ordinary complex polynomial $\psi(z) \in \mathbb{C}[z]$ are *coprime* if there are no common factors $(z - z_0)$ of all polynomial entries $p_{jk}(z)$ $(j = 1, \ldots, p; k = 1, \ldots, q)$ and $\psi(z)$.

Observe that $\mathbb{C}^{n\times n}[z]$, the set of square polynomial matrices, is simultaneously a ring and a $\mathbb{C}[z]$-module.

We also need the notion of a *rational matrix.* $H(z) = [h_{jk}(z)]_{p\times q}$ is a rational matrix if each entry $h_{jk}(z)$ is a rational function of the variable $z \in \mathbb{C}$, $h_{jk}[z] = n_{jk}[z]/d_{jk}[z]$, where $n_{jk}[z]$ and $d_{jk}[z]$ are complex polynomials.

A square polynomial matrix $R(z) \in \mathbb{C}^{p\times p}[z]$ is called *non-singular* if $\det R(z) \not\equiv 0$, that is, its determinant is not identically zero. However, its determinant, which is a complex polynomial, may have finitely many zeros in $\mathbb{C}$. It follows that the inverse $R^{-1}(z)$ exists except for at most finitely many values of z and is a $p \times p$ rational matrix:

$$R^{-1}(z) := \operatorname{adj}(R(z))/\det R(z),$$

where $\operatorname{adj}(R(z))$ denotes adjugate matrix: the transpose of the cofactor matrix.

The extension of the notions of *generator set, linear independence, and basis* from vector spaces to R-modules is straightforward, the only difference is that the elements of the ring R play the role of scalar multipliers.

For example, let us consider $\mathbb{C}^n[z]$, the set of n-dimensional polynomial vectors as a module over the ring $\mathbb{C}[z]$ of polynomials. Then the vectors $x_1(z), \ldots, x_s(z) \in \mathbb{C}^n[z]$ are linearly independent if

$$\sum_{j=1}^{s} r_j(z)x_j(z) \equiv 0 \quad \Leftrightarrow \quad r_1(z) \equiv \cdots \equiv r_s(z) \equiv 0,$$

where $r_j(z)$ are complex polynomials. As a concrete example, the columns of the matrix

$$\begin{bmatrix} x_1(z) & x_2(z) \end{bmatrix} = \begin{bmatrix} z+1 & z+2 \\ (z+1)(z+3) & (z+2)(z+3) \end{bmatrix}$$

are *not* linearly independent, because

$$(z+2)x_1(z) - (z+1)x_2(z) \equiv 0.$$

Correspondingly, the *column rank* of a polynomial matrix $A(z) \in \mathbb{C}^{m\times n}[z]$ is the maximal number of its linearly independent columns over the ring $\mathbb{C}[z]$

of polynomials. It is not difficult to show that the column rank of $A(z)$ is equal to the *row rank* and is equal to the size of its largest nonsingular minor. (A determinant is a polynomial now; it is called nonsingular, if it is not identically 0. Otherwise, it may have finitely many zeros in $\mathbb{C}$.) Thus we can simply call this the *rank of the polynomial matrix* $A(z)$, $\operatorname{rank} A(z) \le \min(m, n)$.

Example D.1. It is easy to find a basis for the $\mathbb{C}[z]$-module $\mathbb{C}^{m\times n}[z]$. For example, the *standard basis*

$$E_{jk} = [e_{rs}]_{m\times n} = [\delta_{rj}\delta_{sk}]_{m\times n} \qquad (j = 1, \dots, m;\ k = 1, \dots, n)$$

for the vector space of ordinary matrices $\mathbb{C}^{m\times n}$ is a basis for this module as well. For, these matrices are clearly linearly independent and any matrix $A(z) = [a_{jk}(z)]_{m\times n} \in \mathbb{C}^{m\times n}[z]$ can be represented as

$$A(z) = \sum_{j=1}^{m} \sum_{k=1}^{n} a_{jk}(z) E_{jk}.$$

The situation is more complex with submodules. The next lemma generalizes a well-known statement from linear algebra to R-submodules.

Lemma D.1. *Let R be a principal ideal domain and M be a left R-module with a finite basis of $m \ge 0$ elements. Then every left R-submodule $N \subset M$ has also a finite basis of $0 \le n \le m$ elements. Similar statement holds for right R-modules as well.*

Proof. If $M = \{0\}$, that is, its basis has 0 elements, then the statement is trivial. Then we prove the theorem by induction over m. Assume that the lemma holds for $(m-1)$ basis elements. Let $B = \{e_1, \dots, e_m\}$ be a basis for M. Then every element $x \in M$ has a unique representation $x = \sum_{j=1}^{m} r_j e_j$, $r_j \in R$. Let $N \subset M$ be a left R-submodule. If N contains only elements of the form $\sum_{j=1}^{m-1} r_j e_j$ then the statement holds by the induction hypothesis. On the other hand, if N contains an element $\sum_{j=1}^{m} r_j e_j$ with $r_m \ne 0$, define $J := \{r_m \in R : \sum_{j=1}^{m} r_j e_j \in N\}$. Clearly, J is an ideal in R: $JR = J$. Since R is a principal ideal domain, it follows that $J = aR$ with some $a \in J$, $a \ne 0$. Thus N has an element

$$y = r_1 e_1 + \dots + r_{m-1} e_{m-1} + a e_m, \quad a \ne 0, \quad r_j \in R.$$

If $\tilde{y} \in N$ is arbitrary then there exists an $r \in R$ such that $\tilde{y} - ry$ belongs to a submodule $\tilde{N}$ of M generated by only $\{e_1, \dots, e_{m-1}\}$. Hence by the induction hypothesis there exists a basis $\{f_1, \dots, f_{n-1}\}$ of $\tilde{N}$ with $n-1 \le m-1$. Clearly, $B' := \{f_1, \dots, f_{n-1}, y\}$ spans N.

We claim that B' is linearly independent. Assume that

$$s_1 f_1 + \dots + s_{n-1} f_{n-1} + s_n y = 0 \quad (s_j \in R).$$

If $s_n = 0$ then, $\{f_1, \dots, f_{n-1}\}$ being linearly independent, it follows that $s_1 = \dots = s_{n-1} = 0$ as well. On the other hand, if $s_n \neq 0$ then it follows that

$$\begin{aligned} & s_1 f_1 + \dots + s_{n-1} f_{n-1} + s_n(r_1 e_1 + \dots + r_{m-1} e_{m-1} + a e_m) \\ & = [s_1 f_1 + \dots + s_{n-1} f_{n-1} + s_n(r_1 e_1 + \dots + r_{m-1} e_{m-1})] + s_n a e_m = 0. \end{aligned}$$

The term $s_n a e_m$ is linearly independent from the term in the brackets. Thus $a = 0$ would follow, which is a contradiction. This finishes the proof of the lemma. □

It can be proved that if R is a principal ideal domain and M is a left R-module with a finite basis of $m \geq 0$ elements, then each basis of M consists of m elements. The number of the elements in a basis of M is called the *rank of the module* M denoted by rank$M$. By Example D.1, rank $\mathbb{C}^{m \times n}[z] = mn$. Clearly, the rank of a polynomial matrix $A(z) \in \mathbb{C}^{m \times n}[z]$ is the same as the rank of the submodule $M := A(z)\mathbb{C}^n[z] \subset \mathbb{C}^m[z]$.

The useful technique of *elementary row (or column) operations* can be extended to polynomial matrices. From now on we discuss mainly row operations, since it is easy to modify the statements to column operations. So the elementary row operations are

1. interchanging the ith and jth rows $(i \neq j)$,
2. multiplying the ith row by a nonzero complex number α,
3. adding a polynomial $p(z) \in \mathbb{C}[z]$ multiple of the ith row to the jth row $(i \neq j)$.

Each of these elementary row operations on a polynomial matrix $A(z) \in \mathbb{C}^{m \times n}[z]$ can be represented by an $m \times m$ *elementary matrix* $E \in \mathbb{C}^{m \times m}[z]$, which multiply $A(z)$ from the left:

$$E_{i \leftrightarrow j} := \begin{bmatrix} 1 & 0 & 0 & 0 \\ 0 & 0 & 1 & 0 \\ 0 & 1 & 0 & 0 \\ 0 & 0 & 0 & 1 \end{bmatrix} \begin{matrix} \\ \leftarrow \; i \\ \leftarrow \; j \\ \\ \end{matrix}$$

$$E_{i,\alpha} := \begin{bmatrix} 1 & 0 & 0 \\ 0 & \alpha & 0 \\ 0 & 0 & 1 \end{bmatrix} \begin{matrix} \\ \leftarrow \; i \\ \\ \end{matrix}$$

$$E_{i,j}(p(z)) := \begin{bmatrix} 1 & 0 & 0 & 0 \\ 0 & 1 & 0 & 0 \\ 0 & p(z) & 1 & 0 \\ 0 & 0 & 0 & 1 \end{bmatrix} \begin{matrix} \\ \leftarrow \; i \\ \leftarrow \; j \\ \\ \end{matrix} \tag{D.1}$$

Clearly, $E_{i\leftrightarrow j}^{-1} = E_{i\leftrightarrow j}$, $E_{i,\alpha}^{-1} = E_{i,1/\alpha}$, and $E_{i,j}(p(z))^{-1} = E_{i,j}(-p(z))$. It implies that all elementary matrices and their products are unimodular. In the case of elementary column operations we should multiply a matrix $A(z)$ by the adjoint matrices E^* of the above elementary matrices from the right. The formula $E_{i,\alpha}^{-1} = E_{i,1/\alpha}$ explains why one cannot multiply a row or column by a polynomial, just by a nonzero complex number if one wants to use only unimodular elementary matrices.

Next we are going to describe a *Hermite form* of a polynomial matrix.

Lemma D.2. *Let $A(z) \in \mathbb{C}^{m\times n}[z]$ and its rank be equal to n, so $A(z)$ has full column rank. Then by elementary row operations, equivalently, by premultiplying by a unimodular matrix $U(z) \in \mathbb{C}^{m\times m}[z]$, it can be transformed to an upper triangular form*

$$\begin{bmatrix} c_{11}(z) & c_{12}(z) & \cdots & c_{1n}(z) \\ 0 & c_{22}(z) & \cdots & c_{2n}(z) \\ \vdots & \vdots & \ddots & \vdots \\ 0 & 0 & \cdots & c_{nn}(z) \\ 0 & 0 & \cdots & 0 \\ \vdots & \vdots & \ddots & \vdots \\ 0 & 0 & \cdots & 0 \end{bmatrix},$$

where

1. *all entries below the main diagonal are identically zero,*
2. *the diagonal entries $c_{kk}(z)$ are monic,*
3. *any entry $c_{jk}(z)$, $j < k$, above the main diagonal has smaller degree than the corresponding diagonal entry $c_{kk}(z)$.*

Proof. By interchanging rows, bring to the $(1,1)$ position a nonzero element of the lowest degree in the first column, call it $a_{11}(z)$. By the Euclidean division algorithm, any entry $a_{j1}(z)$, $j > 1$, can be written as

$$a_{j1}(z) = q_{j1}(z)a_{11}(z) + r_{j1}(z), \quad \deg r_{j1}(z) < \deg a_{11}(z),$$

with some polynomials $q_{j1}(z)$ and $r_{j1}(z)$, where deg denotes the *degree of a polynomial.* (By definition, the identically zero polynomial has degree smaller then zero.) So add $-q_{j1}(z)$ times the first row to the jth; this way the original polynomial entry $a_{j1}(z)$ is replaced by $r_{j1}(z)$, which has smaller degree than $a_{11}(z)$. Perform this for $j = 2, \ldots, m$.

Now repeat the operation choosing a new nonzero $(1,1)$ element with the smallest degree in the first column and continue this until all elements in the first column except the $(1,1)$ element are zero. If the leading coefficient of the $(1,1)$ element is $\alpha \neq 1$, then multiply the first row by $1/\alpha$.

Then consider the second column of the resulting matrix and, temporarily ignoring the first row, repeat the above procedure with the $(2,2)$ element and the rows below, until all entries below the $(2,2)$ element become zero. If the leading coefficient of the $(2,2)$ element is $\alpha \neq 1$, then multiply the second row by $1/\alpha$. Then by a Euclidean division replace the $(1,2)$ entry above the $(2,2)$ entry by a remainder that has smaller degree than the $(2,2)$ entry, if this condition originally did not hold.

Continue this procedure with the 3rd, ..., nth column, and finally we arrive at the desired Hermite form. □

Corollary D.1. *By elementary row operations, equivalently by premultiplying by a unimodular matrix, one may transform an arbitrary polynomial matrix* $A(z) \in \mathbb{C}^{m\times n}[z]$ *to a Hermite form, even if* $\operatorname{rank} A(z) = r < n$. *At the beginning we skip the identically zero columns, if there is any. Then we carry out the algorithm as described in the proof of Lemma D.2, except if after the currently finished* k*th column and* (j,k), $j \leq k$, *'diagonal' element, the next* $(k+1)$*th,* $(k+2)$, ..., *columns have only zeros below the* j*th row. Then we skip these columns. At the end we have a quasi upper triangular matrix called* echelon form, *that still has all entries below the 'main diagonal' identically zero. Also, in the* r *columns that were not skipped in the procedure, properties (2) and (3) described in Lemma D.2 hold as well.*

The next lemma gives a construction for a greatest common right divisor. For simplicity, here we restrict ourselves to the square matrix greatest common right divisors (gcrd's). One can similarly construct a greatest common left divisor.

Lemma D.3. *Given polynomial matrices* $A_j(z) \in \mathbb{C}^{m_j\times n}[z]$ $(j = 1, \ldots, k)$*, there exists a greatest common right divisor* $R(z) \in \mathbb{C}^{n\times n}[z]$ *of them. Moreover, there exist polynomial matrices* $\tilde{U}_j(z) \in \mathbb{C}^{n\times m_j}[z]$, $j = 1, \ldots, k$, *such that*

$$R(z) = \sum_{j=1}^{k} \tilde{U}_j(z) A_j(z). \tag{D.2}$$

Similar statements hold for a greatest common left divisor.

Proof. Let $m := \sum_{j=1}^{k} m_j$ and define the $m \times n$ matrix $A(z)$ as

$$A(z) := \begin{bmatrix} A_1(z) \\ \vdots \\ A_k(z) \end{bmatrix}. \tag{D.3}$$

By Corollary D.1 there exists a unimodular matrix $U(z) \in \mathbb{C}^{m\times m}[z]$ such that $U(z)A(z) = \tilde{R}(z)$,

$$\begin{bmatrix} U_{11}(z) & \cdots & U_{1k}(z) \\ U_{21}(z) & \cdots & U_{2k}(z) \end{bmatrix} \begin{bmatrix} A_1(z) \\ \vdots \\ A_k(z) \end{bmatrix} = \begin{bmatrix} R(z) \\ 0_{(m-n)\times n} \end{bmatrix}, \qquad \text{(D.4)}$$

where $\tilde{R}(z) \in \mathbb{C}^{m\times n}[z]$ is the Hermite form of $A(z)$, $R(z) \in \mathbb{C}^{n\times n}[z]$, $U_{1j}(z) \in \mathbb{C}^{n\times m_j}[z]$, $U_{2j}(z) \in \mathbb{C}^{(m-n)\times m_j}[z]$, $j = 1, \ldots, k$, and $0_{(m-n)\times n}$ is a zero matrix of size $(m-n) \times n$.

Since $U(z)$ is unimodular, its inverse $V(z) \in \mathbb{C}^{m\times m}[z]$ is also a polynomial matrix:

$$U^{-1}(z) = V(z) = \begin{bmatrix} V_{11}(z) & V_{12}(z) \\ \vdots & \vdots \\ V_{k1}(z) & V_{k2}(z) \end{bmatrix},$$

where $V_{j1}(z) \in \mathbb{C}^{m_j\times n}[z]$ and $V_{j2}(z) \in \mathbb{C}^{m_j\times(m-n)}[z]$, $j = 1, \ldots, k$. Then we get

$$A(z) = V(z)\tilde{R}(z), \qquad \begin{bmatrix} A_1(z) \\ \vdots \\ A_k(z) \end{bmatrix} = \begin{bmatrix} V_{11}(z) & V_{12}(z) \\ \vdots & \vdots \\ V_{k1}(z) & V_{k2}(z) \end{bmatrix} \begin{bmatrix} R(z) \\ 0_{(m-n)\times n} \end{bmatrix}.$$

This implies

$$A_j(z) = V_{j1}(z)R(z), \quad j = 1, \ldots, k,$$

so $R(z)$ is a right divisor of each $A_j(z)$, $j = 1, \ldots, k$.

By (D.4) it follows that

$$R(z) = \sum_{j=1}^{k} U_{1j}(z)A_j(z),$$

which proves (D.2). If $S(z) \in \mathbb{C}^{n\times n}[z]$ is another common right divisor:

$$A_j(z) = W_j(z)S(z), \quad W_j(z) \in \mathbb{C}^{m_j\times n}[z], \quad j = 1, \ldots, k,$$

then

$$R(z) = \left\{ \sum_{j=1}^{k} U_{1j}(z)W_j(z) \right\} S(z).$$

This proves that $R(z)$ is a gcrd. □

Remark D.1. The gcrd is not unique, e.g. the product of a unimodular matrix and a gcrd (in this order) is also a gcrd. Any two gcrd's $R_1(z)$ and $R_2(z)$ must be related as

$$R_1(z) = W_2(z)R_2(z), \quad R_2(z) = W_1(z)R_1(z), \quad W_j(z) \in \mathbb{C}^{n\times n}[z],$$

thus

$$R_1(z) = W_2(z)W_1(z)R_1(z).$$

It implies that if $R_1(z)$ is nonsingular, that is, $\det R_1(z) \not\equiv 0$, then $W_1(z)$ and $W_2(z)$ must be unimodular, hence $R_2(z)$ is also nonsingular. So if one gcrd is nonsingular, then all gcrd's must be nonsingular.

Similarly, if one gcrd is unimodular, then all gcrd's must be unimodular.

Moreover, if the matrix $A(z)$ defined in (D.3) has full column rank n, then Lemma D.2 implies that the gcrd $R(z)$ is nonsingular and all other gcrd are also nonsingular, differing from $R(z)$ by a unimodular left factor.

The next *Bézout's identity* characterizes the coprime polynomial matrices.

Lemma D.4. *$P_1(z) \in \mathbb{C}^{m_1 \times n}[z]$ and $P_2(z) \in \mathbb{C}^{m_2 \times n}[z]$ are right coprime if and only if there exist polynomial matrices $X_j(z) \in \mathbb{C}^{n \times m_j}[z]$ $(j = 1, 2)$ such that*

$$X_1(z)P_1(z) + X_2(z)P_2(z) = I_n, \tag{D.5}$$

where I_n is the $n \times n$ identity matrix. Similar statement holds for left coprime matrices.

Proof. (D.2) gives a formula for a gcrd $R(z)$ of $P_1(z)$ and $P_2(z)$:

$$R(z) = \tilde{U}_1(z)P_1(z) + \tilde{U}_2(z)P_2(z)$$

with polynomial matrices $\tilde{U}_1(z)$ and $\tilde{U}_2(z)$. If $P_1(z)$ and $P_2(z)$ are right coprime, then $R(z)$ must be unimodular, so that $R^{-1}(z)$ is also a polynomial matrix. Therefore, we can write

$$I_n = X_1(z)P_1(z) + X_2(z)P_2(z), \quad X_j(z) = R^{-1}(z)\tilde{U}_j(z), \quad j = 1, 2.$$

Conversely, assume that (D.5) holds. Let $R(z)$ be a gcrd of $P_1(z)$ and $P_2(z)$. Then

$$P_j(z) = U_j(z)R(z), \quad j = 1, 2,$$

with some polynomial matrices $U_j(z)$, $j = 1, 2$. These imply that

$$\{X_1(z)U_1(z) + X_2(z)U_2(z)\}\, R(z) = I_n.$$

This shows that $R^{-1}(z) = X_1(z)U_1(z) + X_2(z)U_2(z)$, a polynomial matrix. Thus $R(z)$ is unimodular, so $P_1(z)$ and $P_2(z)$ are right coprime. □

Remark D.2. $P_1(z) \in \mathbb{C}^{m_1 \times n}[z]$ and $P_2(z) \in \mathbb{C}^{m_2 \times n}[z]$ are right coprime if and only if the matrix

$$\begin{bmatrix} P_1(z) \\ P_2(z) \end{bmatrix}$$

has full column rank n for *every* $z \in \mathbb{C}$. For, then Lemmas D.2 and D.3 imply that it is equivalent to the fact that the determinant of the gcrd $R(z)$ of $P_1(z)$ and $P_2(z)$ is a nonzero constant complex number, so $R(z)$ is unimodular.

Lemma D.5. *We have the following properties of submodules.*

(a) $M \subset \mathbb{C}^n[z]$ *is a submodule if and only if it is of the form*

$$M = A(z)\mathbb{C}^k[z], \quad A(z) \in \mathbb{C}^{n\times k}[z],$$

where $A(z)$ *has full column rank* k.

(b) If $A(z) \in \mathbb{C}^{n\times q}[z]$ *and* $B(z) \in \mathbb{C}^{n\times p}[z]$*, then*

$$M_1 := A(z)\mathbb{C}^q[z] \subset M_2 := B(z)\mathbb{C}^p[z] \quad \Leftrightarrow \quad A(z) = B(z)X(z),$$

where $X(z) \in \mathbb{C}^{p\times q}[z]$.

Proof. (a) The 'if' part is obvious. Conversely, Lemma D.1 and Example D.1 imply that a submodule $M \subset \mathbb{C}^n[z]$ has a basis consisting of $k \le n$ elements $a_1(z), \ldots, a_k(z) \in \mathbb{C}^n[z]$. Define $A(z) := [a_1(z) \cdots a_k(z)]_{n\times k}$. Then $M = A(z)\mathbb{C}^k[z]$ and $A(z)$ has rank k.

(b) The implication $\Leftarrow$ is obvious. Conversely, we have to show that if $M_1 \subset M_2$, then the system of equations $B(z)X(z) = A(z)$ has a polynomial matrix solution $X(z)$. Perform elementary row operations on the matrix $[B(z)A(z)]_{n\times(q+p)}$ to obtain its Hermite form. The condition $M_1 \subset M_2$ guarantees that the above system will be solvable and by the back-substitution of the usual Gaussian algorithm one gets a polynomial solution $X(z)$ from the Hermite form. $\square$

Here we extend Lemma D.3 to the non-square divisor case and we establish a connection between the generated submodule and the gcld.

Lemma D.6. *Let* $A_j(z) \in \mathbb{C}^{n\times m_j}[z]$, $j = 1, \ldots, r$*. Then there exists their greatest common left divisor (gcld)* $D(z) \in \mathbb{C}^{n\times q}[z]$*. A polynomial matrix* $D(z) \in \mathbb{C}^{n\times q}[z]$ *is a gcld if and only if for the generated submodule*

$$M := A_1(z)\mathbb{C}^{m_1}[z] + \cdots + A_r(z)\mathbb{C}^{m_r}[z] \subset \mathbb{C}^n[z]$$

we have

$$M = D(z)\mathbb{C}^q[z].$$

Similar statements are true for the greatest common right divisors.

Proof. Define $A(z) := [A_1(z) \cdots A_r(z)]_{n\times m}$, where $m = \sum_{j=1}^r m_j$. A matrix $D(z) \in \mathbb{C}^{n\times q}[z]$ is a common left divisor of the matrices $A_j(z)$, $j = 1, \ldots, r$, if and only if $D(z)$ is a left divisor of $A(z)$:

$$A(z) = D(z)U(z), \quad U(z) \in \mathbb{C}^{q\times m}[z].$$

By Lemma D.5(b) this is equivalent to

$$M := A_1(z)\mathbb{C}^{m_1}[z] + \cdots + A_r(z)\mathbb{C}^{m_r}[z] = A(z)\mathbb{C}^m[z] \subset D(z)\mathbb{C}^q[z]. \quad \text{(D.6)}$$

Since M is a submodule of $\mathbb{C}^n[z]$, by Lemma D.1 it has a finite basis, so there exists a matrix $D'(z) \in \mathbb{C}^{n\times p}[z]$ such that

$$M = D'(z)\mathbb{C}^p[z]. \tag{D.7}$$

Thus $D'(z)$ is also a common left divisor. Assuming that $D(z)$ is a gcld, it follows that

$$D(z) = D'(z)Q(z) \quad \text{for some} \quad Q(z) \in \mathbb{C}^{p\times q}[z].$$

By Lemma D.5(b), this implies that

$$D(z)\mathbb{C}^q[z] \subset D'(z)\mathbb{C}^p[z].$$

In turn, this with (D.6) and (D.7) imply that

$$D(z)\mathbb{C}^q[z] = M = D'(z)\mathbb{C}^p[z].$$

Conversely, assume that

$$M := A_1(z)\mathbb{C}^{m_1}[z] + \cdots + A_r(z)\mathbb{C}^{m_r}[z] = D(z)\mathbb{C}^q[z].$$

By Lemma D.5(a) such a $D(z)$ always exists. Then by (D.6) $D(z)$ is a common left divisor. Moreover,

$$D(z)\mathbb{C}^q[z] = M \subset D'(z)\mathbb{C}^p[z]$$

for any common left divisor $D'(z)$ by (D.6). Then by Lemma D.5(b) this implies that $D(z) = D'(z)Q(z)$ for some $Q(z) \in \mathbb{C}^{p\times q}[z]$. Thus $D(z)$ is a greatest common left divisor. □

Lemma D.7. *Let* $A_j(z) \in \mathbb{C}^{n\times m_j}[z]$, $j = 1,\ldots,r$ *with rank* $q :=$ rank$[A_1(z)\cdots A_r(z)]$.

(a) There exists a gcld $D(z) \in \mathbb{C}^{n\times q}[z]$ *with rank* q*. If* $D'(z) \in \mathbb{C}^{n\times p}[z]$ *is also a gcld with full column rank* p*, then* $p = q$ *and there exists a unique unimodular transformation* $X(z) \in \mathbb{C}^{q\times p}[z]$ *such that* $D'(z) = D(z)X(z)$.

(b) A square gcld $D(z) \in \mathbb{C}^{n\times n}[z]$ *with* $\det D(z) \not\equiv 0$ *exists if and only if* rank$[A_1(z)\cdots A_r(z)] = n$.

Similar statements hold for the greatest common right divisors.

Proof. (a) By Lemma D.1, $M := A_1(z)\mathbb{C}^{m_1}[z]+\cdots+A_r(z)\mathbb{C}^{m_r}[z]$ has a finite basis and rank $M = q \le n$. By Lemma D.5(a), there exists a full column rank matrix $D(z) \in \mathbb{C}^{n\times q}[z]$ such that $M = D(z)\mathbb{C}^q[z]$. By Lemma D.6, then $D(z)$ is a gcld.

Let $D'(z) \in \mathbb{C}^{n\times p}[z]$ be another gcld with full column rank p. Then

$D'(z)\mathbb{C}^p[z] = M = D(z)\mathbb{C}^q[z]$. By Lemma D.5, this implies that $p = q$, $D'(z) = D(z)X(z)$ and $D(z) = D'(z)Y(z)$ with $X(z), Y(z) \in \mathbb{C}^{q\times q}[z]$. Thus

$$D'(z) = D'(z)Y(z)X(z), \quad D(z) = D(z)X(z)Y(z).$$

Since $D(z)$ and $D'(z)$ have full column rank, by the Gaussian algorithm, getting a Hermite form of the systems of equations $[D'(z)D'(z)]$ and $[D(z)D(z)]$ and using back-substitution, we obtain that

$$Y(z)X(z) = I_q = X(z)Y(z).$$

Thus $X(z)$ and $Y(z)$ are unimodular, $Y(z) = X^{-1}(z)$.

(b) If $\operatorname{rank}[A_1(z), \cdots, A_r(z)] = n$, equivalently, rank $M = n$, then by (a) it follows that $D(z) \in \mathbb{C}^{n\times n}[z]$, rank $D(z) = n$, $\det D(z) \not\equiv 0$.

Conversely, if $D(z) \in \mathbb{C}^{n\times n}[z]$, $\det D(z) \not\equiv 0$, then also by (a) it follows that $\operatorname{rank}[A_1(z), \cdots, A_r(z)] = \operatorname{rank} M = n$. □

Theorem D.1. *[21, Theorem 2.29] Assume that $H(z)$ is an $n \times m$ rational matrix, which is not identically zero.*

(a) *Then it has a representation*

$$H(z) = \frac{P(z)}{\psi(z)}, \quad P(z) = [p_{jk}(z)] \in \mathbb{C}^{n\times m}[z], \quad \psi(z) \in \mathbb{C}[z].$$

If we assume that $\psi(z)$ is monic and $P(z)$ and $\psi(z)$ are coprime, then $P(z)$ and $\psi(z)$ are unique.

(b) *Also, $H(z)$ can be represented as a* matrix fraction, *the 'ratio' of two* left coprime *polynomial matrices:*

$$H(z) = \boldsymbol{\alpha}^{-1}(z)\boldsymbol{\beta}(z), \quad \boldsymbol{\alpha}(z) \in \mathbb{C}^{n\times n}[z], \quad \det \boldsymbol{\alpha}(z) \not\equiv 0, \quad \boldsymbol{\beta}(z) \in \mathbb{C}^{n\times m}[z].$$

The polynomial matrices $\boldsymbol{\alpha}(z)$ and $\boldsymbol{\beta}(z)$ are unique up to a unimodular factor.

(c) *Similarly, $H(z)$ can be represented as the 'ratio' of two* right coprime *polynomial matrices:*

$$H(z) = \tilde{\boldsymbol{\beta}}(z)\tilde{\boldsymbol{\alpha}}^{-1}(z), \quad \tilde{\boldsymbol{\alpha}}(z) \in \mathbb{C}^{m\times m}[z], \quad \det \tilde{\boldsymbol{\alpha}}(z) \not\equiv 0, \quad \tilde{\boldsymbol{\beta}}(z) \in \mathbb{C}^{n\times m}[z],$$

which are unique up to a unimodular factor.

Proof. By our assumptions, we can define $\psi(z)$ as the least common multiple of the denominators of the entries of the rational matrix $H(z)$, also assuming that each entry is in lowest terms, that is, its numerator and denominator are coprime. If we assume that $\psi(z)$ has leading coefficient 1 (is monic), then it really gets unique. This proves (a).

By (a) we may factorize as $H(z) = (\psi(z)I_n)^{-1}P(z)$. Consider the submodule

$$M := P(z)\mathbb{C}^m[z] + (\psi(z)I_n)\mathbb{C}^n[z] \subset \mathbb{C}^n[z].$$

M contains the linearly independent elements $\psi(z)\mathbf{e}_j$, $j = 1, \ldots, n$, where $\mathbf{e}_j$ denotes the jth coordinate unit vector in $\mathbb{C}^n$, $j = 1, \ldots, n$. Hence rank $M \geq n$, while M is a submodule of $\mathbb{C}^n[z]$, so rank $M \leq n$. Thus rank $M = n$ and, due to Lemma D.7, there exists a gcld $D(z) \in \mathbb{C}^{n\times n}[z]$, $\det D(z) \not\equiv 0$, of $P(z)$ and $\psi(z)I_p$. Therefore, there exists $\boldsymbol{\alpha}(z) \in \mathbb{C}^{n\times n}[z]$ and $\boldsymbol{\beta}(z) \in \mathbb{C}^{n\times m}[z]$ such that

$$P(z) = D(z)\boldsymbol{\beta}(z), \quad \psi(z)I_n = D(z)\boldsymbol{\alpha}(z).$$

By Lemma D.6,

$$M = D(z)\boldsymbol{\beta}(z)\mathbb{C}^m[z] + D(z)\boldsymbol{\alpha}(z)\mathbb{C}^n[z] = D(z)\mathbb{C}^n[z],$$

and since $D(z)$ is nonsingular, it follows that

$$\boldsymbol{\beta}(z)\mathbb{C}^m[z] + \boldsymbol{\alpha}(z)\mathbb{C}^n[z] = \mathbb{C}^n[z]. \tag{D.8}$$

Let $\mathbf{e}_1, \ldots, \mathbf{e}_n$ be the standard basis in $\mathbb{C}^n[z]$, that is, $\mathbf{e}_j$ is the jth coordinate unit vector in $\mathbb{C}^n$, $j = 1, \ldots, n$. Then by (D.8), there exist elements $\mathbf{x}_1(z), \ldots, \mathbf{x}_p(z) \in \mathbb{C}^m[z]$ and $\mathbf{y}_1(z), \ldots, \mathbf{y}_p(z) \in \mathbb{C}^n[z]$ such that

$$\boldsymbol{\beta}(z)\mathbf{x}_j(z) + \boldsymbol{\alpha}(z)\mathbf{y}_j(z) = \mathbf{e}_j, \quad j = 1, \ldots, n.$$

Consequently, with $X(z) = [\mathbf{x}_1(z), \ldots, \mathbf{x}_n(z)] \in \mathbb{C}^{m\times n}[z]$ and $Y(z) = [\mathbf{y}_1(z), \ldots, \mathbf{y}_n(z)] \in \mathbb{C}^{n\times n}[z]$ we have

$$\boldsymbol{\beta}(z)X(z) + \boldsymbol{\alpha}(z)Y(z) = I_n. \tag{D.9}$$

By Bézout's identity (Lemma D.4) this implies that $\boldsymbol{\alpha}(z)$ and $\boldsymbol{\beta}(z)$ are left coprime. Moreover,

$$H(z) = (\psi(z)I_n)^{-1}P(z) = (D(z)\boldsymbol{\alpha}(z))^{-1}D(z)\boldsymbol{\beta}(z) = \boldsymbol{\alpha}^{-1}(z)\boldsymbol{\beta}(z),$$

which is the stated left coprime factorization. (Since $\psi(z)I_n$ is nonsingular, $\boldsymbol{\alpha}(z)$ must be nonsingular too.)

Assume now that $(N(z), \Delta(z))$ is another left coprime factorization of $H(z)$:

$$\Delta^{-1}(z)N(z) = H(z) = \boldsymbol{\alpha}^{-1}(z)\boldsymbol{\beta}(z).$$

Thus $N(z) = \Delta(z)\boldsymbol{\alpha}^{-1}(z)\boldsymbol{\beta}(z)$. By Bézout's identity (D.9),

$$\boldsymbol{\alpha}^{-1}(z)\boldsymbol{\beta}(z)X(z) + Y(z) = \boldsymbol{\alpha}^{-1}(z).$$

Therefore

$$N(z)X(z) + \Delta(z)Y(z) = \Delta(z)\boldsymbol{\alpha}^{-1}(z),$$
$$\Delta(z) = (N(z)X(z) + \Delta(z)Y(z))\boldsymbol{\alpha}(z) = U(z)\boldsymbol{\alpha}(z),$$

where $U(z) \in \mathbb{C}^{n\times n}[z]$ is a polynomial matrix. Similarly, there exists a polynomial matrix $V(z) \in \mathbb{C}^{n\times n}[z]$ such that $\boldsymbol{\alpha}(z) = V(z)\Delta(z)$. Thus we obtain that

$$\boldsymbol{\alpha}(z) = V(z)U(z)\boldsymbol{\alpha}(z).$$

Since $\boldsymbol{\alpha}(z)$ is nonsingular, it follows that $V(z)U(z) = I_p$, $U(z)$ and $V(z)$ are unimodular, with

$$N(z) = \Delta(z)\boldsymbol{\alpha}^{-1}(z)\boldsymbol{\beta}(z) = U(z)\boldsymbol{\beta}(z), \quad \Delta(z) = U(z)\boldsymbol{\alpha}(z).$$

This completes the proof of (b) of the theorem.

Statement (c) can be proved similarly. □

Remark D.3. Conversely, if a transfer function $H(z)$ can be written in any of the form (a), (b), or (c) in Theorem D.1, then it is a rational matrix.

The case of (a) is obvious. The case of (b) follows from the fact that

$$\boldsymbol{\alpha}^{-1}(z) = \text{adj}\,\boldsymbol{\alpha}(z)/\det\boldsymbol{\alpha}(z),$$

and the case of (c) follows similarly.

Acknowledgment

The research and work underlying the present book and carried out at the Budapest University of Technology and Economics was supported by the National Research Development and Innovation Fund based on the charter of bolster issued by the National Research, Development and Innovation Office under the auspices of the Hungarian Ministry for Innovation and Technology, so was supported by the National Research, Development and Innovation Fund (TUDFO/51757/2019-ITM, Thematic Excellence Program). It was also funded by the project EFOP-3.6.2-16-2017-00015-HU-MATHS-IN: for deepening the activity of the Hungarian Industrial and Innovation Network.

The authors gratefully remember András Krámli for his valuable help with the theory and the related literature. The authors are obliged to Professors Manfred Deistler and György Michaletzky for their valuable help and thank Máté Baranyi for his useful remarks and help in creating some figures.

Bibliography

[1] H. Akaike. Stochastic theory of minimal realization. *IEEE Transactions on Automatic Control*, 19(6):667–674, 1974.

[2] O. Akbilgic, H. Bozdogan, and M. E. Balaban. A novel Hybrid RBF Neural Networks model as a forecaster. *Statistics and Computing*, 24(3):365–375, May 2014.

[3] B. D. O. Anderson, M. Deistler, E. Felsenstein, B. Funovits, L. Koelbl, and M. Zamani. Multivariate AR systems and mixed frequency data: G-identifiability and estimation. *Econometric Theory*, 32(4):793–826, 2016.

[4] M. Bolla. Factor analysis, dynamic. *Wiley StatsRef: Statistics Reference Online*, pages 1–15, 2014.

[5] M. Bolla and A. Kurdyukova. Dynamic factors of macroeconomic data. *Annals of the University of Craiova, Mathematics and Computer Science Series*, 37(4):18–28, 2010.

[6] G. E. P. Box, G. M. Jenkins, G. C. Reinsel, and G. M. Ljung. *Time series analysis: Forecasting and control.* John Wiley & Sons, 2015.

[7] L. Breiman and J. H. Friedman. Estimating optimal transformations for multiple regression and correlation. *Journal of the American Statistical Association*, 80(391):580–598, 1985.

[8] D. R. Brillinger. The canonical analysis of stationary time series. In P. R. Krishnaiah, editor, *Multivariate analysis II*, pages 331–350. Academic Press, New York, 1969.

[9] D. R. Brillinger. *Time series: Data analysis and theory*, volume 36. SIAM, 1981.

[10] P. J. Brockwell and R. A. Davis. *Introduction to time series and forecasting.* Springer, 2016.

[11] P. J. Brockwell, R. A. Davis, and S. E. Fienberg. *Time series: Theory and methods.* Springer Science & Business Media, 1991.

[12] H. Cramér. *Mathematical methods of statistics.* Princeton University Press, 1946.

[13] M. Deistler, B. D. O. Anderson, A. Filler, C. Zinner, and W. Chen. Generalized linear dynamic factor models: An approach via singular autoregressions. *European Journal of Control*, 16(3):211–224, 2010.

[14] M. Deistler and W. Scherrer. *Modelle der Zeitreihenanalyse.* Springer, 2018.

[15] M. Deistler, C. Zinner, et al. Modelling high-dimensional time series by generalized linear dynamic factor models: An introductory survey. *Communications in Information & Systems*, 7(2):153–166, 2007.

[16] J. L. Doob. *Stochastic processes*, volume 101. Wiley, 1953.

[17] L. Fejér. Über trigonometrische polynome. *Journal für die Reine und Angewandte Mathematik*, 146:53–82, 1916.

[18] M. Forni and M. Lippi. The general dynamic factor model: One-sided representation results. *Journal of Econometrics*, 163(1):23–28, 2011.

[19] B. Friedman. Eigenvalues of composite matrices. *Mathematical Proceedings of the Cambridge Philosophical Society*, 57(1):37–49, 1961.

[20] P. A. Fuhrmann. *Linear systems and operators in Hilbert space.* Courier Corporation, 2014.

[21] P. A. Fuhrmann and U. Helmke. *The mathematics of networks of linear systems.* Springer, 2015.

[22] L. Gerencsér, Z. Vágó, and B. Gerencsér. *Financial time series.* Pázmány Péter Catholic University, 2013.

[23] I. I. Gikhman and A. V. Skorokhod. *The theory of stochastic processes I.* Springer, 1974.

[24] I. I. Gikhman and A. V. Skorokhod. *The theory of stochastic processes III.* Springer, 1979.

[25] I. I. Gikhman and A. V. Skorokhod. *The theory of stochastic processes II.* Springer Science & Business Media, 2004.

[26] G. H. Golub and C. F. Van Loan. *Matrix computations*, volume 3. JHU Press, 2012.

[27] L. Györfi, M. Kohler, A. Krzyzak, and H. Walk. *A distribution-free theory of nonparametric regression.* Springer Science & Business Media, 2006.

[28] E. J. Hannan. *Multiple time series*, volume 38. John Wiley & Sons, 2009.

[29] E. J. Hannan and M. Deistler. *The statistical theory of linear systems.* SIAM, 2012.

[30] G. H. Hardy, J. E. Littlewood, and G. Pólya. *Inequalities.* Cambridge University Press, 1952.

[31] B. L. Ho and R. E. Kálmán. Effective construction of linear state-variable models from input/output functions. *Regelungstechnik*, 14(12):545–548, 1966.

[32] R. E. Kálmán. A new approach to linear filtering and prediction problems. *Journal of Basic Engineering*, 82(1):35–45, 1960.

[33] R. E. Kálmán and R. S. Bucy. New results in linear filtering and prediction theory. *Journal of Basic Engineering*, 83(1):95–108, 1961.

[34] A. N. Kolmogorov. Stationary sequences in Hilbert space. *Moscow University Mathematics Bulletin (in Russian)*, 2(6):1–40, 1941. Translated into English: Selected works of A. N. Kolmogorov, Vol. II, Probability theory and mathematical statistics, ed. A. N. Shiryayev, 228–271, Springer, 1986.

[35] A. Krámli. On factorization of spectral matrices (in Hungarian). *MTA III. Osztály Közleményei*, 18:183–186, 1968.

[36] A. Krámli. Regularity and singularity of stationary stochastic processes (in Hungarian). *MTA III. Osztály Közleményei*, 18:155–168, 1968.

[37] J. Lamperti. *Probability: A survey of the mathematical theory.* Wiley, second edition, 1996.

[38] J. Lamperti. *Stochastic processes: A survey of the mathematical theory*, volume 23. Springer Science & Business Media, 2012.

[39] A. Lindquist and G. Picci. *Linear stochastic systems: A geometric approach to modeling, estimation and identification*, volume 1. Springer, 2015.

[40] P. Loubaton. Non-full-rank causal approximations of full-rank multivariate stationary processes with rational spectrum. *Systems & Control Letters*, 15(3):265–272, 1990.

[41] H. Lütkepohl. *New introduction to multiple time series analysis.* Springer Science & Business Media, 2005.

[42] P. Mörters and Y. Peres. *Brownian motion*, volume 30. Cambridge University Press, 2010.

[43] N. Nikolski. *Hardy spaces*, volume 179. Cambridge University Press, 2019.

[44] C. R. Rao. *Linear statistical inference and its applications*, volume 2. Wiley, 1973.

[45] C. R. Rao. Separation theorems for singular values of matrices and their applications in multivariate analysis. *Journal of Multivariate Analysis*, 9(3):362–377, 1979.

[46] A. Rényi. On measures of dependence. *Acta Mathematica Academiae Scientiarum Hungarica*, 10(3–4):441–451, 1959.

[47] Yu. A. Rozanov. *Stationary random processes.* Holden-Day, 1967.

[48] P. Rózsa. *Linear algebra and its applications (in Hungarian).* Műszaki Kiadó, Budapest, third edition, 1991.

[49] W. Rudin. *Functional analysis.* McGraw-Hill, 1991.

[50] W. Rudin. *Real and complex analysis.* Tata McGraw-Hill education, 2006.

[51] W. Scherrer and M. Deistler. Vector autoregressive moving average models. In *Conceptual Econometrics Using R*, page 145. Elsevier, 2019.

[52] T. Szabados and B. Székely. Stochastic integration based on simple, symmetric random walks. *Journal of Theoretical Probability*, 22(1):203–219, 2009.

[53] G. J. Tee. Eigenvectors of block circulant and alternating circulant matrices. *New Zealand Journal of Mathematics*, 36(8):195–211, 2007.

[54] R. S. Tsay. *Multivariate time series analysis: With R and financial applications.* John Wiley & Sons, 2013.

[55] G. Tusnády and M. Ziermann, editors. *Analysis of time series (in Hungarian).* Műszaki Könyvkiadó, Budapest, 1986.

[56] R. von Sachs. Nonparametric spectral analysis of multivariate time series. *Annual Review of Statistics and Its Application*, 7:361–386, 2020.

[57] N. Wiener. *Time series.* MIT Press, 1949.

[58] N. Wiener. *Extrapolation, interpolation, and smoothing of stationary time series: With engineering applications*, volume 8. MIT Press, 1964.

[59] N. Wiener and P. Masani. The prediction theory of multivariate stochastic processes, I. The regularity condition. *Acta Mathematica*, 98:111–150, 1957.

[60] N. Wiener and P. Masani. The prediction theory of multivariate stochastic processes, II. The linear predictor. *Acta Mathematica*, 99:93–137, 1958.

[61] J. H. Wilkinson. *The algebraic eigenvalue problem*, volume 662. Oxford Clarendon, 1965.

[62] H. Wold. *A study in the analysis of stationary time series.* PhD thesis, Almqvist & Wiksell, 1938.

Index